Microbial Applications for Environmental Sustainability

Arun Karnwal •
Abdel Rahman Mohammad Said Al-Tawaha
Editors

Microbial Applications for Environmental Sustainability

Springer

Editors
Arun Karnwal
Department of Microbiology
Lovely Professional University
Punjab, India

Abdel Rahman Mohammad Said Al-Tawaha
Department of Biological Sciences
Al Hussein Bin Talal University
Ma'an, Jordan

ISBN 978-981-97-0675-4 ISBN 978-981-97-0676-1 (eBook)
https://doi.org/10.1007/978-981-97-0676-1

This Springer imprint is published by the registered company Springer Nature Singapore Pte Ltd.
The registered company address is: 152 Beach Road, #21-01/04 Gateway East, Singapore 189721, Singapore

Preface

In a world grappling with complex environmental challenges, the role of microorganisms has emerged as a beacon of hope and innovation. Welcome to *Microbial Applications for Environmental Sustainability*, a comprehensive exploration of the remarkable potential that tiny, often overlooked, life forms hold in shaping the future of our planet. As the pressures of climate change, pollution, resource depletion, and habitat destruction continue to mount, our understanding of the intricate web of life that sustains us has deepened. Within this intricate tapestry, microorganisms stand as unsung heroes, operating at scales beyond our perception but with impacts that resonate globally.

This book seeks to unveil the captivating world of microbial life, showcasing the manifold ways in which these microscopic powerhouses can be harnessed for the betterment of our environment. From the lush soils beneath our feet to the depths of our oceans, from waste treatment plants to bioremediation efforts, microorganisms are quietly working to maintain ecological balance and restore our ecosystems. Through the pages of this volume, readers will embark on a journey that delves into the realms of microbiology, ecology, biotechnology, and environmental science. We will explore the innovative strategies developed by scientists, engineers, and researchers worldwide to employ microorganisms as sustainable tools to address pressing environmental issues. These applications extend far beyond traditional methods, offering eco-friendly solutions that have the potential to transform industries, conserve natural resources, and protect the delicate balance of our biosphere.

While each chapter of this book offers insights into specific microbial applications, they collectively paint a picture of hope and possibility. From harnessing microbial communities to clean up contaminated sites, to leveraging their metabolic prowess to produce renewable energy, and even manipulating their genetic code to mitigate pollution, the potential of microorganisms is awe-inspiring.

As we journey through the pages ahead, we invite readers from diverse backgrounds—scientists, students, environmentalists, policymakers, and curious minds alike—to join us in this exploration. *Microbial Applications for Environmental Sustainability* is not merely a compilation of scientific findings; it is a call to action. It invites us to reevaluate our relationship with the microbial world, appreciate the power it holds, and engage in collaborative efforts to safeguard the planet we call home.

This book is the culmination of the dedication and passion of numerous experts who have dedicated their careers to understanding and harnessing the potential of microorganisms for environmental sustainability. We extend our heartfelt gratitude to them for their invaluable contributions.

Ultimately, this preface marks the beginning of an exciting journey—one that promises to expand our horizons and redefine our approach to environmental stewardship. The future of our planet lies in the hands of not just governments and institutions but also in the profound influence of microorganisms and the collective actions of individuals and communities. Let us embark on this journey together and discover the incredible world of microbial applications for environmental sustainability.

Punjab, India Arun Karnwal
Ma'an, Jordan Abdel Rahman Mohammad Said Al-Tawaha

Contents

Chapter 1
Microbes for Sustainable Development of Environmental Issues

Kiranjot Kaur and Pratibha Vyas

Abstract The most hazardous environmental issues are the depleting soil quality, food insecurities, plastic waste, threat to marine life including flora and fauna. Agriculture is completely dependent on the quality of soil and various climatic factors which are deteriorating day by day, ultimately impacting word's ecosystem and daily life of living beings because food and nutrition heavily relies on the agriculture. These problems are interlinked to each other and can probably be solved by the augmentation and engagement of beneficial microorganisms. Because of the present trend of excessive use of chemical fertilizers and pesticides, sustainability in the agriculture industry has become a challenging job. In addition, anthropogenic activities such as urbanization and industrialization are producing more garbage and endangering ecosystem. The incredible diversity of microorganisms on earth plays a critical function in the ecological processes of elements influencing soil fertility, organic matter breakdown, and carbon storage. Microorganisms mainly plant growth-promoting bacteria are now being utilized worldwide as bioenhancers for promoting growth of plant and development under various environmental stresses like heavy metals, etc. They are critical in a number of environmental friendly practices in addition to new advancements, from biogeochemical cycles to diverse industrial productions. Plastic is becoming a pest to environment and is a major threat to world in upcoming years. Microorganisms ensure the capability of degrading plastic and other harmful substances which causes damage to the environment. Microbes are very helpful in combatting bioterrorism which is a serious hazard to ecosystem as they can be used against pathogen and help the environment. This chapter focuses on the diverse role played by microorganisms in combating various environmental issues for the sustainable development.

K. Kaur · P. Vyas (✉)
Department of Microbiology, College of Basic Science and Humanities, Punjab Agricultural University, Ludhiana, Punjab, India

A. Karnwal, A. R. Mohammad Said Al-Tawaha (eds.), *Microbial Applications for Environmental Sustainability*, https://doi.org/10.1007/978-981-97-0676-1_1

Keywords Soil Quality · Food Insecurity · Plastic Waste · Marine Life · Agriculture Sustainability · Microorganisms · Chemical Fertilizers · Urbanization · Industrialization · Bioterrorism

1.1 Introduction

The sustainability of the environment is seriously threatened by current practices, which include the indiscriminate use of chemicals, greater reliance on nonrenewable energy resources, and unchecked waste production in all conceivable industrial processes. The Green revolution, technologies though assured higher crop yields have made crops more dependent on the chemical fertilizers. One of the main challenges for agriculture in the twenty-first century is the sustainable crop production. The continuous use of these fertilizers leading to the environmental pollution and their fluctuating cost has necessitated finding out the cost-effective and eco-friendly agricultural practices to improve crop productivity in a sustainable manner (Kaur et al. 2022). The usage of cleaner production and green technologies is the major goal looking into the unhealthy environment which is the reason behind the deaths of millions of people each year in the world (Omer 2008). This enlightens how infections are caused by microorganisms that influence people's ability to maintain adequate cleanliness, food access and health sector, and quality of life (McCall and Irwin 2006). Additionally, phytopathogens, a serious threat for both to the growth and yield of various types of crops cause huge yield loss in crops every year. To mitigate harmful effects of different phytopathogens, a variety of chemical pesticides and fungicides are relied upon, but these agrochemicals have deleterious effects on soil, ecosystem and human health (Latz et al. 2018). The usage of agrochemicals affects flora and fauna on earth, ecosystem, and soil health. Additionally, chemical fungicides cause resistance and adaptability in pathogens.

Microorganisms play vital functions in the conservation of resources and also the existence of life on earth. There are many definite traits of microorganisms with which they wisely contribute incredibly to sustainable development (Fig. 1.1). For an approach to sustainable developmental, microbes can get exploited to attain green technology through agriculture. By adopting endophytes inoculation in agriculture practices as biocontrol agent and biofertilizers, harmful effects of agrochemicals can be mitigated (Hardoim et al. 2015; Bora et al. 2019; Vyas and Singh 2023). Endophytic microorganisms produced various types of secondary metabolites, lytic enzymes and VOCs which are ultimately enhances plant immunity and induce resistance of host plant against various pathogens. Endophytic microorganisms as biocontrol agent is the most sustainable way to cope up with various plant pathogens. Beside numerous benefits of endophytes, limited work has been done on endophytes. There are still various unexplored mechanisms and novel compounds to be found from endophyte, from where more benefits can be drawn. Microbial

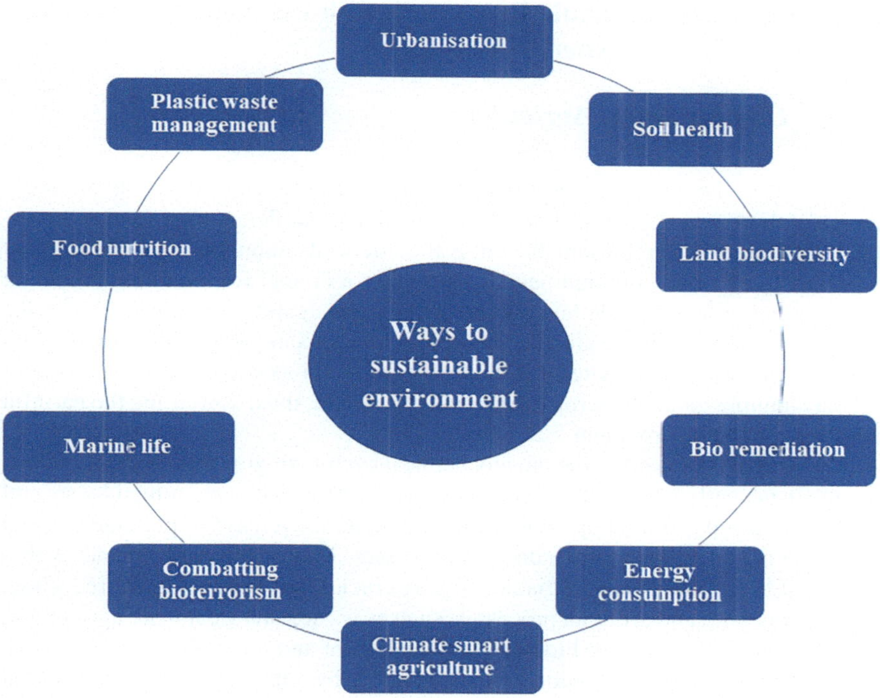

Fig. 1.1 Role of microorganisms for sustainable environment

biotechnology has the potential to enhance industrial setups, job rates, and total GDP of nations. Promoting the utilization of microbes as biofertilizers can help to improve soil health and boost crop productivity. Furthermore, scientists are encouraged to focus not just on the disease and health elements of microbiology, but also on the science's other key uses, such as fermented food products, microbial power, microorganisms as a source of food, and biocontrol agents. Microorganisms have greater capacity in increasing agricultural yield, fertility of soil and inhibiting the growth of phytopathogens. However, the overpowering benefits of microbes in the several perspective of sustainable ecosystem are often underestimated in the dissertation of operationalizing the sustainable development goals.

1.2 Ways to Sustainable Ecosystem with the Action of Microorganisms

1.2.1 Climate Smart Agriculture

The quality of the soil resources is depleting because of several anthropogenic activities as it decreases the capacity for the production of sufficient food for an increasing human population. Sustainable cultivation depends upon the functions of plant associated microbial communities, as these play a crucial role in nutrient cycling and combating many biotic and abiotic stresses. Instigating microbial roles in the agriculture may lead to sustainable environment, for example, exploitation of plant growth-promoting rhizobacteria as biofertilizers and biopesticides can play a major role in attaining the higher crop productivity and yield, thus, decreasing the harmful influence on the environment.

Microorganisms control the biogeochemical cycling of greenhouse gases including methane, carbon dioxide, and nitrous oxide. These gases are emitted by several daily activities performed by humans, for example, the release of pollutants, fossil fuel combustion, and various industrial processes. Microorganisms interact with a variety of biotic and abiotic variables to play crucial part in the element recycling. For many natural and artificial processes, such as wastewater treatment, agriculture, remediation, the creation of biofuels, the breakdown and mineralization of organic matter, the production of secondary metabolites by microorganisms is essential (Bodelier 2011; Akinsemolu 2018). Precipitation, carbon dioxide concentrations and temperature might change because of the responses provided by microorganisms towards vital processes such as denitrification and respiration. Contrarily, changing global conditions affect the ecosystem's microbiota, creating a vicious loop that makes things worsen. Due to the activity of microbes, the carbon levels in the atmosphere are sinking and soil microbiota provide vital functions in maintaining it. The rise in C and N cycles where only CO_2 alone or in combination with warming, and the decrease in N cycling processes under warming, is the potential feedback responses of the soil microbiome to several climate change factors. A slight change in the soil microbiota has a significant impact on this process. The organic carbon is present in soil is higher than atmosphere and vegetation.

Biosensors are diagnostic tools used to find samples such metabolites, medicines, microbial load, pollutants, control parameters, etc. by combining biological components with transducers. By transforming biochemical events into quantifiable physiochemical signals, such as electrical impulses, they are able to determine the amount of sample needed to determine the concentration of the sample which can be useful tool in detecting the climate changes (Nag et al. 2018). In addition to using biosensors to develop climate smart agriculture, a smart system is capable of carrying out a comprehensive approach, through detecting to responding, to carry out the best online monitoring for productivity, biosensor future technologies has the potential to increase agricultural productivity. Hence it is necessary to know the roles and natural science of soil microbes to increase carbon sequestration. New evolving technologies such as high throughput screenings are needed to understand the soil microbiota and its prospects in alleviating changes in climate (Singh et al. 2010).

1.2.2 Food Security

Excellent health begins with good nutrition. A person who is deprived of daily dietary needs tends to be unhealthy and is more susceptible to sickness. The goal of the agro-ecological system is to maximize the agricultural food production for the future population growth whilst protecting the environment and the economic and environmental performances of beneficial ecosystem services. In this regard, the goal should not be just for production of more food, but also to assure its quality and safety in order to preserve the correct well-being of those who consume it. As previously noted, the preservation of food with technology that include fermentation may be a helpful technique for assuring quality of food and waste reduction. Some fermented alcoholic drinks, on the other hand, can be utilized as a functional diet, preventing cardiovascular disease, infections, and other medical disorders (Tamang et al. 2020). Probiotics are components of microorganisms which are beneficial for host's health. Lactic acid bacteria are most commonly used probiotics. Dietary supplements such as probiotics are added to diet that increases life of humans and immunity to fight diseases. They are the utilizable substitutes to antibiotics for cattle and may also be used for animal farming. Furthermore, immunological traits are found in probiotics which increases the microflora of gut and promote abscrption of nutrients (Huerta-Rábago et al. 2019).

Probiotics influence the function and conformation of the gut microbial diversity which enhances immune system (Swain and Ray 2016). Microbes play an important role in both health and sickness. Host health is also greatly influenced by commensals, implicating pathogenic microbes as they reach distant organs, thereby interfering with the normal body functions. As a result, understanding the microbial transcriptome is critical for better human defence against species that are pathogenic. It is critical to preserve healthy flora consisting in gastrointestinal area because it has a significant impact on the T-helper (Th) cell-mediated immune response. The recent rise in Th-2-mediated illnesses can be related to the lack of immune system exposure to microbial antigens throughout development. Additionally, microorganisms are important for both maternal and newborn health. It is now clear that interactions between host and microorganism impact a child's immunity, body, and immunological phenotype. Bacterial transfer from the mother to her child trains the neonatal immune system to know particular pattern of bacteria's molecular system. This assists the progeny respond appropriately to pathogens and commensals. Nutrition and microorganism go hand in hand as the number of deaths is increasing due to inadequate nutrition and basic needs of life which is putting negative impact on the environment.

1.2.3 Marine Life

The Earth's ecosystem consist of seas and oceans which are life's cradle whereas these are considered as an essential component and a necessary resource and are important to long-term development. The demand for marine resources has increased as the world's population has grown. Overutilization of the marine resources has resulted in the collapse of highly productive fisheries, and global climate changes have had an impact on marine life. Wild-capture fishing and aquaculture generate over 141 million tonnes of fish each year, with the seas accounting for former's majority percent. Increase in fishing has lowered food output and affected the ecology severely. These circumstances have heightened the importance of using green and sustainable techniques in maritime activities.

Furthermore, global temperature change promotes cyanobacteria blooms in bodies of water, resulting in eutrophication. According to research, rise in the concentration of atmosphere and the growth of cyanobacteria in water is accelerated by greenhouse gas carbon dioxide, which may potentially result in a genetic alteration. However, blue-green algae provide a solid substitute to future generations' food security. These organisms are simple to cultivate, can grow in dirty water, have a high biomass production, and may be utilized to manufacture a variety of goods. They may be used to improve soil fertility and reduce greenhouse gas emissions, so contributing to a more sustainable future. Additionally, in the production of antibiotics and biofuels, marine microorganisms are considered as a major source (Peralta-Yahya et al. 2012; Nikapitiya 2012; Song and Lee 2015). Algae are also considered to be a powerful adsorbing agent that may be utilized to remediate different water surfaces. Traditional processes such as flocculation and filtration are costly. In turn, algae such as *Chlorella*, *Azolla,* and *Nostoc*, open the door to an economical and green feasible source of bioremediation that may then be utilized to produce biofuels.

Furthermore, as a component of marine ecosystems, microbial communities are blamable for nutrient uptaking which serve as a food supply for macro-organisms. The microorganisms including bacteria, algae and viruses play an important role in deciding the microbiota of ocean ecosystem. Viruses govern bacterial blooms and alter their genetic makeup, having a profound impact on ocean health. As a result, the accessibility of water which should be clean and fresh, as well as pollution-free air, are the two fundamental concerns confronting humans in the twenty-first century (Thijssen-van Zuylen et al. 1998; Simonovic 2002). These issues may be addressed if society is willing to be responsible by collaborating with the rules and regulations and put contribution in the establishment of a sustainable living. The use of necessary needs, as well as the limited utilization of ecologically dangerous or polluting items, is needed at the consumer level (Connor 2015). *Candidatus Pelagibacter communis*, the bacterium which can be found in almost all marine biodiversity and play an important function in marine geochemical cycles; it is considered to be essential as it play role in recycling of the organic matter in the waterbodies. Therefore, to isolate useful microbes various methods do exist from the natural environment, but they remain quite labour-intensive.

1.2.4 Soil Health

The animal and plant biodiversity loss has been a hot subject in recent years. Nothing has been said concerning the disappearance of unseen creatures: soil microbes. One gram of soil contains about ten million microorganisms, making it an important component of the food web and biogeochemical cycles. Due to an extreme usage of inorganic fertilizers and pesticides, a widespread burden is imposed on the agriculture. A sustainable crop production is required in agriculture for feeding ever-rising human population. Therefore, to lessen the utilization of synthetic enrichers, efficient and safe substitutes are needed. The free-living bacteria colonizing the rhizosphere and roots of plant aggressively which whilst pertained to seed, soil, and seedlings increase the progress and crop yield are called plant growth-promoting rhizobacteria. These PGPR with numerous plant growth-promoting activities may further be utilized as biofertilizers, bioinoculants, or bioenhancers. These PGPR aid in solubilization of insoluble phosphates, fix atmospheric nitrogen, produce phytohormones, and help plants to resist various environmental stresses and pathogens (Fig. 1.2). These are omnipresent, most of the plant is colonized, and are isolated from almost all plants examined till yet. They last in a variety of tissues within extensive range of plants, colonizing the plant analytically with bacterial/

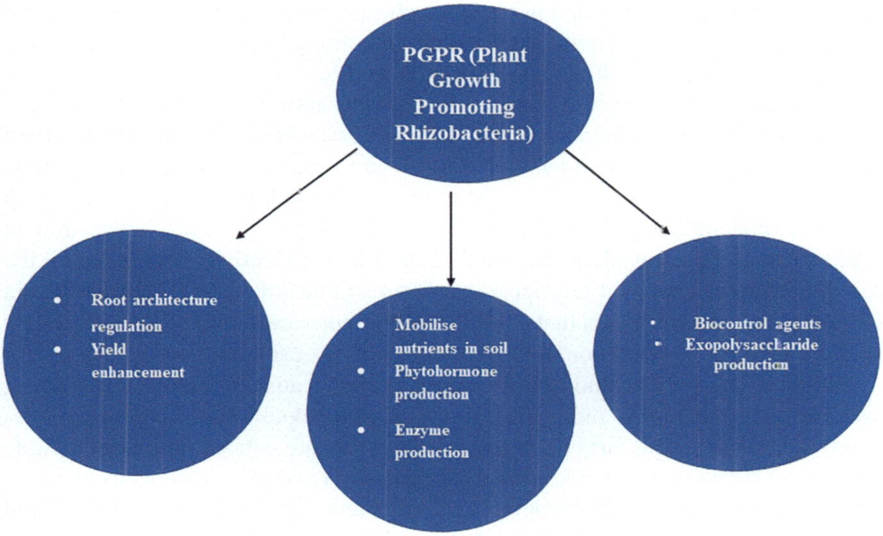

Fig. 1.2 Role of microorganisms in promoting plant growth and improving soil health

fungal colonies and biofilms. Whilst growing inside the plant, the endophytes show no visible symptoms of infection and disease (Bacon et al. 2000). They are important as they sustain plants by enabling the transfer of nutrients and perform many biological activities (Table 1.1). Amongst microorganisms, bacteria have acquired concern, primarily due to their crucial role in plant growth stimulation through mobilizing nutrients in soil, producing plethora of plant growth promoters, improving soil structure, providing protection to plants from phytopathogens. The production of antibiotics, by the production of enzymes and fungicidal compounds against harmful microorganisms is also an important feature of these PGPR. These bioinoculants have potential to increase seedling emergence rate (Hafeez et al. 2004). The microbial symbionts which reside in the healthy plant tissue, called as endophytes are gaining importance at commercial and scientific level for leguminous and non-leguminous crops. In comparison to rhizospheric and epiphytic microbes, these microorganisms are protected from adverse environmental stress (Senthilkumar et al. 2011; Vyas 2021; Rana et al. 2023). They help to maintain the soil structure, allow plants to absorb nutrients, manage pests and diseases, digest organic matter, and breakdown toxic substances, as well as serve as an indication of soil health. Currently an immense amount of interest regarding eco-friendly and sustainable agriculture has surfaced emphasizing the utilization of beneficial microorganisms. Consequently, it is important to opt for alternative fertilizers and pesticides like bioinoculants which can help in the production of crops on sustainable basis without affecting the environment. In current years, microorganisms have gained the absolute attention, as they appear to be playing a key role in promoting better plant environment for growth (Bacon et al. 2000; Rana et al. 2023; Vyas and Singh 2023). Deterioration in soil health has a direct impact on all living forms, since poor soil structures have a negative impact on groundwater, surface water, and air. Microbes, as major participants in regulating nitrogen, phosphorus, and sulphur levels, as well as the breakdown of decaying organic waste, are critical for maintaining soil health. They help in nutrient uptaking, allowing them to be consumed by higher species.

Furthermore, bacteria breakdown contaminants introduced into the soil as a result of agricultural, industrial, and building processes. Microbes play an important role in agricultural operations by regulating biogeochemical cycles. They contribute to soil fertility by delivering carbon, nitrogen, and phosphorus to crops for growth and development. They fix nitrogen, solubilize phosphorus, and chelate iron to make these nutrients available to plants, increasing productivity and reducing the use of hazardous chemical fertilizers that harm soil health and may accumulate in food. Another green process that can be used to improve soil health and ultimately land life is the bioremediation of hazardous chemicals using microorganisms.

Additionally, as pests and illnesses are a primary cause of significant losses in agricultural productivity, there is currently a tremendous effort to develop low-input, more sustainable farming practices. Scientists are concentrating on possible bio control bacteria as possible changes for the pests and plant pathogens management due to the negative effects of pesticides on human health, the environment, and living organisms. Additionally, the pathogens also develop resistance to these chemicals on continuous exposure. Biological control with the help of microorganisms is

Table 1.1 Role of microorganism in combatting environmental issues

Environmental issue	Microorganism involved	Functions of microorganism	References
Food and nutrition	*Lactobacillus*, *Streptococcus*	Fermentation Pickles, soy sauces, vinegar, bread, cheese, milk	Kim et al. (2012)
Climate smart agriculture	*Pseudomonas* and *Bacillus*	Biogeochemical cycles Nutrient uptaking which affects the carbon and nitrogenous gas	Augùeres (2015), Aguilar-Paredes et al. (2020)
Energy consumption	*Escherichia coli*, *Methanobacterium formicicium*	Production of biofuel Methanogenesis (gobar gas)	Hannon et al. (2010)
Combatting biodiversity	*Yersinia pestis; Pseudomonas syringae*	Microbes act as biocontrol and attack the pathogen	Innerebner et al. 2011; Sohrabi et al. 2012; D'allesandaro et al. 2014
Marine life	*Chlorella*, *Azolla*, and *Nostoc*	Increase in the number of fish by improving water quality (e.g., nitrogen fertilizers) on biodiversity loss and perform functions in the ecosystem	Fimbres-Olivarría et al. (2016)
Soil health	*Bacillus*, *Psuedomonas*, *Advenella*, *brevendimonas*, and *Serratia*	Fixation of nitrogen, solubilization of phosphorus, and chelation of iron Production of biofertilizer Biopesticides Biocontrol agents	Pérez-Montaño et al. (2014) Vyas and Kaur (2019), Kaur et al. (2022)
Land biodiversity	*Pleurotus pulmonarius*, *Microbispora*, *Actinobifida*	Composting of solid waste	Wachira et al. (2022)
Plastic waste management	*Phormidiumlucidum*, *Oscillatoria subbrevis*, *Xanthomonas* sp., *Sphingobacterium* sp., and *Bacillus* sp.	Degrading the artificial plastics	Gougoulias et al. (2014), Bilal et al. (2021)

(continued)

Table 1.1 (continued)

Environmental issue	Microorganism involved	Functions of microorganism	References
Urbanization	*Bradyrhizobium,* and *Azospirillum*	Facilitation cycling of nutrients Carbon shortage processes	Jang et al. (2017)
Bioremediation	*Burkholderia, Arthrobacter, Chromobacterium, Micrococcus, Candida, Pseudomonas*	Removal of pollutant Biofuel production Survival of aquatic life	Forgacs et al. (2004), Singh et al. (2010)
	Pseudomonas sp.	Atrazine degradation	James et al. (2018)
	Rhizospheric bacterial community	Atrazine degradation	Cao et al. (2018)
	Bacillus and Halobacillus	Zinc, Aluminium and Lead	Chandra Shekharaiah et al. (2020)
	Enterobacter aerogenes	Cadmium	Pramanik et al. (2018)
	Enterobacter sp.	Cadmium	Mitra et al. (2018)
	Aeromonas sp.	Cadmium	Itusha et al. (2019)
	Rhodococcus and *Acinetobacter*	Degradation of naphthalene, anthracene and phenanthrene	Belovezhets et al. (2020)

an emerging alternative to control phytopathogens. A growing body of research shows how phyllospheric and root-associated microbes may improve plant productivity and yield in cropping systems. Understanding the function of these microbes in fostering growth and preventing disease, as well as their use as bioinoculants and biopesticides whose effectiveness in the field is still uneven, is crucial. Plant defence mechanisms are modified by biocontrol microorganisms, which also carry out biocontrol actions in plants and provide novel methods for eradicating plant diseases. There are currently initiatives to enhance, facilitate, and maintain long-term plant colonization in addition to merely administering particular biocontrol bacteria. With the novel strategies of exploiting "plant-optimized microbiomes" (microbial engineering) and identifying the genetic underpinnings of advantageous interactions of plants with microorganism to enable the breeding of "microbe-optimized crops", there are particularly high hopes.

1.2.5 Energy Production/Biofuels

Sustainable consumption and production frameworks must be fully integrated into sectoral and national strategies. It involves green techniques, liable behaviour to consume, ecological products and services, and effective waste and hazardous chemical treatment. The sufficient consumption of natural occurring resources, usage of substitute energy sources, reduction of global food waste, effective recycling, expansion of scientific study into sustainable green practices are important components in ensuring sustainable development. Archaebacteria have a significant impact on the generation of biofuels and the global carbon cycle. Biological sequestration process in the form of methanogenesis and formation of methane from carbon dioxide is one of the processes which produce a valuable fuel (methane) that can be used for the generation of electricity and for other applications. This method is sustainable and eco-friendly. It involves anaerobic digestion, climate management, and model organisms for harsh conditions. Ancient organisms known as methanogens play a crucial role in carbon-neutralization process. They perform a vital function in the earth's nutrient cycling of carbon and act as carbon dioxide scrubber.

Bioenergy is energy derived from natural source materials, such as biomass. Bioenergy is a scientifically and economically possible alternative for societal development that is sustainable for clean and green earth (Banos et al. 2021). New advancements in science and technology have demonstrated that technology that involves microorganism may significantly contribution in the creation of green energy. Methanogenic archaea is one of the simplest and oldest organisms on our earth. They do not require organic nutrients and they are non-photosynthetic, which indicates they could exist in subsurface environments. It is clear that they have a mechanism in place to deal with salt or osmostress by the presence of a methanogenic archaea habitat with highly diverse salinities. Biogas is created by microbial fermentation of solid and liquid biomass, which may be used for translocation or heat generation and power. The fermentation of sugar, which is mostly produced from grains. is important for the production of first-generation biomass/biofuels (Goldemberg and Guardabassi 2010). Due to the rapidly expanding worry that both rising global energy consumption and environmental pollution are major problems in both wealthy and emerging countries, particularly India, it has received more attention to date. It is a potential technique for both recycling the gas and creating an alternate fuel source. Thus, the general population has a positive image for the environment. Methanogenic biofuel is the most efficient alternative strategy for meeting our world's energy needs when compared to other types of biofuels. These concerns may be solved if society takes responsibility by working with the government to develop a sustainable way of life. At the consumer level, the usage of basic necessities, as well as the limited use of environmentally harmful or polluting commodities, is required.

1.2.6 Land Biodiversity

Human population growth has raised the need of land as their habitat, contributing in the devastation of natural forests and the damage of biodiversity on land. It is projected that the world's forest acreage decreased from 31.6% to 30.6% between 1900 and 2015 which impacts heavily on the ecosystem (Nowak and Greenfield 2018). The two primary causes of this process have been identified as infrastructure development and agriculture whereas shortage of food is greatly linked to societal consequences (Slopen et al. 2010). Even though there is no direct relationship of bacteria to peace, they may play an important role in preserving a peaceful society. Thus, microbes may be employed to promote biodiversity of land organism, enhances biodiversity of plant, increase production of food, and decrease land pollution. Microbes, since they are so effective and play essential roles in the soil ecosystem, have the potential to help achieve sustainable ecosystem. Human population growth has raised the need for land for their habitat, contributing in the forest damage and biodiversity's land loss.

1.2.7 Combatting Bioterrorism

Microbes, as stated in earlier sections, may afford a method to earn a living through small-scale setups. Furthermore, looking into the limited tolerance of nations for one another, bioterrorism, or the utilization of live creature (pathogens) or their products (toxins), has become a big concern. Nonetheless, microbial forensics can be a significant weapon in fighting against bioterrorism. The use of standard investigative methods in conjunction with biotechnological tools can aid in locating the offenders and origins of the bio-hazard (Budowle et al. 2005; Akinsemolu 2018). Unluckily, microbial forensics is a relatively new field, and more study and training are needed to prepare officials to use these techniques to prevent bioterrorism and maintain peace. The majority of industrialized and developing countries believe that biological warfare is a faraway concern. However, there is cause for concern over the potential of bioweapons, in which violent extremists employ biological agents as weapons targeting local populations. The ability to selectively develop some of the most lethal viruses and use them as agents in terrorist attacks is possessed by nations and dissident groups. Despite the difficulty in foreseeing a bioterrorist strike, the effects of a major attack could be disastrous and cannot be disregarded. Almost any pathogenic microbe might be utilized to carry out a bioterrorist strike. Microbes such as vaccinia virus plays an important role in fighting with small pox as it is one of the bioterrorist agent, the utilization of live virus in vaccine helps to get rid of small pox.

The public health care system will be asked to handle the impacts of bioterrorism, which might place a significant burden on it. To prevent any overt or covert bioterrorist attack, a strong public health care system with quick laboratory and epidemiological inquiry, efficient medical management, and information, education, and contact (IEC) will be necessary. Microorganisms may contribute to the achievement of sustainable environment by combatting bioterrorism, serving as a nutritional supplement, improving environmental conditions, and adopting green technologies.

1.2.8 Urbanization

The environment has suffered greatly as a result of society's growing urbanization. In 2015, cities were home to a rough estimation of 54% of the people. As the population grows, so will the need for basic services and infrastructure, result in an increased number of slum inhabitants, pollution in air, and uncontrolled urban expansion. These circumstances will render the planet more vulnerable to calamities. Furthermore, from 2000 to 2015, metropolitan city expansion surpassed urban population growth. Advancements and the rationalization of wasteful fossil fuel fundings are one of the aims to accomplish the goal of development of natural environment. One of the main causes of the fast disappearing biodiversity is expanding urbanization. It has been suggested that the rarity of contact with various ambient bacteria in urbanized civilizations has a negative effect on immunity and bodily functions which eventually increases the risk for allergens and other immunological-mediated diseases. Furthermore, one of the goals is to deploy responsible and cleaner manufacturing technologies that include the utilization of microorganisms. When evaluating the effects of environmental perturbation, the structure of the microbial community is crucial. Involvement of beneficial bacteria in the surroundings can make a great positive impact in the ecosystem.

The environment has suffered as a result of society's growing urbanization. Surprisingly, little research has been done to test the fundamental premise that urbanization lowers exposure to ambient microbiota and its transfer indoors. The microorganisms are omnipresent, they play several roles in ecological cycles whereas, they put great contribution in the various antibiotic production and increased crop productivity, their potential as a cause of alternative fossil fuels, and their bioremediation characters make it critical to promote consciousness, nearly to the effective organisms. Because bacteria are crucial to biogenic hazardous gases such as methane, nitrous oxide, and carbon dioxide, it is critical to thoroughly investigate and comprehend their responsibilities, as well as act in accordance, in order to fully utilize them to modifying changes in the climate (Singh et al. 2010).

1.2.9 Plastic Waste Management

Plastic garbage buildup has become a threat to environment and is a great issue in the society. There is a critical need for the development of innovative disposal methods for plastic waste. Recent studies on the microbial or enzyme biodegradation of synthetic plastics suggests a potential to progress the biological linked waste treatment technology. The bacteria and enzymes that can break down a number of commonly utilized artificial plastics which includes polyvinyl chloride (PVC), polyurethane (PUR) polypropylene (PP), polyethylene (PE), polyethylene terephthalate (PET), and polystyrene (PS) (Bilal et al. 2021). Plastic garbage production worldwide is rising quickly in tandem with the world's increasing plastic usage. By 2050, up to 26 billion tonnes of plastic garbage are expected to be created, of which more than half will end up in landfills before eventually making their way into lakes and oceans, posing a severe threat to the ecosystem (Jambeck et al. 2016; Silva-Cavalcanti et al. 2017). As a result, plastic garbage has come to represent our wasteful society negatively. It was believed that the combined effects of thermo or photo-oxidation and biotic microbial activity may degrade long-chain polyethylene in the environment (Hakkarainen and Albertsson 2004). A range of settings, involve marine water, mulch films, soil contaminated by sewage sludge landfills and crude oil have yielded strains which has capacity of breaking down un-pretreated polyethylene.

Phormidium lucidum and *Oscillatoria subbrevis*, two cyanobacteria, demonstrated the ability to degrade initial weight of the PE up to 30% (Sarmah and Rout 2018). Polyethylene has been found to degrade after coming into touch with the homogenate of the waxworm *Galleria mellonella* or after passing through the stomach of the lesser waxworm *Achroia grisella* (Mehnaz and Javaid 2020; Kundungal et al. 2019). *Sphingobacterium* sp., *Bacillus* sp., and *Xanthomonas* sp. are three soil bacteria that may break down PS (Mor and Sivan 2008) discovered that *Rhodococcus ruber*, an actinomycete, could use PS as the only carbon source for growth, which resulted in a weight reduction of polymers.

1.2.10 Bioremediation

Bioremediation is the process utilized for the cleanup of contaminated places, such as water which is wasteful, surface waters or ground, sediments, soils, and the gases in the atmosphere. In the course of bioremediation, waste is transformed into inorganic substances including carbon dioxide, water, and methane through detoxification and mineralization (Olguín and Sánchez-Galván 2012). Microbial remediation and phytoremediation are both included under the umbrella term "bioremediation". In turn, microbial remediation involves using bacterial, fungal, and algal components for cleanup resolutions, including numerous phases of different enzymatic reactions (Table 1.1). The removal of toxic substances from the environment has

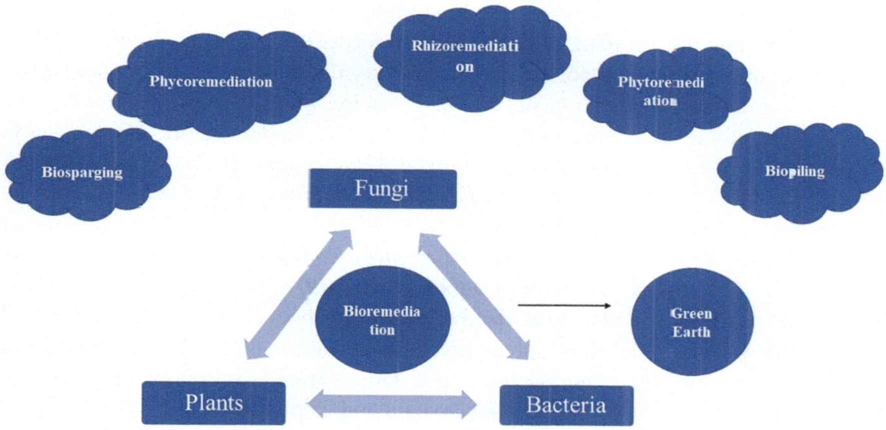

Fig. 1.3 Role of microorganisms in bioremediation

become more sustainable in recent years, and one of the finest methods is phycoremediation (Fig. 1.3). The economically and environmentally sustainable handling of commercial sludge is a serious challenge across the world. Globally accepted industrial sludge disposal procedures include properly disposing, elevated drying, slurry dispersion on land surface, lime added stabilization, burning, and composting. Due to high sludge treatment costs, the bulk of textile manufacturers in India discharged their wastewater effluents onto traditional farming, open dumps, fallow land, ineffectively regulated sanitary landfills, and beside railway lines. It contaminates groundwater further, posing major human health risks. Meanwhile, due to a scarcity of landfill space, a strict national waste disposal policy, and a lack of public knowledge, open dumping and land expansion have become prohibitively expensive and unfeasible. Most poor nations' sludge management techniques are underdeveloped. Several research groups around the world are now working to increase the dye breakdown capabilities of bacteria in order to speed up the bioremediation process. Numerous azo-dyes can be degraded aerobically and anaerobically by various microbial consortiums. Bacterial consortiums are thought to be more advantageous for full dye breakdown. Toxic intermediates produced by one bacterium's activities are destroyed by other bacteria in the consortium in mixed culture (Ajaz et al. 2020). Thus, dye degradation will be affected by the origin and chemical composition of the dyes as well as the microorganisms included in the consortium. Bioremediation of hydrocarbon-polluted groundwater and soils using bacteria has lately received a lot of attention. Bacteria may digest a wide range of hazardous compounds in both anaerobic and aerobic environments.

The primary benefit of growing algae photoautotrophically is the capture of CO_2 during cell development. Utilizing macroalgae and cyanobacteria to remove minerals and toxins from wastewater as well as carbon dioxide from the air is known as phycoremediation. The utilization of photoautotrophic microorganisms is advantageous because of their non-polluting nature as long as the biomass produced is recycled. Through photosynthetic reactions, light is used as a source of energy by photoautotrophic metabolism, which then transforms it into chemical energy. This releases O_2 into the atmosphere by consuming a significant amount of CO_2 and turning it into biomass (Pacheco et al. 2020). By using a process known as intrinsic remediation, microorganisms such as, *Arthrobacter*, *Burkholderia*, *Micrococcus*, *Chromobacterium*, *Pseudomonas*, *Candida*, and *Bacillus* which degrades hydrocarbons and crude oil without the use of any synthetic augmentation. *Chlamydomonas*, *Phormidium*, *Cyanothece*, *Botryococcus*, *Chlorella*, *Oscillatoria*, *Spirulina*, *Desmodesmus*, *Scenedesmus*, *Nostoc Arthrospira*, *Nodularia*, etc. are amongst the species used for phycoremediation (Dubey et al. 2011). The microbes which are algal based products can also be utilized to create fertilizers, cosmetics, and food or feed additives. Due to their widespread presence in waterbodies and soil ecosystems and flexible metabolism, cyanobacteria have demonstrated exceptional abilities for bioaccumulation and biosorption (Zinicovscaia and Cepoi 2016).

1.3 Conclusion

Considering the growing use of nonrenewable energy sources and the deterioration of global climatic quality, with rising footprints of carbon and widespread poverty throughout the world, finding ways to achieve a cleaner environment and sustainable development has become an essential necessity. Moreover, significant development in this manner, operative approaches utilizing such resources which include microbes are still absent. Microorganisms can make a substantial contribution in this by limiting changes in climate, boosting production with cleaner and green technologies, increasing agricultural productivity, and providing a means of livelihood generating a living. It is still not early or too late to take action. Using green microbial technologies, we can achieve the objective of cleaner industrial methods and sustainable development (Table 1.2).

Despite boosting outcomes of microbial technologies, the expense of setups of industrial processes and production processes remains a key barrier to using technology which include microorganism. The development of expense technology for integrating microbiological production methods might be a game changer, helping in the "microbial revolution" and, eventually, increasing the challenges in the

Table 1.2 Initiatives taken by different countries for sustainable ecosystem

Location	Description	Reference
African countries	Clean water management practices	Bansal et al. (2017)
India	Food and beverages	Tamang et al. (2020)
Denmark	Bioenergy	Lund et al. (2007)
South Africa	Public health	Chola et al. (2015)
Portugal	Public health	Mafra et al. (2015)
India	Clean water and sanitation	Surie (2017)

environmental. Despite it is impossible until national blockages are crossed and national infrastructure and policies are improved. Greater international efforts and cooperation are necessary, and reasonable efforts will provide cleaner production technologies, sustainable outcomes, and a balanced and sustainable ecology. To effectively fulfil the goals of sustainable development, the governments must take the most substantial initiatives. Governance, laws, and regulations are desperately needed to put the ideals of responsible growth and long-term viability into action. Better infrastructure for education and awareness campaigns, increased funding for research and development, deployment and enhancement of current green and clean manufacturing technologies and the development of microbiota startups in the industries are some of the ways that can contribute to achievement of the goals for sustainable development of ecosystem.

References

Aguilar-Paredes A, Valdés G, Nuti M (2020) Ecosystem functions of microbial consortia in sustainable agriculture. Agronomy 10(12):1902

Ajaz M, Shakeel S, Rehman A (2020) Microbial use for azo dye degradation—a strategy for dye bioremediation. Internat Microbiol 23:149–159

Akinsemolu AA (2018) The role of microorganisms in achieving the sustainable development goals. J Clean Prod 182:139–155

Auguères AS (2015) Biotic regulation of global biogeochemical cycles: a theoretical perspective (Doctoral dissertation, Université Paul Sabatier-Toulouse III)

Bacon CW, Stone JK, White JF (2000) Microbial endophytes. CRC Pess, Boca Raton, pp 17–44

Bansal V, Tumwesige V, Smith JU (2017) Water for small-scale biogas digesters in sub-Saharan Africa. GCB Bioenergy 9(2):339–357

Banos V, Deuffic P, Brahic E (2021) Engaging or resisting? How forest–based industry and private forest owners respond to bioenergy policies in Aquitaine (Southwestern France). Forest Policy Econ 2022 Nov 1;144:102843

Belovezhets LA, Tretyakova MS, Markova YA, Levchuk AA (2020) Use of rhizosphere microorganisms for bioremediation of oil contaminated soils. In: IOP conference series: earth and environmental science 2020 Vol, 408, No. 1, 012086. IOP Publishing

Bilal H, Raza H, Bibi H, Bibi T (2021) Plastic biodegradation through insects and their symbionts microbes: a review. J Bioresour Manag 8:7

Bodelier PL (2011) Interactions between nitrogenous fertilizers and methane cycling in wetland and upland soils. Curr Opin Env Sustain 3:379–388

Bora P, Saikia K, Hazarkia H, Ragesh G (2019) Exploring potential of bacterial endophytes in disease management of horticultural crops. Curr Hort 7:32–37

Budowle B, Wilson MR, Burans JP, Breeze RG, Chakraborty R (2005) Microbial forensics. In: Microbial forensics 2005 Jan 1. Academic Press, pp 1–25

Cao Y, Fatemi V, Fang S, Watanabe K, Taniguchi T, Kaxiras E, Jarillo-Herrero P (2018) Unconventional superconductivity in magic-angle graphene superlattices. Nature 556:43–50

Chandra Shekharaiah PS, Sanyal D, Dasgupta S, Sapre A, Banik A (2020) Heavy metal mitigation with special reference to bioremediation by mixotrophic algae-bacterial protocooperation. Cellular and Molecular Phytotoxicity of Heavy Metals 2020:305–334

Chola L, Michalow J, Tugendhaft A, Hofman K (2015) Reducing diarrhoea deaths in South Africa: costs and effects of scaling up essential interventions to prevent and treat diarrhoea in underfive children. BMC Public Health 15:1–10

Connor R (2015) The United Nations world water development report 2015: water for a sustainable world. UNESCO publishing, pp 121–125

D'allesandaro M, Erb M, Ton J, Brandenburg A, Karlen D, Zopfi J, Turlings TC (2014) Volatiles produced by soil-borne endophytic bacteria increase plant pathogen resistance and affect tritrophic interactions. Plant Cell Environ 37:813–826

Dubey SK, Dubey J, Mehra S, Tiwari P, Bishwas AJ (2011) Potential use of cyanobacterial species in bioremediation of industrial effluents. Afr J Biotechnol 10(7):1125–1132

Fimbres-Olivarría D, López-Elías JA, Carvajal-Millán E, Márquez-Escalante JA, Martínez-Córdova LR, Miranda-Baeza A, Enríquez-Ocaña F, Valdéz-Holguín JE, Brown-Bojórquez F (2016) Navicula sp. sulfated polysaccharide gels induced by Fe (III): rheology and microstructure. Int J Mol Sci 17(8):1238

Goldemberg J, Guardabassi P (2010) The potential for first-generation ethanol production from sugarcane. Biofuels, Bioproducts and Biorefining: Innovation for a sustainable economy 2010 Jan;4(1):17–24

Gougoulias C, Clark JM, Shaw LJ (2014) The role of soil microbes in the global carbon cycle: tracking the below-ground microbial processing of plant-derived carbon for manipulating carbon dynamics in agricultural systems. J Sci Food Agric 94:2362–2371

Hafeez FY, Safdar ME, Chaudhry AU, Malik KA (2004) Rhizobial inoculation improves seedling emergence, nutrient uptake and growth of cotton. Aust J Exp Agric 44:617–622

Hakkarainen M, Albertsson AC (2004) Environmental degradation of polyethylene long term properties of polyolefins. Microb Ecol 10:177–200

Hannon M, Gimpel J, Tran M, Rasala B, Mayfield S (2010) Biofuels from algae: challenges and potential. Biofuels 1:763–784

Hardoim PR, Van Overbeek LS, Berg G, Pirttilä AM, Compant S, Campisano AM, Döring, Sessitsch A (2015) The hidden world within plants ecological and evolutionary considerations for defining functioning of microbial endophytes. Microbiol Mol Boil Rev 79:293–320

Huerta-Rábago JA, Martínez-Porchas M, Miranda-Baeza A, Nieves-Soto M, Rivas-Vega ME, Martínez-Córdova LR (2019) Addition of commercial probiotic in a biofloc shrimp farm of Litopenaeus vannamei during the nursery phase: effect on bacterial diversity using massive sequencing 16S rRNA. Aquaculture 502:391–399

Innerebner G, Knief C, Vorholt JA (2011) Protection of *Arabidopsis thaliana* against leaf-pathogenic *Pseudomonas syringae* by *Sphingomonas* strains in a controlled model system. Appl Environ Microb 77:3202–3210

Itusha A, Osborne WJ, Vaithilingam M (2019) Enhanced uptake of cd by biofilm forming Cd resistant plant growth promoting bacteria bioaugmented to the rhizosphere of *Vetiveria zizanioides*. Int J Phytoremediation 21:487–495

Jambeck JR, Geyer R, Wilcox C, Siegler TR, Perryman M, Andrady A, Narayan R, Law KL (2016) Plastic waste inputs from land into the ocean. Science 347(6223):768–771

James SL, Abate D, Abate KH, Abay SM, Abbafati C, Abbasi N, Abbastabar H, Abd-Allah F, Abdela J, Abdelalim A, Abdollahpour I (2018) Global, regional, and national incidence, prevalence, and years lived with disability for 354 diseases and injuries for 195 countries and territories 1990–2017: a systematic analysis for the Global Burden of Disease Study. Lancet 392:1789–1858

Jang J, Hur HG, Sadowsky MJ, Byappanahalli MN, Yan T, Ishii S (2017) Environmental Escherichia coli: ecology and public health implications: a review. J Appl Microbiol 123:570–581

Kaur M, Vyas P, Rahi P, Sharma S (2022) Chlorpyrifos-and carbofuran-tolerant phosphate-solubilising *Arthrobacter oxydans* and *Bacillus flexus* improved growth and phosphorus content in potato in pesticide-amended soils. Potato Res 65:213–231

Kim J, Bang J, Beuchat LR, Kim H, Ryu JH (2012) Controlled fermentation of kimchi using naturally occurring antimicrobial agents. Food Microbiol 32(1):20–31

Kundungal H, Gangarapu M, Sarangapani S, Patchaiyappan A, Devipriya SP (2019) Efficient biodegradation of polyethylene (HDPE) waste by the plastic-eating lesser waxworm (*Achroia grisella*). Environ Sci Pollut Res 26:18509–18519

Latz MA, Jensen B, Collinge DB, Jørgensen HJ (2018) Endophytic fungi as biocontrol agents: elucidating mechanisms in disease suppression. Plant Ecol Divers 11:555–567

Lund L, Jacobsen J, Borre M, Larsen EH, Graversen PH, Sørensen HT, Notheren Danish Cancer Quality Assesment Group (2007) Improved survival of patients with prostate cancer in northern Denmark 1985–2004 Scandinavian. J Urol Nephrol 41:308–313

Mafra P, Lima N, Carvalho GS (2015) Experimental activities in primary school to learn about microbes in an oral health education context. J Biolog Educ 49(2):190–203

McCall AC, Irwin RE (2006) Florivory: the intersection of pollination and herbivory. Ecol Lett 9:1351–1365

Mehnaz S, Javaid A (2020) Microbes and plastic waste management. Environ Sust 3:337–339

Mitra S, Pramanik K, Sarkar A, Ghosh PK, Soren T, Maiti TK (2018) Bioaccumulation of cadmium by Enterobacter sp. and enhancement of rice seedling growth under cadmium stress. Ecotoxicol Environ Safety 156:183–196

Mor R, Sivan A (2008) Biofilm formation and partial biodegradation of polystyrene by the actinomycete *Rhodococcus ruber*. Biodegradation 19:851–858

Nag A, Mitra A, Mukhopadhyay SC (2018) Graphene and its sensor-based applications: a review. Sens Actuators A Phys 270:177–194

Nikapitiya C (2012) Bioactive secondary metabolites from marine microbes for drug discovery. Adv Food Nutr Res 65:363–387

Nowak DJ, Greenfield EJ (2018) US urban forest statistics, values, and projections. J Forestry 116(2):164–177

Olguín EJ, Sánchez-Galván G (2012) Heavy metal removal in phytofiltration and phycoremediation the need to differentiate between bioadsorption and bioaccumulation. New Biotechnol 30:3–8

Omer AM (2008) Energy, environment and sustainable development. Renew SusEnergy Rev 12(9):2265–2300

Pacheco D, Rocha AC, Pereira L, Verdelhos T (2020) Microalgae water bioremediation: trends and hot topics. Applied Sci 10(5):1886

Peralta-Yahya PP, Zhang F, Del Cardayre SB, Keasling JD (2012) Microbial engineering for the production of advanced biofuels. Nature 488(7411):320–328

Pérez-Montaño F, Alías-Villegas C, Bellogín RA, Del Cerro P, Espuny MR, Jiménez-Guerrero I, Cubo T (2014) Plant growth promotion in cereal and leguminous agricultural important plants: from microorganism capacities to crop production. Microbiol Res 169:325–336

Pramanik K, Mitra S, Sarkar A, Maiti TK (2018) Alleviation of phytotoxic effects of cadmium on rice seedlings by cadmium resistant PGPR strain Enterobacter aerogenes MCC 3092. J Hazar Mat 351:317–329

Rana AK, Vyas P, Sharma S, Sardana V (2023) Groundnut harbours non-nodulating non-rhizobial plant growthpromoting bacterial endophytes. 3 Biotech 13(12):1–4

Sarmah P, Rout J (2018) Efficient biodegradation of low-density polyethylene by cyanobacteria isolated from submerged polyethylene surface in domestic sewage water. Environ Sci Pollut Res 25:33508–33520

Senthilkumar M, Anandham R, Madhaiyan M, Venkateswaran V, Sa T (2011) Endophytic bacteria: perspectives and applications in agricultural crop production. In: Bacteria in agrobiology: crop ecosystems, vol 2011, pp 61–96

Silva-Cavalcanti JS, Silva JD, de França EJ, de Araújo MC, Gusmao F (2017) Microplastics ingestion by a common tropical freshwater fishing resource. Environ Pollut 221:218–226

Simonovic SP (2002) World water dynamics: global modeling of water resources. J Environ Manag 66(3):249–267

Singh A, Agrawal M, Marshall FM (2010) The role of organic vs. inorganic fertilizers in reducing phytoavailability of heavy metals in a wastewater-irrigated area. Ecol Eng 36:1733–1740

Slopen N, Fitzmaurice G, Williams DR, Gilman SE (2010) Poverty, food insecurity, and the behavior for childhood internalizing and externalizing disorders. J American Acad Child Adolescent Psyc 49(5):444–452

Sohrabi M, Alizade Naini M, Rasekhi A, Oloomi M, Moradhaseli F, Ayoub A, Bazargani A, Hashemizadeh Z, Shahcheraghi F, Badmasti F (2012) Emergence of K1 ST23 and K2 ST65 hypervirulent Klebsiella pneumoniae as true pathogens with specific virulence genes in cryptogenic pyogenic liver abscesses shiraz Iran. Front Cell Infec Microbiol 12:964290

Song M, Lee OM (2015) A study of newly recorded genera and species of filamentous blue-green algae (Cyanophyceae, cyanobacteria) in Korea. J Ecol Environ 38(4):619–627

Surie G (2017) Creating the innovation ecosystem for renewable energy via social entrepreneurship: insights from India. Technol Forecast Soc Change 121:184–195

Swain MR, Ray RC (2016) Nutritional values and bioactive compounds in lactic acid fermented vegetables and fruits lactic acid fermentation of fruits and vegetables. In: Paramithiotis S (ed) Lactic acid fermentation of fruits and vegetables. CRC Press, pp 37–52

Tamang JP, Cotter PD, Endo A, Han NS, Kort R, Liu SQ, Mayo B, Westerik N, Hutkins R (2020) Fermented foods in a global age: east meets west. Compr Rev Food Sci Food Saf 19:184–217

Thijssen-van Zuylen CW, de Beer T, Leeflang BR, Boelens R, Kaptein R, Kamerling JP, Vliegenthart JF (1998) Mobilities of the inner three core residues and the man ($\alpha 1 \rightarrow 6$) branch of the glycan at Asn78 of the α-subunit of human chorionic gonadotropin are restricted by the protein. Biochemistry 37(7):1933–1940

Vyas P, Kaur R (2019) Culturable stress-tolerant plant growth-promoting bacterial endophytes associated with *Adhatoda vasica*. J Soil Sci Plant Nutr 19:290–298

Vyas P (2021) Endophytic microbes: an array of organic volatiles and secondary metabolites. In: Volatiles and metabolites of microbes 2021 Jan 1. Academic Press, pp 105–141

Vyas P, Singh B (2023) Plant growth-promoting endophytes from Stevia rebaudiana identified to possess biocontrol potential against maize sheath blight pathogen Rhizoctonia solani. Europ J Plant Pathol 23:1–21

Wachira JW, Nguluu S, Kimatu J (2022) Differential growth and productivity of oyster mushroom (Pleurotus pulmonarius) on agro-waste substrates in semi-arid regions of Kenya. Internat J Recycling Org Waste Agri 11(3):375–383

Zinicovscaia I, Cepoi L (eds) (2016) Cyanobacteria for bioremediation of wastewaters Basel. Springer International Publishing, Switzerland, pp 1–124

Chapter 2
Harnessing Microbial Solutions for Sustainable Food and Environmental Security

Priyanka Adhikari, Pooja Thathola, Kuldeep Joshi, Abdel Rahman Mohammad Said Al-Tawaha, and Abdel Razzaq Mohammad Al-Tawaha

Abstract How to sustainably provide food and fodder to a mounting worldwide resident is a perquisite interrogation which is still lacking a response from the scientific world. Additional difficulties are due to industrial development and other human actions are responsible for upsurging the dissimilar contaminants, which gather in soils and marine ecosystems. Which leads not only the social security at danger, but then again also a major threat to ecological healthiness. For sustainable remedies for different environmental issues there are a numerous methods and tactics which are different industries and NGOs are using to reduce contamination from the environments, but the only techniques which we can use reduce pollutants without conciliatory environmental healthiness and well-beings are microbial products. Microbial methods can efficaciously be castoff for sustainable farming expansion. Additionally, these microbial solutions have positive touches on the treatment of bioaugmentation, solid waste managements, remediation of environmental pollutants, and detoxification of heavy metals. Microorganism are a gem in itself, if they uncertainty oppressed sensibly, they can subsidize to the sustainable development. In the current scenario the foremost tribunals are climate change, food safety,

P. Adhikari
Centre for GMP Extraction Facility (Department of Biotechnology), National Institute of Pharmaceutical Education and Research (NIPER), Guwahati, Assam, India

P. Thathola · K. Joshi
G.B. Pant National Institute of Himalayan Environment, Almora, Uttarakhand, India

A. R. Mohammad Said Al-Tawaha (✉)
Department of Biological Sciences, Al Hussein Bin Talal University, Ma'an, Jordan
e-mail: Abdel-al-tawaha@ahu.edu.jo

A. R. M. Al-Tawaha
Department of Crop Science, Faculty of Agriculture, Universiti Putra Malaysia, Serdang, Selangor, Malaysia

National Agricultural Research Center, Baqa, Jordan

A. Karnwal, A. R. Mohammad Said Al-Tawaha (eds.), *Microbial Applications for Environmental Sustainability*, https://doi.org/10.1007/978-981-97-0676-1_2

and increase in pollutants, and microbes are effectively sufficient in solving out these encounters alone. Microbes are ubiquitous and repeatedly in actual great figures. Therefore, microbes can be utilized for solving and resolving the numerous environmental issues in the world which needs a sustainable approach to minimize pollutants in the ecosystems.

Keywords Water pollution · Environmental issues · Bioremediation · Endophytes · Solid waste management

2.1 Introduction

Uncontrolled growth on the population worldwide have a great impact on the different environmental issues, i.e., agriculture, solid waste accumulation, aquaculture, soil and water pollutants, etc. (Ahmad et al. 2018). Through increase in population a significant impact on the industrialization and urbanization, that directly or indirectly gives burden on worldwide environment and numerous habitats (Ellis et al. 2010). By the way of day-by-day industrialization increasing worldwide, simultaneously many environmental pollutants and industrial contaminants are growing delinquent and unreal problems in the twenty-first century (Ahmad et al. 2018). Our existing techniques involves the undifferentiating usage of chemicals and unrestrained production of solid waste stuffs in every single probable manufacturing progression that directly or indirectly has stood as a huge menace to the sustainability of the environment (Ferronato and Torretta 2019).

How to sustainably provide food and fodder to a mounting worldwide resident is a perquisite interrogation which is still lacking a response from the scientific world. Predominantly agriculturalists and workers, to upsurge the crops and cereals yields, incline to spread over more fertilizers and pesticides, which is a great thereat for the environments (Zhang et al. 2022). For sustainable remedies for different environmental issues there are a numerous of methods and tactics which are different industries and NGOs are using to reduce contamination from the environments, but the only techniques which we can use reduce pollutants without conciliatory environmental healthiness and well beings are microbial products (Timmusk et al. 2017). Microbial methods can effectively be a catalyst for sustainable farming expansion (Adhikari et al. 2021; Zhang et al. 2022).These microorganisms can improve plant growth by refining soil nutrient accessibility for small crop to big plants by numerous plant growth and promotion (PGP) activities (Pan et al. 2020; Yu et al. 2022). At the moment, reprocessing, land-filling, ignition, and pyrolysis are some methods that are in usage to decrease the quantities of noxious contaminants from polluted locations. But then again, they have adversative possessions on the atmosphere, creating even further resilient and extremely poisonous midway compounds. Furthermore, these approaches are exclusive and problematic to perform for soil, water, and air contaminants (Ferronato and Torretta 2019).

Based on the above discussed problems, this chapter will deliver an outline on the existing studies on microbial solutions as a method to either substitute or decrease the practice of harmful chemicals and at the same time how microbes are reducing the environments pollutants, i.e., heavy metals and pesticides. Microbial solutions can also be utilized to redevelop polluted places by decreasing the accumulation of insecticides, heavy metals, polyaromatic hydrocarbons, and other toxic sewage waste. The possible purposes of microbes and their significant procedures and involvement in sustainable development and ecological supervision will be deliberated in this chapter, followed by their future prospects.

2.2 Sustainable Development in Environmental Issue

As day-by-day world population are increasing, simultaneously numerous pollutions are also increasing and affecting the environment raising the many environmental issues. The only beneficial partner to cope up with this problem is microorganism, i.e., bacteria and fungi. Microorganisms are ubiquitous in nature and play a significant role in the improvement and degradation of different environmental issues, i.e., agriculture, bioaugmentation, waste water treatment, waste management, environmental pollutants, etc. In this chapter, we will discuss wastewater problems in brief.

2.2.1 Role of Microbes in Sustainable Agriculture

Agriculture plays a substantial role in global economies, making it a potential contributor to key continental objectives such as poverty eradication, hunger alleviation, rapid industrialization, economic diversification, sustainable resource management, investment promotion, and environmental stewardship (Sudheer et al. 2020). Microbial technology is an expanding field that exhibits significant growth and encompasses a wide range of applications within the realm of sustainable agriculture (Adhikari and Pandey 2020). A comprehensive comprehension of the behavior of microbial communities in soil and their collaborative functioning in adapting to fluctuating environmental circumstances is of utmost importance (Shen et al. 2019). The presence of microorganisms in soil has a crucial role in enhancing soil fertility, promoting its sustainability, facilitating the decomposition of agricultural waste, increasing plant resilience to environmental stressors, safeguarding plants against soil-borne diseases, and facilitating the remediation of heavy metal contamination in soil.

According to Sudheer et al. (2020), arbuscular mycorrhizal fungi (AMF), i.e., *Glomus mosseae* and *Funneliformis mosseae,* have synergetic effects on root morphology, soil nutrient availability, and plant growth, respectively (Pan et al. 2020; Yu et al. 2022). Endophyte *Curtobacterium citreum* isolated from *Manihot esculenta* was reported for nitrogen fixation and, at the same time, to increase the biomass of *Cassava* (Zhang et al. 2022). *Aneurinibacillus aneurinilyticus* and *Paenibacillus* sp. isolated from rhizosphere of *Allium sativum* have the ability to produce ACC deaminase (Gupta and Pandey 2019). *Pseudomonas* sp. isolated from cold dessert soil were reported as a potential phosphate solubilizer in the broad temperature range of 5–35 °C (Adhikari et al. 2021).

Endophytes both bacteria (*Burkholderia contaminants,* and *Enterobacter* sp.) and fungi (*Aspergillus* sp. and *Penicillium* sp.) isolated from the roots of *Taxus wallichiana* were reported to solubilize phosphate in the broad range of temperature, i.e., 5–35 °C. Additionally, endophytic bacteria were reported to increase soil nutrients, PGPR activity, i.e., ACC deaminase, siderophores, and ammonia production. They are also increasing biomass of rice (*Oryza sativa*) and black soybean (*Glycine* max) (Adhikari and Pandey 2019, 2020). Details on beneficial bacteria and fungi and their role in agriculture are given in Table 2.1.

2.2.2 Role of Microbes in Waste or Sewage Water Treatment

As population and industrialization is increasing, side by side water bodies are also getting contaminated due to which worldwide water scarcity is becoming a great problem (Liu et al. 2017; Vliet et al. 2021). Addressing the issue of water scarcity, adopting a long-term and sustainable approach involves recycling and reprocessing waste and polluted water. In this monotonous progression, i.e., bioaugmentation, the major step is biodegradation of solid waste and pollutants through connecting microbes exist in activated sludge is single of the greatest imperative approaches to eliminate organic impurities on or after sewer water (Biswas and Tortajada 2018; Vanham et al. 2018). In bioaugmentation, the significant and imperative role is played by indigenous or previously existing microorganism in the polluted site, but additionally for the fast and plane degradation of pollutants or xenobiotic compounds, sewage water is augmented by adding special microorganisms for the backing deprivation process. This technique is very operative when the pre-existing microbes are unable to reduce the contaminants or xenobiotic compounds in the sewage water (Goswami et al. 2018).

Previously many workers reported the use of bioaugmentation for the removal of chlorinated and fluorinated compounds, e.g., bacteria *Comamonas testosterone* and *Acinetobacter* sp. was reported for the degradation of 3-chloroaniline and 4-fluoroaniline, respectively (Boon et al. 2000; Wang et al. 2013). Another class of compounds, i.e., lignin's are there which are harmful compounds produced by paper and pulp industries in this sequence consortia of lignocellulose degrading bacteria isolated from degraded bamboo slips was reported for cleaning of sludge waste water (Zheng et al. 2013). Some bacterial strains, i.e., *Paracoccus* sp., Shinella zoogloeoides, *Pseudomonas* sp., are reported for the removal contaminants (N-heterocyclic) like pyridine and quinoline from coking wastewater (Bai et al.

Table 2.1 Beneficial microbes and their role in agriculture sustainability

S. no.	Microorganisms	Plants/source	Application	References
1	*Curtobacterium citreum*	*Manihot esculenta*/Plant root endophytes	Endophytes have the potential to solubilize nitrogen while at the same time helping to increase the Cassava biomass.	Zhang et al. (2022)
2	*Funneliformis mosseae*	Red clover	AMF have effects that are synergistic on the shape of roots, the availability of nutrients in the soil, and plant development.	Yu et al. (2022)
3	*Pseudomonas* sp.	Soil bacteria	Four different *Pseudomonas* bacteria have the potential of phosphate solubilization, and phytase, phosphatase enzyme production.	Adhikari et al. (2021)
4	*Burkholderia contaminants*, and *Enterobacter* sp.	*Taxus wallichiana*/ Plant root endophytes	Both the endophytes have positive response on increasing the soil fertility and increasing the rice and black soybean growth and developments (Phosphate solubilization potential)	Adhikari and Pandey (2020)
5	*Glomus mosseae*	*Populus euphratica/ rhizosphere soil*	Root development was aided by fungi, which also altered root architecture and enhanced soil enzyme activity.	Pan et al. (2020)
6	*Azotobacter vinelandii*, *Bacillus megaterium*, and *B. subtilis*	–	Under alkaline circumstances, these three bacterial strains show promise for siderophore synthesis and chlorosis correction.	Ferreira et al. (2019)
7	*Aneurinibacillus aneurinilyticus* and *Paenibacillus* sp.	*Allium sativum*/ rhizosphere soil	ACC deaminase producing ability	Gupta and Pandey (2019)
8	*Aspergillus* sp. and *Penicillium* sp.	*Taxus wallichiana*/ Plant root endophytes	Endophytic fungi have potential of phosphate solubilization, and phytase, phosphatase enzyme production.	Adhikari and Pandey (2019)
9	*Bacillus*, *Caballeronia*, *Paenibacillus*, and *Pseudomonas*	Soil bacteria	Isolated bacteria have potential to increase the soil macro and micro nutrients, same time have ability of nitrogen fixation	Puri et al. (2017)

2010). *Shewanella* sp. was reported for the degradation or removal of azo-dyes, acid orange from aerated biofilm reactor (Wang et al. 2012a, b). Additional assemblage of frequent found contaminants in manufacturing factories wastewater are polycyclic aromatic hydrocarbons (PAHs). These PAHs are disobedient to biodegradation; therefore, they keep on extensive in the nature, with the connected significances on noxiousness to the nature e.g., naphthalene, one worker used strain of *Streptomyces* sp., for naphthalene degradation from coal gasification wastewater (Xu et al. 2015).

In another study consortia of five *Pseudomonas* sp. and *Paracoccus denitrificans* were reported for bioaugmentation and removal of PAHS, i.e., naphthalene, phenol, pyridine, quinoline, and carbazole from the coking wastewater (Zhu et al. 2015). One important point to noted out is that currently all these reported studies are lab scale procedures only. Consequently, full-scale wastewater treatment plant for contaminants and xenobiotic compound removal with the usage of biodegrading microbes are perquisite.

2.2.3 Role of Microbes in Solid Waste Managements

In current times, solid waste clearance and its management are serious problems, and they are rising rapidly with the increase in population. Incineration and open dumping are widespread practices for solid waste clearance in developing countries (Jain et al. 2021).According to the findings of Mani et al. (2020), global population growth has resulted in the production of around 13 billion tons of waste in 2011, with projections indicating that that amount may rise to 18 billion tons by the conclusion of 2020. The substances or wastes materials are available in the form of liquid, solid, and gaseous, and are usually discard as sewage sludge, household waste, electronic waste, industrial waste waters, and nuclear waste, etc. (Hassan et al. 2008; Angulo et al. 2012). Over a period of time after disposal these wastes transform in diverse toxic or hazardous forms and harmful to environment. Solid wastes consist of domestic wastes, sewage sludge, industrial waste, agricultural wastes, construction wastes, food wastes, mining wastes, petroleum wastes, etc. (Demirbas 2011). The modern era management practices of solid waste promote material reduction, recycling, stabilization of solid waste and energy recovery (Ionescu et al. 2015; Bong et al. 2017). Therefore, it becomes very crucial to reduce and eliminate the toxic and hazardous waste materials from the environment for the protection of the human health and ecosystem. However, for solid waste management several sustainable methods are used, viz., vermin composting, composting, sanitary land-filling, anaerobic digestions, and ethanol production (Jain et al. 2021). The utilization of solid waste management strategies aimed to mitigate the volume of solid waste materials being deposited in open landfills by enhancing the extraction of valuable resources and energy from these waste materials (Giusti 2009; Tchobanoglous 2009). The management and exploitation of degradable wastes include the implementation of three key strategies, sometimes referred to as the three Rs: reduction, recycling, and reuse. The commonly employed methods for managing solid waste include landfills, open dumping, burning, and usage of biological processes (Abdel-Shafy and Mansour 2018; Shekdar 2009). Physicochemical variables encompass a range of characteristics, including redox potential (Eh), pH, ionic strength, solubility, temperature, and half-life of organic compounds. The aforementioned criteria are of significant importance in the processes of deterioration. To reduce and solve the solid waste problems the microbes are widely used in different biological processes, for example, bioremediation, bioaugmentation, decomposition, aerobic decomposition, anaerobic decomposition, and recycling (Chang et al. 2017).

The majority of microorganisms, particularly bacteria and fungi, possess a remarkable capacity for adaptation in environments characterized by both aerobic and anaerobic conditions. They are capable of breaking down a wide range of manufactured chemicals as well as natural substances, making them very versatile in their degradation capabilities (Ellis and Wackett 2012).

Bioremediation procedures are commonly employed to address the removal, purification, or conversion of highly polluted or contaminated substances through the utilization of biological agents (Chaturvedi et al. 2015; Wang et al. 2012a, b). Aquatic ecosystem is one of the highly polluted ecosystems worldwide through any point or nonpoint source of pollution but discharge of such pollutants also effects aquatic ecosystem, terrestrial animals, and human beings (Barik et al. 2011). Hence, it is vital to address the treatment of effluents that are being released into the aquatic ecosystem. Microorganisms have been identified as highly sustainable and economically efficient catalysts for the bioremediation of hazardous chemicals or contaminants within ecosystems (Pillay 1992; Bharagava et al. 2017). In the bioremediation process, the efficacy of microbial decomposition is enhanced by the supplementation of external nutrients, including carbon and nitrogen sources, as well as substrates such as phenol. Additionally, electron acceptors such as oxygen, toluene, methane, and others have been shown to contribute to this process (Ma et al. 2007; Baldwin et al. 2008). The microorganisms often employed in bioremediation include *Arthrobacter, Alcaligenes, Achromobacter, Corynebacterium, Acinetobacter, Flavobacterium, Micrococcus, Mycobacterium, Nocardia, Pseudomonas, Sphingomonas, and Vibrio species* (Gupta et al. 2001; Kim and Aga 2007; Jayashree et al. 2012).

In a comparable manner, bioaugmentation refers to the utilization of microorganisms to breakdown recalcitrant molecules within polluted environments or ecosystems. This approach is both environmentally sustainable and economically advantageous when compared to physicochemical techniques (Boon et al. 2003; Semrany et al. 2012; Curiel-Alegre et al. 2022). Decomposition is a supplementary procedure involving the utilization of chemical and biological agents to facilitate the breakdown of intricate organic substances, ultimately transforming them into compost. During the process of decomposition, many microorganisms, including bacteria, fungus, actinomycetes, protozoa, and other saprophytic organisms, engage in the consumption of deceased solid waste materials, subsequently transforming them into composting substances (Pandey et al. 2018; Kallapiran and Kannan 2022). Generally, decomposition process is based on aerobic and anaerobic two different pathways where organism depends on availability and non-availability of oxygen to decompose the organic matters (Insam et al. 2010). During the recycling process, the breakdown of organic solid waste materials and the recycling of nutrients derived from different organic sources usually occurs by microorganisms. Subsequently, the aforementioned recycled nutrients are discharged into various ecosystems with the purpose of providing nourishment to algae or plants (Rebollido et al. 2008; Mani et al. 2020). *Alcanivorax borkumensis*, a microorganism known for its ability to consume hydrocarbons, is commonly employed in the remediation of oil spills occurring in deep-sea environments. This particular microbe utilizes enzymes and oxygen present in seawater to facilitate the breakdown of hydrocarbon ring structures within the oil (Warr et al. 2018).

Waste materials, which lack significance to both producers and consumers, are generated in vast quantities due to human activities. Consequently, the management of these waste materials has become a pressing concern in contemporary times, necessitating the utilization of conventional and biological approaches. Chemical and physical processes often exhibit detrimental effects due to their elevated expenses and maintenance requirements. Hence, biological methodologies are often favored due to their utilization of microorganisms and the environmentally sustainable outcomes they provide.

2.2.4 Role of Microbes in Remediation of Environmental Pollutants

Microorganisms are pervasive throughout the biosphere and are capable of thriving in diverse environmental conditions thanks to their metabolic prowess. Their remarkable nutritional adaptability also makes them valuable for biodegrading pollutants, a process known as bioremediation. Microorganisms possess the unique ability to transform and utilize toxic pollutants as their primary source of carbon and energy in metabolic processes (Tang et al. 2007). Microbial degradation involves the breakdown or transformation of contaminants into less or non-toxic forms. A wide array of microorganisms, including bacteria, archaea, and fungi, have been reported for their roles in bioremediation processes (Strong and Burgess 2008).

In the realm of environmental remediation, various methods have been proposed, with microbial approaches offering promising solutions. One such approach is the use of microbial fuel cells (MFCs), which exhibit efficiency in reducing resistance to metals. Bioremediation via biofilm is also employed to detoxify environmental pollutants present in polluted ecosystems. Microbial enzymes represent another avenue for breaking down complex molecules into simpler nutrients, mobilizing metal ions, promoting pollution uptake by plants, and enhancing overall bioremediation processes.

Due to the overuse of agricultural pesticides, mulching, and fertilizers, which mostly lead to heavy metal pollution in the soil (Wang et al. 2018), environmental pollution from new contaminants is a big problem that hurts crop yield and food quality. Currently, industrial effluents are a major source of environmental contamination, containing substantial quantities of persistent and non-degradable emerging contaminants (Iloms et al. 2020). While physical and chemical techniques for treating these pollutants are time-consuming, expensive, and often ineffective, the use of microbes for detoxifying environmental pollutants emerges as an efficient alternative (Conesa et al. 2012). This method is recognized as a viable means of pollution control and a novel approach for detoxifying environmental pollutants using biological systems (Wan et al. 2012).

Microorganisms also play a pivotal role in promoting plant growth and enhancing the absorption of toxic pollutants. Several hormonal processes in plants, including the production of siderophores, ACC, IAA, and deaminase, help the plants grow faster and have lower levels of ethylene. Among various types of pollutants, heavy metals tend to have limited mobility within the soil matrix, posing challenges for direct absorption by plant roots. The interaction between plant roots and soil microorganisms is essential for enhancing the absorption of metals by plant roots (Saravanan et al. 2007). Nevertheless, it is important to note that various environmental factors can influence the rate of biodegradation in bioremediation processes.

2.2.4.1 Factors Affecting Microbial Bioremediation

Microorganisms can effectively combat pollutants only when they have access to a variety of materials and compounds that enable them to generate energy and nutrients for cell growth. The rate of microbial degradation is influenced by several factors, including the chemical nature and concentration of pollutants (Uqab et al. 2015). One key reason for variations in degradation rates is the limited contact between affected bacteria and pollutants. Both pollutants and microorganisms are not uniformly distributed in the environment. Consequently, optimizing process parameters during degradation is crucial, owing to a multitude of factors (He et al. 2013; Dharni et al. 2014). These factors encompass the microbial population's capacity to degrade environmental pollutants, the accessibility of contaminants to the microbial population, and environmental conditions such as soil type, temperature, pH, and nutrient availability.

2.2.4.2 Physicochemical Factors Affecting Bioremediation

Physicochemical factors encompass a range of variables, including redox potential (Eh), pH, ionic strength, solubility, temperature, and the half-life of organic compounds. These characteristics play a critical role in degradation processes. Biodegradation techniques involve the conversion of metal ions into insoluble forms by microorganisms, transitioning them from a higher to a lower oxidation state. This process is primarily pH-dependent and can only occur through enzymatic reduction. The solubility of metals or emerging contaminants is contingent on the higher oxidation states of the metals; when metals have a higher oxidation state, their solubility is correspondingly higher (Wołejko et al. 2020). Solubility is an important factor in the breakdown of new pollutants because compounds that are hydrophobic or sparsely soluble stay in the environment for a long time and are not bioavailable (Babu et al. 2015).

2.2.4.3 Biological Factors Affecting Bioremediation Processes

Biological factors may not always be readily apparent, but their significance becomes evident when implementing these techniques in pilot studies for biodegradation processes. Several inherent characteristics of microorganisms undergo alterations as substrate degradation takes place. For instance, plasmid-encoded genes confer specificity for particular substrates and encode the corresponding enzymes (proteins). The nature of microbial cells has diversified in terms of specificity for various substrates. In the context of microbial recalcitrance, bacterial chemotaxis stands out as an advantageous behavior Ingham et al. (2007) during biodegradation processes.

2.3 Role of Microbes in Detoxification of Heavy Metals

The ubiquitous presence of microbes in nature plays a paramount role in addressing heavy metal pollution, enabling them to crucially contribute to the conversion of toxic heavy metals into non-toxic forms. Microbes employ several processes to mineralize organic pollutants, transforming end-products into either CO_2 and H_2O or non-toxic intermediates that serve as primary substrates for cell growth during bioremediation. Microorganisms maintain a dual defense mechanism: (i) they produce enzymes responsible for breaking down the target contaminants, and (ii) they resist relevant heavy metals. In the biodegradation process (as depicted in Fig. 2.1), microorganisms employ various methods, including biodeterioration, assimilation, bio-fragmentation, and mineralization. The microbial approach enhances the efficiency of bioremediation processes by comprehending the mechanisms governing microbial growth in polluted sites, their metabolic potential, and their response to modifying environmental parameters (Álvarez-Barragán et al. 2016).

Microorganisms play a pivotal role in the reduction and oxidation of transition metals rather than their dissolution. Their tolerance to contaminants also holds significance in bioremediation processes. Elevated compound concentrations can disrupt microbial cell membranes, resulting in microorganism mortality. However, some microorganisms activate defense mechanisms through the production of organic solvents, with hydrophobic or solvent efflux pumps forming an outer protective cell membrane material in such cases (Dixit et al. 2015).

Bioremediation, dependent on microbes, hinges on biologically encoded oxidation state modifications that are instrumental in remediating environmental sediments, soil, and water contaminated with heavy metals. While some microorganisms cannot convert toxic inorganic compounds like heavy metals into non-toxic forms, they are employed based on their specificity toward particular contaminants. Consequently, the bioremediation of heavy metals relies on actively metabolizing microbes. Utilizing microorganisms for contaminant removal emerges as the most effective, straightforward, and safest technology. Anthropogenic activities, including fuel combustion, industrial processes, ore mining, organic solvent use, gas

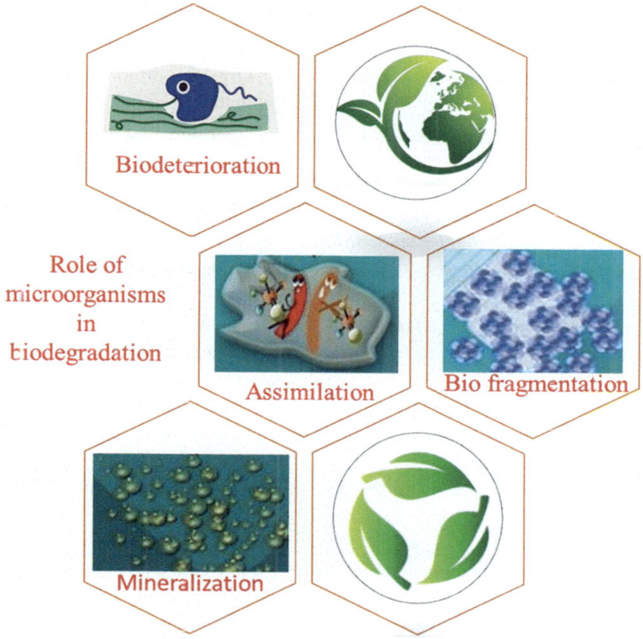

Fig. 2.1 Different processes followed by the bacteria during biodegradation

extraction, pigment production, pesticide application, and excessive plastic consumption, have introduced various toxic contaminants into the environment, necessitating detoxification or complete removal through native microorganisms (Garbisu and Alkorta 2001).

2.4 Role of Microbes in Heavy Metals Phytoremediation

Endophytic microbes are known to colonize the internal tissues of plants, namely dwelling in the regions underneath the layers of epidermal cells. The bacteria inside these interior plant tissues exhibit adaptability to different host lifestyles, as seen in Table 2.2. The relationship between endophytes and plants covers a range of interactions, including symbiotic, communalistic, mutualistic, and trophobiotic relationships (Schulz and Boyle 2006). Many different kinds of plants have arbuscular mycorrhizal fungi (AMF), which are mostly found in the roots and become less common as they grow toward the leaves, as shown by Perpetuo et al. (2011). Many epiphytic bacterial communities live with endophytes in different parts of plants, like the rhizosphere and phyllosphere. However, these communities can sometimes be moved from seeds or damaged foliar tissues (Bacon et al. 2012). The enduring mutualistic association between plants and endophytic bacteria has resulted in a

Table 2.2 Endophytes and their host plants for the degradation of heavy metals

S.no.	Host plant	Microorganisms	Target compound	References
1	*Lupinus luteus*	*Consortia of bacteria*	Organic pollutants and metals	Leroy et al. (2017)
2	*Brassica napus*	*Alternaria* sp. *CBSF68, Penicillium* sp. *CBRF65, Fusarium* sp. *CBRF44,*	Cadmium and lead	Chen et al. (2016)
3	*Brassica juncea*	*Genetically engineered Enterobacter* sp. *CBSB1*	Cadmium and lead	Persoh (2015)
4	*Oryza sativa L.*	*Phomopsis liquidambar*	Phenanthrene, phenolic acids	Wang et al. (2017)
5	*Brassica campestris L.*	*Mucor* sp. *MHR-7*	The elements zinc (Zn), manganese (Mn), chromium (Cr), copper (Cu), and cobalt (Co)	Wang et al. (2018)
6	*Trifolium arvense*	*Pseudomonas azotoformans ASS1*	Copper (Cu), zinc (Zn), and nickel (Ni)	Toju et al. (2018)
7	*Leptochloa fusca (L.) Kunth*	*Enterobacter* sp. *HU38,*	The elements uranium (U) and lead (Pb)	Lemaire et al. (2012)
8	*Alnus firma*	*Bacillus thuringiensis GDB-1*	The elements copper (Cu), cadmium (Cd), nickel (Ni), lead (Pb), and zinc (Zn)	Babu et al. (2013)
9	*Pelargonium graveolens*	*P. monteilii PsF84, P. plecoglossicida PsF610*	Cr	Dharni et al. (2014)
10	*Polygonum pubescens*	*Rahnella* sp. *JN6*	Cd, Pb and Zn	He et al. (2013)
11	*Lupinus luteus L*	*Burkholderia cepacia L.S.2.4, Herbaspirillum seropedicae LMG2284*	The elements copper (Cu), cadmium (Cd), cobalt (Co), nickel (Ni), lead (Pb), and zinc (Zn).	Lodewyckx et al. (2001)
12	*Alyssum serpyllifolium*	*Pseudomonas* sp. *A3R3*	Ni	Ma et al. (2011)
13	*Noccaea caerulescens*	*Microbacterium* sp. *NCr-8, Arthrobacter* sp. *NCr-1, Bacillus* sp. *NCr-5, Bacillus* sp. *NCr-9 and Kocuria* sp. *NCr-3*	Ni	Visioli et al. (2012)
14	*Solanum nigrum L*	*Serratia nematodiphila LRE07*	Cd	Wan et al. (2012)
15	*Barley*	*P. indica*	AS	Baghaie and Jabari (2019)

close-knit ecology that facilitates plants' ability to acclimate to and endure unfavorable circumstances (Ryan et al. 2008). Endophytic microorganisms, particularly in the setting of metal stress, have developed several methods to mitigate the toxicity of metal ions by converting them into less complex forms. The mechanisms described in the study by Rajkumar et al. (2013) involve the expulsion of metal ions from the cell, the conversion of metal ions into less harmful forms, the containment of metals on the surface of the cell or within internal polymers, and various processes such as precipitation, adsorption/desorption, or biomethylation. Hyperaccumulator plants, which possess the ability to collect significant quantities of metals, have provided evidence that the introduction of metal-resistant endophytic bacteria into soils, seeds, or seedlings can augment plant development and accelerate the process of phytoremediation in soil polluted with metals. The enhancement mentioned is achieved by many mechanisms, such as enhanced nutrient acquisition, elongation of cells, accumulation or stability of metals, and mitigation of metal stress in plants (Luo et al. 2011; Wei et al. 2014; Zhu et al. 2014; Ma et al. 2015; Chen et al. 2016). In addition, the proliferation and establishment of microorganisms that facilitate the development of plants are frequently denoted as plant growth-promoting endophytic bacteria (PGPE). People know that these bacteria can make the soil more fertile and help host plants grow by making growth regulators (Phetcharat and Duangpaeng 2012) and vital nutrients that plants can use (Persoh 2015). Adding endophytic microbes that help plants grow in many ways, like those that can detoxify and change things, to bioaugmentation strategies might help plants deal with the bad effects of metals and change how easy it is for metals to get into polluted soils. According to the previous research by Weyens et al. (2009) and Ma et al. (2011), the aforementioned methodology offers an ideal choice for investigations involving the use of microorganisms in the field of phytoremediation.

2.5 Conclusion and Future Perspective

Environment or nature or we can say our surrounding plays a very crucial role in our mental and physical health. As day-by-day worldwide population is increasing, simultaneously pressure on nature is also increasing, i.e., more and extra pressure is applying on our industrialization and urbanization, which is leading numerous environmental issues. These issues are countless agrochemicals, seepages and drain from industries and automobiles, which are polluting our significant resources, i.e., soil, water, and atmosphere. Directly or indirectly through these xenobiotic compounds and pollutants are introducing in our food chains and ecosystems, and simultaneously creating insalubrious circumstances for social lifespan. This chapter overviewed approaches to re-establish and withstand the environment with the usage of microorganisms and their role for spot cleansing. Through all the compiled data we can established that microorganisms are operative and real in the deprivation of agrochemicals, industrial effluents, and other xenobiotic compounds. Microbes have inordinate possibilities to purify or clean contaminated spots though their imperative part in the deprivation of organic impurities and reclamation of inorganic mixtures. One of the best approaches for removal of pollutants is bioaugmentation, it is a striking approach for the elimination of recalcitrant contaminants and impurities from wastewater. This approach has showed positive results in the lab scale inquiries, but the major research is still needed, particularly for scaling up the bioaugmentation methods in large scale. In the future research more focus is needed to enhance the biological activity of the microbes for the sustainable solutions for the environmental issues. Additionally, there is a necessity to examine site-specific bacterial groups further down a widespread series of environmental conditions. Alternative extent of interest is the preparation and construction of appropriate inoculums and the challenging and examine their ecological influence. If we talk about the beneficial microbe's individual, then only 2–4% has been recognized, while the massive commodity of beneficial microbes and their probable extant yet to be explored.

Acknowledgement Authors are grateful to Centre for GMP Extraction Facility (Department of Biotechnology) NIPER Guwahati, Assam-India, and NMHS (National Mission on Himalayan Studies) Ministry of Environment, Forest & Climate Change, Govt. of India, New Delhi for support.

References

Abdel-Shafy HI, Mansour MS (2018) Solid waste issue: sources, composition, disposal, recycling, and valorization. Egypt J Pet 27(4):1275–1290

Adhikari P, Jain R, Sharma A, Pandey V (2021) Plant growth promotion at low temperature by phosphate solubilizing *Pseudomonas* spp. isolated from high-altitude Himalayan soil. Microb Ecol 82:677–687

Adhikari P, Pandey A (2019) Phosphate solubilization potential of endophytic fungi isolated from *Taxus wallichiana* Zucc. roots. Rhizosphere 9:2–9

Adhikari P, Pandey A (2020) Bioprospecting plant growth promoting activities of endophytic bacteria isolated from Himalayan yew (*Taxus wallichiana* Zucc.). Microbiol Res 239(2):126536

Ahmad M, Pataczek L, Hilger TH, Zahir ZA, Hussain A, Rasche F, Schafleitner R, Solberg S (2018) Perspectives of microbial inoculation for sustainable development and environmental management. Front Microbiol 9:2992

Álvarez-Barragán J, Domínguez-Malfavón L, Vargas-Suárez M, González-Hernández R, Aguilar-Osorio G, Loza-Tavera H (2016) Biodegradative activities of selected environmental fungi on a polyester polyurethane varnish and polyether polyurethane foams. Appl Environ Microbiol 82:5225–5235

Angulo J, Mahecha L, Yepes A, Yepes AM, Bustamante G, Jaramillo H, Valencia E, Villamil T, Gallo J (2012) Quantitative and nutritional characterization of fruit and vegetable waste from marketplace: a potential use as bovine feedstuff? J Environ Manage 95:S203–S209

Babu AG, Kim JD, Oh BT (2013) Enhancement of heavy metal phytoremediation by Alnus firma with endophytic Bacillus thuringiensis GDB-1. J hazard mater 250:477–483

Babu AG, Shea PJ, Sudhakar D, Jung IB, Oh BT (2015) Potential use of Pseudomonas koreensis AGB-1 in association with Miscanthus sinensis to remediate heavy metal(loid)-contaminated mining site soil. J Environ Manag 151:160–166

Bacon CW, Hinton DM, Mitchell TR, Snook ME, Olubajo B (2012) Characterization of endophytic strains of *Bacillus mojavensis* and their production of surfactin isomers. Biol Control 62:1–9

Baghaie AH, Jabari AG (2019) Effect of Nano Fe-oxide and endophytic fungus (*P. indica*) on petroleum hydrocarbons degradation in an arsenic contaminated soil under barley cultivation 05 environmental sciences 0503 soil sciences. J Environ Health Sci Eng 17:853–861

Bai Y, Sun Q, Zhao C, Wen D, Tang X (2010) Bioaugmentation treatment for coking wastewater containing pyridine and quinoline in a sequencing batch reactor. Appl Microbiol Biotechnol 87:1943–1951

Baldwin AN, Shen LY, Poon CS, Austin SA, Wong I (2008) Modeling design information to evaluate pre-fabricated and pre-cast design solutions for reducing construction waste in high rise residential buildings. Autom constr 17(3):333–341

Barik P, Vardia HK, Gupta SB (2011) Bioremediation of ammonia and nitrite in polluted water. Int J Fish Aquac 3(7):135–141

Bharagava RN, Chowdhary P, Saxena G (2017) Bioremediation an eco-sustainable green technology, its applications and limitations. In: Bharagava RN (ed) Environmental pollutants and their bioremediation approaches. CRC Press, Taylor & Francis Group, USA, pp 1–22

Biswas AK, Tortajada C (2018) In: Biswas AK, Tortajada C, Rohner P (eds) Assessing global water megatrends assessing global water megatrends. Springer Singapore, Singapore, pp 1–26

Bong CPC, Ho WS, Hashim H, Lim JS, Ho CS, Tan WSP, Lee CT (2017) Review on the renewable energy and solid waste management policies towards biogas development in Malaysia. Renew Sustain Energy Rev 70:988–998

Boon N, Goris J, De-Vos P, Verstraete W, Top EM (2000) Bioaugmentation of activated sludge by an indigenous 3-chloroaniline-degrading *Comamonas testosteroni* strain, I2gfp. Appl Environ Microbiol 66:2906–2913

Boon N, Top EM, Verstraete W, Siciliano SD (2003) Bioaugmentation as a tool to protect the structure and function of an activated-sludge microbial community against a 3-chloroaniline shock load. Appl Environ Microbiol 69(3):1511–1520

Chang CH, Yang HY, Hung JM, Lu CJ, Liu MH (2017) Simulation of combined anaerobic/aerobic bioremediation of tetrachloroethylene in groundwater by a column system. Int Biodeterior Biodegradation 117:150–157

Chaturvedi AD, Pal D, Penta S, Kumar A (2015) Ecotoxic heavy metals transformation by bacteria and fungi in aquatic ecosystem. World J Microbiol Biotechnol 31(10):1595–1603

Chen F, Ren CG, Zhou T, Wei YJ, Dai CC (2016) A novel exopolysaccharide elicitor from endophytic fungus *Gilmaniella* sp. AL12 on volatile oils accumulation in *Atractylodes lancea*. Sci Rep 6:1–17

Conesa HM, Evangelou MWH, Robinson BH, Schulin R (2012) A critical view of current state of phytotechnologies to remediate soils: still a promising tool? Sci World J 2012:173829

Curiel-Alegre S, Velasco-Arroyo B, Rumbo C, Khan AHA, Tamayo-Ramos JA, Rad C, Gallego JLR, Barros R (2022) Evaluation of biostimulation, bioaugmentation, and organic amendments application on the bioremediation of recalcitrant hydrocarbons of soil. Chemosphere 307:135638

Demirbas A (2011) Waste management, waste resource facilities and waste conversion processes. Energy Conv Manag 52(2):1280–1287

Dharni S, Srivastava AK, Samad A, Patra DD (2014) Impact of plant growth promoting *Pseudomonas monteilii* PsF84 and *Pseudomonas plecoglossicida* PsF610 on metal uptake and production of secondary metabolite (monoterpenes) by rose-scented geranium (*Pelargonium graveolens* cv. bourbon) grown on tannery sludge amended soil. Chemosphere 117:433–439

Dixit R, Wasiullah MD, Pandiyan K, Singh UB, Sahu A, Shukla R, Singh BP, Rai JP, Sharma PK, Lade H, Paul D (2015) Bioremediation of heavy metals from soil and aquatic environment: an overview of principles and criteria of fundamental processes. Sustain For 7:2189–2212

Ellis EC, Klein GK, Siebert S, Lightman D, Ramankutty N (2010) Anthropogenic transformation of the biomes, 1700 to 2000. Glob Ecol Biogeogr 19:589–606

Ellis LB, Wackett LP (2012) Use of the university of Minnesota biocatalysis/biodegradation database for study of microbial degradation. Microb Inform Exp 2:1–10

Ferreira CMH, Vilas-Boas A, Sousa CA, Soares HMVM, Soares EV (2019) Comparison of five bacterial strains producing siderophores with ability to chelate iron under alkaline conditions. AMB Express 9:78

Ferronato N, Torretta V (2019) Waste mismanagement in developing countries: a review of global issues. Int J Environ Res Public Health 16(6):1060

Garbisu C, Alkorta I (2001) Phytoextraction: a cost-effective plant-based technology for the removal of metals from the environment. Bioresour Technol 77:229–236

Giusti L (2009) A review of waste management practices and their impact on human health. Waste Manag 29(8):2227–2239

Goswami M, Chakraborty P, Mukherjee K, Mitra G, Bhattacharya P, Dey S, Tribedi P (2018) Bioaugmentation and biostimulation: a potential strategy for environmental remediation. J Microbiol Exp 6(5):223–231

Gupta S, Pandey S (2019) ACC deaminase producing bacteria with multifarious plant growth promoting traits alleviates salinity stress in french bean (*Phaseolus vulgaris*) plants. Front Microbiol 10:1506

Gupta VK, Shrivastava AK, Jain N (2001) Biosorption of chromium (VI) from aqueous solutions by green algae Spirogyra species. Water Res 35:4079–4085

Hassan MM, Ahmed SA, Rahman KA, Biswas TK (2008) Pattern of medical waste management: existing scenario in Dhaka City, Bangladesh. BMC public health 8:1–10

He H, Ye Z, Yang D, Yan J, Xiao L, Zhong T, Yuan M, Cai X, Fang Z, Jing Y (2013) Characterization of endophytic *Rahnella* sp. JN6 from *Polygonum pubescens* and its potential in promoting growth and Cd, Pb, Zn uptake by Brassica napus. Chemosphere 90:1960–1965

Iloms E, Ololade OO, Ogola HJO, Selvarajan R (2020) Investigating industrial effluent impact on municipal wastewater treatment plant in vaal, South Africa. Int J Environ Res Public Health 17:109

Ingham CJ, Sprenkels A, Bomer J, Molenaar D, van den Berg A, van Hylckama Vlieg JET, de Vos WM (2007) The micro-Petri dish, a million-well growth chip for the culture and high-troughput screening of microorganisms. Proc Natl Acad Sci USA 104:18217–18222

Insam H, Franke-Whittle I, Goberna M (2010) Microbes in aerobic and anaerobic waste treatment. Microbes at Work: From Wastes to Resources, 1–34

Ionescu G, Rada EC, Cioca LI (2015) Municipal solid waste sorting and treatment schemes for the maximization of material and energy recovery in a latest EU member. Environ Eng Manag J 14(11)

Jain R, Pattanaik L, Padhi SK, Naik SN (2021) Role of microbes and microbial consortium in solid waste management. Environ Agric Microbiol Appl Sustain:383–422

Jayashree R, Nithy SE, Rajesh PP, Krishnaraju M (2012) Biodegradation capability of bacterial species isolated from oil contaminated soil. J Academia Indust Res 1(3):127–135

Kallapiran KA, Kannan R (2022) A review on the agricultural waste residues management by different microbes. J Agric Ecol Res Int:93–113

Kim S, Aga DS (2007) Potential ecological and human health impacts of antibiotics and antibiotic-resistant bacteria from wastewater treatment plants. J Toxicol Environ Health Part B10(8):559–573

Lemaire B, Van Oevelen S, De Block P, Verstraete B, Smets E, Prinsen E, Dessein S (2012) Identification of the bacterial endosymbionts in leaf nodules of Pavetta (Rubiaceae). Int J Syst Evol Microbiol. 62:202–209

Leroy C, Jauneau A, Martinez Y, Cabin-Flaman A, Gibouin D, Orivel J, Sejalon-Delmas N (2017) Exploring fungus-plant N transfer in a tripartite ant-plant-fungus mutualism. Ann Bot. 120(3):417–426

Liu J, Yang H, Gosling SN, Kummu M, Florke M, Pfister S, Hanasaki N, Wada Y, Zhnag X, Zheng C, Alcamo J, Oki T (2017) Water scarcity assessments in the past, present and future. Earth's Future 5:545–559

Lodewyckx C, Taghavi S, Mergeay M, Vangronsveld J, Clijsters H, Lelie DV (2001) The effect of recombinant heavy metal-resistant endophytic bacteria on heavy metal uptake by their host plant. Int J Phytoremediat. 3(2):173–187

Luo S, Wan Y, Xiao X, Guo H, Chen L, Xi Q, Zeng G, Liu C, Chen J (2011) Isolation and characterization of endophytic bacterium LRE07 from cadmium hyperaccumulator *Solanum nigrum* L. and its potential for remediation. Appl Microbiol Biotechnol 89:1637–1644

Ma X, Novak PJ, Ferguson J, Sadowsky M, LaPara TM, Semmens MJ, Hozalski RM (2007) The impact of H2 addition on dechlorinating microbial communities. Bioremediat J 11(2):45–55

Ma Y, Oliveira RS, Nai F, Rajkumar M, Luo Y, Rocha I, Freitas H (2015) The hyperaccumulator Sedum plumbizincicola harbors metal-resistant endophytic bacteria that improve its phytoextraction capacity in multi-metal contaminated soil. J Environ Manag 155:62–69

Ma Y, Prasad MNV, Rajkumar M, Freitas H (2011) Plant growth promoting rhizobacteria and endophytes accelerate phytoremediation of metalliferous soils. Biotechnol Adv 29:248–258

Mani S, Chowdhary P, Zainith S (2020) Microbes mediated approaches for environmental waste management. In: Microorganisms for sustainable environment and health. Elsevier, pp 17–36

Pan J, Huang C, Peng F, Zhang W, Luo J, Ma S, Xue X (2020) Effect of arbuscular mycorrhizal fungi (AMF) and plant growth promoting bacteria (PGPR) inoculations on *Elaeagnus angustifolia* L. in saline soil. Appl Sci 10(3):945

Pandey SN, Abid M, Abid AKMM (2018) Diversity, functions, and stress responses of soil microorganisms. Plant Microbiome: Stress Response, 1–19

Perpetuo EA, Souza CB, Nascimento CAO, Perpetuo EA, Souza CB, Nascimento CAO (2011) Engineering bacteria for bioremediation. In: Carpi A (ed) Progress in molecular and environmental bioengineering—from analysis and modeling to technology applications, pp 605–632

Persoh D (2015) Plant-associated fungal communities in the light of meta'omics. Fungal Divers 75:1–25

Phetcharat P, Duangpaeng A (2012) Screening of endophytic bacteria from organic rice tissue for indole acetic acid production. Procedia Eng 32:177–183

Pillay TVR (1992) Aquaculture and the environment. Fishing New Books, England

Puri A, Padda KP, Chanway CP (2017) Plant Growth Promotion by Endophytic Bacteria in Nonnative Crop Hosts. In: Maheshwari D, Annapurna K (eds) Endophytes: Crop Productivity and Protection. Sustainable Development and Biodiversity, vol 16. Springer, Cham. https://doi.org/10.1007/978-3-319-66544-3_2

Rajkumar M, Prasad MNV, Swaminathan S, Freitas H (2013) Climate change driven plant–metal–microbe interactions. Environ Int 53:74–86

Rebollido ROCIO, Martinez JORGE, Aguilera YURI, Melchor KENIA, Koerner I, Stegmann RAIGNER (2008) Microbial populations during composting process of organic fraction of municipal solid waste. Appl Ecol Environ Res 6(3):61–67

Ryan RP, Germaine K, Franks A, Ryan DJ, Dowling DN (2008) Bacterial endophytes: recent developments and applications. FEMS Microbiol Lett 278:1–9

Saravanan VS, Madhaiyan M, Thangaraju M (2007) Solubilization of zinc compounds by the diazotrophic, plant growth promoting bacterium *Gluconacetobacter diazotrophicus*. Chemosphere 66:1794–1798

Schulz B, Boyle C (2006) What are endophytes? In: Schulz BJE, Boyle CJC, Sieber TN (eds) Microbial root endophytes, pp 1–13

Semrany S, Favier L, Djelal H, Taha S, Amrane A (2012) Bioaugmentation: possible solution in the treatment of bio-refractory organic compounds (Bio-ROCs). Biochem Eng J 69:75–86

Shekdar AV (2009) Sustainable solid waste management: an integrated approach for Asian countries. Waste Manag 29(4):1438–1448

Shen FT, Yen JH, Liao CS, Chen WC, Chao YT (2019) Screening of rice endophytic biofertilizers with fungicide tolerance and plant growth promoting characteristics. Sustain For 11(4):1133

Strong PJ, Burgess JE (2008) Treatment methods for wine-related and distillery wastewaters: *a review*. Biorem J 12:70–87

Sudheer S, Bai RG, Usmani Z, Sharma M (2020) Insights on engineered microbes in sustainable agriculture: biotechnological developments and future prospects. Curr Genomics 21(5):321–333

Tang CY, Fu QS, Criddle CS, Leckie JO (2007) Effect of flux (transmembrane pressure) and membrane properties on fouling and rejection of reverse osmosis and nanofiltration membranes treating perfluorooctane sulfonate containing wastewater. Environ Sci Technol 41:2008–2014

Tchobanoglous G (2009) Solid waste management. Environmental engineering: environmental health and safety for municipal infrastructure, land use and planning, and industry. Wiley, New Jersey, pp 177–307

Timmusk S, Behers L, Muthoni J, Muraya A, Aronsson AC (2017) Perspectives and challenges of microbial application for crop improvement. Front Plant Sci 8:49

Toju H, Peay KG, Yamamichi M, Narisawa K, Hiruma K, Naito K, Fukuda S, Ushio M, Nakaoka S, Onoda Y, Yoshida K (2018) Core microbiomes for sustainable agroecosystems. Nat plant. 4(5):247–257

Uqab A, Syeed M, Saleemm F, Ruqeya N (2015) Factors affecting bioremediation. J Res Dev 15:102–109

Vanham D, Hoekstra AY, Wada Y, Bouraoui F, Roo D, Mokennen MM, Bund WJVD, Batelaan O, Pavelic P, Bastiaanssen WGM, Kummu M, Rockstrom J, Liu J, Bisselink B, Pistocchi A, Bidoglio G (2018) Physical water scarcity metrics for monitoring progress towards sdg target 6.4: an evaluation of indicator 6.4.2 "level of water stress". Sci Total Environ 612:218–232

Visioli G, D'Egidio S, Vamerali T, Mattarozzi M, Sanangelantoni AM (2012) Culturable endophytic bacteria enhance Ni translocation in the hyperaccumulator Noccaea caerulescens. Chemosphere. 1(117):538–544

Vliet MTH, Jones ER, Florke M, Franssen WHP, Hanasaki N, Wada Y, Yearsley JR (2021) Global water scarcity including surface water quality and expansions of clean water technologies. Environ Res Lett 16:024020

Wan Y, Luo S, Chen J, Xiao X, Chen L, Zeng G, Liu C, He Y (2012) Effect of endophyte-infection on growth parameters and Cd-induced phytotoxicity of Cd-hyperaccumulator *Solanum nigrum* L. Chemosphere 89:743–750

Wang J, Liu GF, Lu H, Jin RF, Zhou JT, Lei TM (2012a) Biodegradation of acid orange 7 and its auto-oxidative decolorization product in membrane-aerated biofilm reactor. Int Biodeterior Biodegrad 67:73–77

Wang J, Liu XD, Lu J (2012b) Urban river pollution control and remediation. Procedia Environ Sci 13:1856–1862

Wang L, Ji B, Hu Y, Liu R, Sun W (2017) A review on in situ phytoremediation of mine tailings. Chemosphere. 184:594–600

Wang LP, Jia ZB, Liu Y, Gao Q, Cheng SJ, Jin D, Ma L, Yin XH (2018) Inhibitory effect of wild-type P53 gene transfer on graft coronary artery disease. Transpl Immunol 48:1–9

Wang M, Xu J, Wang J, Wang S, Feng H, Shentu J, Shen D (2013) Differences between 4-fluoroaniline degradation and autoinducer release by *Acinetobacter* sp. TW: implications for operating conditions in bacterial bioaugmentation. Environ Sci Pollut Res Int 20:6201–6209

Warr LN, Schlüter M, Schauer F, Olson GM, Basirico LM, Portier RJ (2018) Nontronite-enhanced biodegradation of Deepwater Horizon crude oil by *Alcanivorax borkumensis*. Appl Clay Sci 158:11–20

Wei Y, Hou H, ShangGuan YX, Li JN, Li FS (2014) Genetic diversity of endophytic bacteria of the manganese-hyperaccumulating plant Phytolacca americana growing at a manganese mine. Eur J Soil Biol 62:15–21

Weyens N, van der Lelie D, Taghavi S, Newman L, Vangronsveld J (2009) Exploiting plant–microbe partnerships to improve biomass production and remediation. Trends Biotechnol 27:591–598

Wołejko E, Jabłońska-Trypuć A, Wydro U, Butarewicz A, Łozowicka B (2020) Soil biological activity as an indicator of soil pollution with pesticides—a review. Appl Soil Ecol 147:103356

Xu P, Ma W, Han H, Jia S, Hou B (2015) Isolation of a naphthalene-degrading strain from activated sludge and bioaugmentation with it in a MBR treating coal gasification wastewater. Bull Environ Contam Toxicol 94:358–364

Yu L, Zhang H, Zhang W, Liu K, Liu M, Shao X (2022) Cooperation between arbuscular mycorrhizal fungi and plant growth-promoting bacteria and their effects on plant growth and soil quality. Peer J 10:e13080

Zhang X, Tong J, Dong M, Akhtar K, He B (2022) Isolation, identification and characterization of nitrogen fixing endophytic bacteria and their effects on *cassava* production. Peer J 10:e12677

Zheng Y, Chai LY, Yang ZH, Tang CJ, Chen YH, Shi Y (2013) Enhanced remediation of black liquor by activated sludge bioaugmented with a novel exogenous microorganism culture. Appl Microbiol Biotechnol 97:6525–6535

Zhu LJ, Guan DX, Luo J, Rathinasabapathi B, Ma LQ (2014) Characterization of arsenic-resistant endophytic bacteria from hyperaccumulators Pteris vittata and Pteris multifida. Chemosphere 113:9–16

Zhu X, Liu R, Liu C, Chen L (2015) Bioaugmentation with isolated strains for the removal of toxic and refractory organics from coking wastewater in a membrane bioreactor. Biodegradation 26:465–474

Chapter 3
Microbial Technology: Tools for Waste Management; Environmental Sustainability and Environmental Safety

Anis Mirza, Arun Kumar, Gursharan Singh, Shailendra Kumar Arya, Aditya Bhalla, and Jatinder Singh

Abstract In the current scenario, land, water, and air are highly polluted with diverse types of the contaminants that deteriorate the human and animal health directly or indirectly. Although most of the pollution is self-generated by mankind due to the establishment of various industries without appropriate norms, unmanaged waste of livestock, municipal solid waste, and excess use of chemicals in agriculture and daily life. The best method to rid off from all types of the pollution and harmful effects are avoid or imposing ban on such type of industrial units and activities. But in modern era it's not possible to because of always scare of sudden increase in unemployment and huge economical loss. Second alternate is taking advantage of highly efficient diverse types of the microbes such as prokaryotes and eukaryotes. They are excellent bio-degraders of organic waste and recycled the waste to further value-added products. In literature there are so many reports available in which researchers also bio-remediate the soils from high contaminants like heavy metals and toxic pesticides. Additionally, the use of genetically modified microorganisms (GMOs) can help in the conversion of highly toxic chemicals to less harmful. This chapter provides the proper updated information on the state of the art of different waste types and shows how the microorganisms can be effectively used for the waste management (WM) and more environmentally sustainable

---

A. Mirza · A. Kumar · J. Singh (✉)
Department of Horticulture, SAGR, Lovely Professional University, Phagwara, Punjab, India
e-mail: jatinder.19305@lpu.co.in

G. Singh
Department of Medical Laboratory Sciences, Lovely Professional University, Phagwara, Punjab, India

S. K. Arya
Department of Biotechnology, University Institute of Engineering and Technology, Punjab University (UIET), Chandigarh, India

A. Bhalla
Department of Biochemistry and Molecular Biology, Michigan State University, East Lansing, MI, USA

A. Karnwal, A. R. Mohammad Said Al-Tawaha (eds.), *Microbial Applications for Environmental Sustainability*, https://doi.org/10.1007/978-981-97-0676-1_3

practices. It also confers the interpretation to the readers about the areas of highly explored and the field where there is still much left to be done.

Keywords Pollution · Contaminants · Industrial pollution · Waste management · Microbes · Biodegradation · Genetically modified microorganisms (GMOs) · Soil remediation · Environmental sustainability · Pollution control

3.1 Introduction

The development of microbiology as a science has impacted the history of human adoption of biotechnological methods. Ancient civilizations like those in China and Mesopotamia utilized microbes for fermentation and alcohol production, as evidenced by archeological findings. Genetic data from the Neolithic hamlet of Jiahu in China and the Zagros Mountains in northern Mesopotamia support the use of microbes for cereal grain fermentation. Discoveries of tartaric acid in antique jars and grape juice remnants further indicate the early production of wine. These techniques enabled large-scale production of wine around 5000 BC. In Egypt, yeast was employed for beer production, and samples dating from 2000 to 1200 BC have been found. These historical practices laid the foundation for Louis Pasteur's later research on fermentation and the connection between microbes and diseases. Pasteur's discoveries not only advanced microbiology but also led to the development of industrial processes based on microorganisms. The rise of technological microbiology occurred during the post-World War II economic boom, driven by a growing understanding of microbial genetics. The authorization of patenting microorganisms for pollutant digestion further propelled the field in the 1980s. Today, the need for environmental protection and sustainability is paramount, with improper waste management posing significant threats to ecosystems and natural resources. Microorganisms offer potential solutions for waste biodegradation and bioremediation, including the innovative nano-bioremediation technique for radioactive waste cleanup. Genetically modified organisms also hold promise for addressing pollution challenges. This chapter explores various waste types and highlights the effective use of microorganisms for waste management and environmentally sustainable practices.

3.2 Sustainable Waste Management

At present, the waste materials produced by humanity are having a significant impact on the land we inhabit, the water we consume, and the air we breathe. According to the United Nations, people are responsible for a large portion of the approximately 11.2 billion tonnes of solid waste generated globally. However, outdated methods of waste disposal, such as burning, flushing, or discarding, are no longer sustainable practices for our civilization. Hawken et al. (1999) reported that the USA alone produces over 115 billion tonnes of waste annually, with 80% of that being wastewater. Shockingly, less than 2% of this waste is recycled. The consequences of such practices are evident, with over 45% of Americans residing in areas with occasional hazardous air quality due to high levels of air pollutants (USEPA 2002), and approximately 40% of all US waters being too polluted to fulfill their intended functions. The waste management (WM) situation varies greatly between developed and developing countries, with inadequate waste collection and disposal systems posing significant challenges in poorer nations. As a result, there is a growing global emphasis on waste management due to concerns about environmental degradation and sustainability (Brewer 2001). It is crucial not only to manage waste but also to develop sustainable strategies for its handling. Waste management involves the collection, transportation, processing, and disposal of various waste products, while ensuring proper monitoring and control. To effectively manage all types of waste and avoid indiscriminate landfilling, sustainability must be at the forefront of waste management practices. The concept of "sustainable waste management" encompasses the comprehensive approach of collecting, handling, recycling, and disposing of waste without jeopardizing the environment, the well-being of future generations, or human health. It encompasses all stages of waste management, from its generation to its final disposal.

It is crucial to understand that waste encompasses different categories such as municipal waste, electronic waste (e-waste), and radioactive waste, among others. Sustainable waste management focuses on minimizing resource consumption and reducing waste generation (Fig. 3.1).

Protecting sustainability is our responsibility for the well-being of our environment and future generations. An effective and adaptable sustainable waste management system incorporates feedback loops and prioritizes practices like waste avoidance. It plays a vital role in the circular economy, offering numerous opportunities and benefits for the environment, society, and the economy. Sustainable waste management involves collection, sorting, processing, recycling, and, when appropriate, the generation of energy and resources. This improves waste management

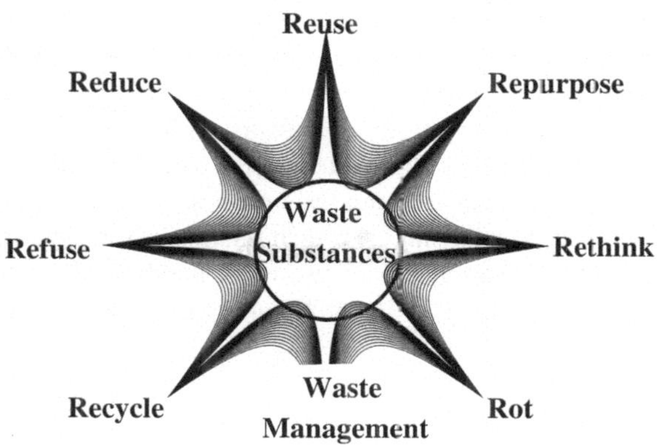

Fig. 3.1 Different types of waste management (WM)

practices, creates employment opportunities, mitigates the adverse impacts of human activities on the environment, and enhances air and water quality. Additionally, it reduces food waste, prevents significant environmental costs, and safeguards public health, thereby enhancing overall human well-being.

Given the current circumstances, there is a pressing need for an economically and environmentally sustainable waste management (WM) system. Several widely used WM and therapy modalities are regularly used, including the following:

1. **Incineration technique**
   Garbage and wastes are heated up and further treated.
2. **Sanitary landfills**
   This technique is more effective and is used all around the world. The locations where wastes are maintained and insulated from the environment are considered to be practically hygienic landfills.
3. **Recycling**
   Material transformation to be used in a new form.
4. **Avoidance and reduction**
   In recent years, the significance of microorganisms in waste management (WM) and the degradation of pollutants has gained recognition. Microorganism-based biotechnological techniques, such as biodegradation, bio composting, and biotransformation, have proven effective in the collection and removal of various contaminants (Banerjee et al. 2018). Additionally, certain microorganisms have demonstrated unique capabilities in waste management. For example, Cladophora sp., a type of green algae, has shown a remarkable capacity for accumulating toxic metals and can serve as an efficient alternative for waste management (Maghraby and Hassan 2018).

The microbial ecology plays a crucial role in the proper functioning of biological processes in wastewater treatment systems. Bacteria and archaea dominate the bioreactors of wastewater treatment systems in Finland's polar arctic region, influencing the efficiency of the treatment process (Martinez et al. 2018). Furthermore, innovative approaches such as nano-bioremediation have successfully utilized nanoparticles to enhance bacterial activity. For instance, the extremophile bacterium *Deinococcus radiodurans,* known for its radiation resistance, has been employed in the cleanup of radioactive waste in the USA (Brim et al. 2000; Varma et al. 2017).

By combining microorganisms with various biotechnologies, an efficient and environmentally sustainable approach can be achieved for treating different types of waste. This approach not only proves to be economical and efficient but also promotes responsible and practical methods for mitigating pollution through the use of microbiological agents. Thus, the utilization of microorganisms in waste management presents a promising and responsible solution for addressing pollution challenges.

3.3 Benefits of Waste Management and Environmental Sustainability

Effective waste management is of utmost importance for a variety of reasons. Firstly, it plays a vital role in reducing greenhouse gas emissions and improving air and water quality, resulting in a healthier environment. Secondly, it helps to minimize the need for resource extraction and reduces pollution and energy consumption associated with the production of new materials. Waste management also contributes significantly to the overall well-being of the population and the preservation of natural resources such as wood, water, and minerals. Recycling waste materials allows for their transformation into useful products, promoting a more sustainable and circular economy. Additionally, proper waste disposal helps prevent the proliferation of diseases by eliminating breeding grounds for harmful germs and bacteria. Furthermore, techniques like composting not only enrich the soil but also save money by reducing the reliance on landfill space. By prioritizing effective waste management, we can make a positive impact on the environment, public health, and resource conservation.

Other benefits can be grouped under:

3.3.1 It Provides Space

In the absence of effective waste management, waste accumulates and finds its way to landfills, which can be dispersed or centralized. However, landfills consume valuable space and pose challenges for land resources. To address this issue and optimize land utilization, it becomes crucial to adopt sustainable waste management practices that prioritize waste reduction and resource recovery. Singapore serves as an exemplary case, as the country, with its limited land area and dense population, recognizes the importance of recycling and proper waste disposal.

3.3.2 It Saves and Makes Money

Reusing and recycling items eliminates the need for unnecessary purchases, resulting in cost savings. By adopting these practices, individuals and businesses can avoid unnecessary expenses by utilizing resources that can be recycled or repurposed. Moreover, with a reduction in waste generation, waste management organizations can minimize their operational frequency, leading to cost savings in disposal and management. This approach promotes financial efficiency while encouraging organizations to be more environmentally conscious and less wasteful.

3.3.3 It Enhances Sustainability

The principle of sustainability focuses on the effective management of resources such as waste, energy, and water. Embracing sustainability as individuals, businesses, governments, or organizations can lead to an improved reputation, attracting reputable tenants, clients, and patrons. Moreover, it fosters active participation from volunteers, staff members, and citizens, promoting a sense of shared responsibility and collaborative efforts toward sustainable practices.

3.3.4 It Controls Pollution

Every bit of garbage we discard has a different effect on the ecosystem. For example, leftover food draws mosquitoes and flies, while drug waste contaminates our water. Sustainable waste management enables us to comprehend our garbage and determine the best way to handle it. Because of this, food waste can be composted, plastics can be recycled, and trash from pharmaceutical products should be returned to the company that made them so that it can be properly disposed of, such as by incineration. Pollution may be reduced thanks to all of these actions. Thus, plastic garbage won't block marine life, food waste won't attract mice, and pharmaceutical waste won't poison the water.

3.3.5 It Is the Central Point of Environmental Conservation

Human actions have a detrimental impact on the environment, particularly due to inadequate waste management practices despite the alarming rate of waste generation. Therefore, it is crucial to prioritize sustainable waste management for the sake of environmental conservation. By implementing sustainable waste management strategies, we can safeguard the environment and improve it for future generations, as well as for the well-being of other species. This includes preserving valuable resources such as trees, metals, and water, reducing greenhouse gas emissions that contribute to global warming, and promoting the efficient utilization of existing resources. Additionally, sustainable waste management involves the utilization of biodegradable waste to enrich the soil, fostering a more sustainable and balanced ecosystem.

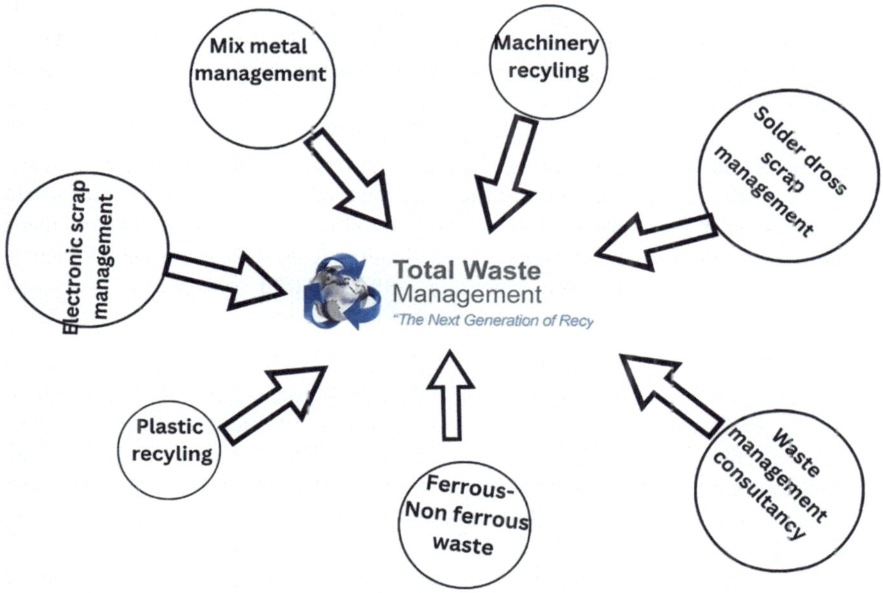

Fig. 3.2 Total waste management

3.3.6 *Betterment of Ourselves and Responsible Inhabitants of the Earth*

Sustainable waste management is vital for our existence as humans. It enables us to fulfill our responsibility as accountable members of society by effectively and responsibly managing our waste. By implementing intelligent and efficient waste management strategies, we can develop innovative technologies, employ improved waste management techniques, and explore suitable alternatives for different types of waste (refer to Fig. 3.2). Instead of indiscriminately sending all waste to landfills, we can focus on composting food scraps and organic materials, as well as recycling plastics and other recyclable.

3.4 Challenges to Waste Management

Sustainable waste management (SWM) poses significant challenges in India due to the country's rapid population growth and the emergence of sprawling cities. The existing waste management practices predominantly involve disposal methods and informal economy participation, while the proper waste infrastructure is lacking. One of the key hurdles in achieving effective waste management is the lack of public engagement, as individuals often fail to take responsibility for their waste. It is crucial to raise community awareness about waste issues and foster a shift in people's attitudes toward establishing efficient and sustainable waste management systems. To ensure sustainable and financially viable waste management, it is imperative to emphasize maximum resource recovery from waste while securely disposing of residual garbage through the establishment of engineered landfills and waste-to-energy plants. However, India faces challenges in terms of waste regulations, technological options, and the availability of trained professionals in the waste management sector. Failure to address these critical factors will perpetuate the inadequate waste management practices in India, leading to adverse impacts on the environment and public health. When dealing with industrial solid waste, cities and municipalities frequently experience the following issues.

1. The lack of designated sanitary landfill sites poses a challenge for businesses in urban areas when it comes to waste disposal. This issue is particularly prominent among small-scale businesses that operate within city limits and may not have obtained the necessary permissions from regulatory bodies such as State Pollution Control Boards (SPCBs) or Pollution Control Committees (PCCs). Additionally, the presence of businesses in non-conforming areas contributes to water and air pollution concerns.
2. Urban local bodies and SPCBs/PCCs do not frequently communicate to resolve matters pertaining to the disposal and treatment of waste and the granting of licenses in non-conforming areas. There are not enough facilities in industrial estates inside city borders for businesses to organize the collection, handling, and disposal of their liquid and solid waste.

3.5 Various Microorganisms Are Employed in the Process of Waste Management

Microorganisms have been helping humanity by solving several issues related to maintaining environmental quality. They have been successfully used in the fields of environmental preservation, waste management in both urban and industrial locations, and human and animal health. Microorganisms have been used to create solutions that were before unaffordable and practical utilizing chemical or physical engineering methods (Satyanarayana et al. 2012). Microbial technologies have emerged as effective solutions for addressing environmental issues, particularly in the realm of waste management. Human activities frequently involve the synthesis of diverse chemicals, leading to the production of waste materials that ultimately find their way into the environment. To tackle this challenge, integrating these waste materials back into the environment through microbial processes has been identified as a viable approach (Ezeonu et al. 2012) (Fig. 3.3).

Microbial biotechnology utilizes microorganisms such as yeasts, bacteria, and fungi, as well as their by-products, to facilitate the production of various industrial products and perform important functions. These microorganisms are added to

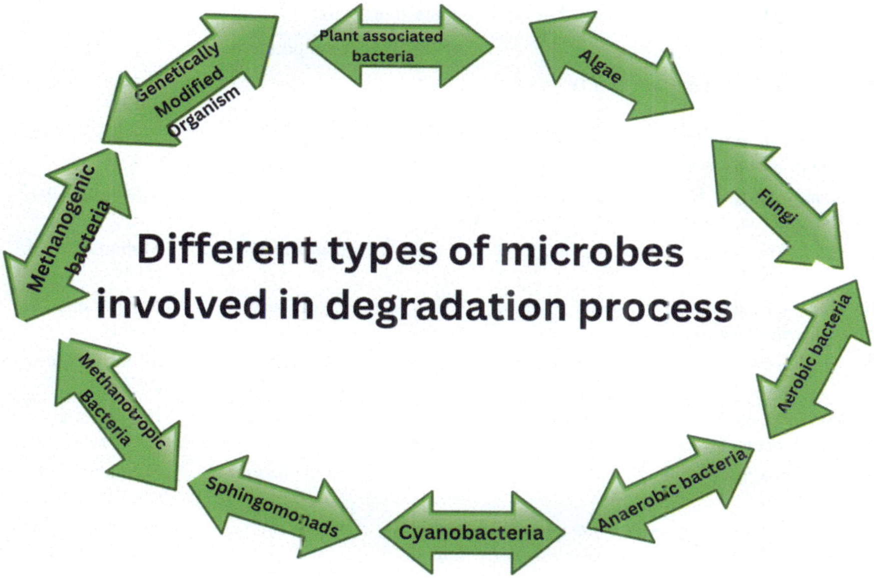

Fig. 3.3 Distinct microbes included in degradation process

substrates to achieve specific outcomes, including the production of biocatalysts, biomass fuel, biomonitoring, bioleaching, bio-detergents, biotreatment of pulp and wastes, biofiltration, aquaculture treatments, textile biotreatment, and more. Microorganisms also play a crucial role in nitrogen and carbon cycles and contribute to waste recycling and decomposition, making them vital to both humans and the environment. In symbiotic relationships, microorganisms are significant in the functioning of many multicellular organisms. Microbial biotechnology involves the application of scientific methodologies and controlled conditions to manage waste without disrupting the ecosystem. Common waste management practices include composting, biodegradation, bioremediation, and biotransformation. Various microorganisms such as *Bacillus* sp., *Corynebacterium* sp., *Staphylococcus* sp., *Streptococcus* sp., *Scenedesmus platydiscus, Chlorella vulgaris, Selenastrum capricornutum, Scenedesmus quadricauda*, and others have been successfully employed in waste management applications (Table 3.1).

3.5.1 *Bacteria*

Bacteria play a vital role in aerobic waste treatment systems by breaking down organic compounds found in industrial wastes. These systems consist of a diverse range of microorganisms, including facultative and aerobic bacteria. The competitiveness of a particular species determines its ability to utilize the organic materials

Table 3.1 Microorganism involved in biodegradation process

	Name of Microorganism	Compound name	References
1	*Marinobacter, Shewanella, Pseudomonas*	–	Gerdes et al. (2005)
2	*Sphingomonas, Halanaerobium, Roseobacter Pseudomonas, and Afipia*	–	Bordenave et al. (2007)
3	*Methylophaga, Marinobacter Stappia, Martelella, Alkanivorax, Paracoccus, Roseovarius,* etc.	–	Vila et al. (2015)
4	*Staphylococcus, Bacillus*	Endosulfan	Mohamed et al. (2011)
5	*Trametes versicolor Gloeophyllum trabeum*	Hydrocarbons	Karigar and Rao (2011)
6	*Microbacterium* sp., Acinetobacter sp.	Aromatic hydrocarbons	Simarro et al. (2013)
7	*Nocardia atlantica, Listeria denitrificans Micrococcus luteus*	Textile dyes	Hassan et al. (2013)
8	*Pseudomonas putida*	Benzene and xylene	Safiyanu et al. (2015)
9	*Bacillus cereus, Pseudomonas cepacia, Serratia ficaria, Citrobacter koseri, Bacillus coagulans*	Crude oil	Kehinde and Isaac (2016)

present in the system, leading to reproduction and population growth. Bacterial predomination can be observed in two primary groups: those that directly utilize the organic waste compounds and those that rely on the by-products produced by the first group. The characteristics of the treatment system are determined by the dominant species that can efficiently utilize the available organic resources and proliferate rapidly. As the organic substrate is consumed, a period of scarcity occurs, leading to the prevalence of secondary bacterial species. Eventually, most of the bacteria die and lyse, releasing their biological components and creating opportunities for other microorganisms to thrive. Flocculation, the process of separating microorganisms from the liquid phase, is crucial for the complete stabilization of aerobic waste treatment systems. While a bacterial species called *Zoogloea ramigera* was initially believed to be responsible for flocculation, further research has revealed that several other bacterial species are also capable of this process. Flocculation is influenced by factors such as the energy level and surface charges of the bacteria. When the surface charges are neutralized, "Van der Waal forces of attraction" come into play, causing bacteria to aggregate. Autoagglutination, also known as flocculation, occurs when bacteria lack motility and cannot withstand the Van der Waal forces. During flocculation, some bacteria lie and die, while a fraction of the bacterial cell, mainly composed of polysaccharides, remains insoluble. Over time, more polysaccharides accumulate, resulting in the aging of the floc with fewer active bacteria.

3.5.2 Fungi

Fungi are essential for the stabilization of organic waste products. Fungi can decompose almost every type of organic compound found in industrial waste, just like bacteria can. Fungi can outweigh bacteria, although this only occurs in extreme environmental conditions. The majority of fungi present in industrial wastes are filamentous, which renders them undesirable since they are prone to settling and find it challenging to build a tight, compact floc. Due to the latter of these two factors, major efforts are made to alter the environment to reduce the likelihood that filamentous fungus will predominate over bacteria. Filamentous fungi thrive in environments with low nitrogen contents, low pHs, and low oxygen tensions. Low oxygen tension is the result of either limited oxygen supply or high organic load that increases demand over supply. Low oxygen levels cause metabolism to stop, producing organic alcohols, aldehydes, and acids instead of carbon dioxide and water. If the system does not have enough buffering capacity, the organic acids reduce the pH to a level that is more favorable to fungi. Therefore, there might be a connection between low oxygen tension and pH. While few bacteria can grow well enough to compete, many fungi thrive in a pH range of 4–5. In comparison to bacteria, fungi need less nitrogen per unit of protoplasm. Fungi flourish in nitrogen-deficient wastes because they can generate more active masses of protoplasm from the wastes than bacteria can. Fungi normally contain 5–6% nitrogen, whereas bacteria typically contain 10–12% nitrogen. Fungi will typically be present and will help stabilize the organic waste in a typical environment. The fungus, though, is not necessary and won't take over.

3.5.3 Algae

Algae are the third category of biological plant that helps to stabilize organic wastes overall. Opposite to bacteria and fungi, algae don't need to ingest organic components because they get their energy for synthesis from sunlight. To produce protoplasm, the algae predominantly use the inorganic waste components including carbon dioxide, ammonia, magnesium, phosphate, potassium, iron, sulfate, calcium, sodium, and other ions. Due to the fact that they don't use the same waste products, bacteria and algae may co-dominate. The waste's organic components are broken down by bacteria, and some of the inorganic elements that algae require are also released. The bacteria entirely aerobically stabilize the organic materials by using the oxygen that the algae produce during protoplasm formation. Algae must use the energy generated by the metabolism of organic matter, just like bacteria and fungi, to survive when there is no sunlight. The majority of algal species obtain this organic material from food that has been retained within the cell, but some may also obtain it from organic waste.

3.5.4 *Viruses*

In prokaryotic or eukaryotic cells that are still alive, virus particles consisting of biopolymers can reproduce and combine to form new virus particles (Ivanov 2010). Pathogenic viruses often referred to as bacteriophages, must be eliminated, reserved, or devastated during wastewater treatment due to their capacity to infect and damage bacterial populations in the environment. Bacteriophages can also be employed to identify specific microbiological pollution of environmental waste.

3.5.5 *Protozoa*

Protozoa are the most basic organisms found in waste disposal systems. It was not until recently that pure culture protozoa research and natural discoveries from various biological treatment systems were combined to understand the role of protozoa in stabilizing organic wastes. This study showed that the main purification process was not carried out by protozoa, but rather by reducing the amount of free-swimming bacteria, which helped to produce a cleansed effluent. The succession of protozoa in biological waste disposal systems has long been observed, but its origins are unknown. The same conditions that determine the dominance of any biological species also affect the succession of protozoa. The primary elements that affect protozoa's predominance are intensity, type of food, and competition.

3.6 Applications of Microbes for the Environmental Sustainability

Various researchers, organizations, and legislators are interested in using microbial tools to boost agricultural production sustainably. Various initiatives over the past 10 years have hastened the development of microbial products and market access:

1. Consumer and policymaker pressure to increase crop productivity while reducing use of agrochemicals.
2. Frequent occurrence of biotic and abiotic stress threats.
3. Loss of efficacy of some agrochemicals.
4. Although this sector is still in its infancy, the path to successful microbial goods is taking shape and the global market for these products has grown faster than that of agrochemicals.

Microbes, or microorganisms, have the remarkable ability to metabolize a wide range of chemicals, which can have profound environmental consequences. One notable example is their capacity to effectively cleanup hazardous oil spills by consuming the pollutants and converting them into harmless by-products.

The microbial communities present in the rhizosphere, the soil surrounding plant roots, play a vital role in indirectly benefiting human health through their influence on food production. With climate change predicted to bring about significant alterations in the water cycle, agricultural regions will be affected differently, experiencing increased flooding or heightened aridity. Consequently, the composition of rhizosphere microbial communities is expected to change, but the extent of their adaptation is still being explored. Fundamental research is essential to enhance our understanding of cultivating healthy soil microbiomes and promoting food security.

An essential step toward enabling more sustainable precision agriculture lies in comprehensively understanding how microbial communities respond to climatic stressors. This predictive understanding is crucial for designing interventions that can facilitate effective adaptation and, ultimately, contribute to sustainable agricultural practices.

Considering phosphate as an example, plants require this nutrient for growth, but it is primarily sourced from depleted mines. Exploring the potential of rhizosphere bacteria to detect and mobilize phosphate in the soil, including identifying when and where it is needed, holds promise for sustainable agricultural practices. Harnessing the potential of microbial communities has the power to revolutionize agriculture and make a significant positive impact on our planet. The key lies in advancing our understanding of these communities in different settings and effectively applying this knowledge, particularly in the context of climate change.

Now consider some of the following examples:

3.6.1 Utilizing Microorganisms to Remediate Municipal Sewage

Aerobic biological treatment systems house a wide range of microorganisms, including bacteria, fungi, algae, protozoa, and even higher organisms. The specific bacteria that thrive in these systems depend on factors such as the composition of the waste, environmental conditions, and the biochemical traits of the microorganisms. Each microorganism present in an industrial waste treatment system plays a role in shaping the system's overall characteristics, which can be both advantageous and detrimental. Therefore, comprehending the contributions of different organisms to the stabilization of organic waste is essential for designing and operating effective waste treatment systems.

3.6.2 Utilizing Microorganisms to Treat Soil

Depending on the kind of soil, different microbial populations exist in the soil. The most numerous and diverse soil residents are bacteria, which may be found in all soil levels and frequently forms biofilms, particularly around plant roots. Because they can easily adapt to changing circumstances, bacteria are common in most soils. The next largest group is fungi. Only in topsoil can both free-living and symbiotic fungi be found, where they can grow to create enormous mycelia that may cover several acres. In comparison to bacteria and fungi, algae and protozoa are also found in soil, though in smaller concentrations. Soil algae can be found on or close to the soil surface since they are photoautotrophs. Protozoans can migrate across the soil and may eat other microorganisms, particularly bacteria. Even while viruses are located inside bacteria, they are typically not discovered in free form.

Rejuvenation of soil using microbes like bacteria and fungi is a type of environmental remediation. The removal of pollutants and toxins from soil is the goal of microbial remediation. Anaerobes aerobes and facultative anaerobes have all contributed to soil development over billions of years, although microorganisms have only been used industrially for eliminating toxins for the past three decades. They aid in the soil's decomposition of heavy metals, pesticides, and hydrocarbons as well as the fixation of nitrogen and prevention of growth of plant diseases. The microbial flora consumes the pollutants and breaks them down for food and reproduction. Three levels of microbial remediation are distinguishable:

- **Natural attenuation**: Local soil microorganisms carry out the process spontaneously.
- **Bio stimulation**: The process is externally assisted with nutrients, moisture, and a pH that is optimal for the bacteria.
- **Bioaugmentation**: In circumstances like oil spills, where the naturally present bacteria may disappear due to the severity of the contamination, this technique entails using microorganisms that have been introduced externally.

To combat soil contamination, genetically modified microorganisms (GEMs) are becoming more and more significant. Microbes with the necessary evolution can break down naturally occurring compounds. However, naturally occurring bacteria cannot break down the man-made compounds present in paints, herbicides, insecticides, cleansers, and cosmetics into less hazardous substances. POPs are man-made compounds that may be persistent, such as dioxin and DDT (POPs). They are referred to be persistent because they are hard to eliminate and have a propensity to build up inside of creatures, including people, where they can cause cancer and reproductive system problems.

Of great interest to geneticists is the soil bacterium *Deinococcus radiodurans,* which exhibits an exceptional ability to survive radiation doses that would be fatal to humans. This bacterium has also demonstrated its capability to degrade toluene and radioactive mercury. Consequently, there is potential for its application in the degradation of radioactive waste generated in industries such as medicine, energy, and the military. This requires the bacterium to be engineered to thrive in radioactive environments.

When it comes to microbial remediation, there are two primary approaches: in situ and ex situ. The key distinction between these strategies lies in the location where the remediation takes place.

3.6.3 *In Situ Remediation: Advantages, Disadvantages, and Environmental Impact*

In situ remediation refers to the process that takes place at the site of contamination without removing the contaminated soil from the substrate. This technique has the advantage of being simple to adopt and causing little disruption to the local flora and fauna. No danger exists that contamination will move from one place to another. Humans are protected from exposure to hazardous materials because there is no handling of contaminated soil required. When pollution levels are low and extensive, in situ remediation—a gradual process—is chosen. The disadvantage of this approach is that the region may not be uniformly decontaminated at the end of the operation.

3.6.4 *Ex Situ Microbial Remediation: A Swift Solution for Contaminated Soil Treatment*

Ex situ microbial remediation involves removing the soil from the substrate and transporting it for treatment. Cleaning occurs in ideal circumstances that are simpler to control and observe. As a result, treating polluted soil off-site is a quicker option. This method works well in situations when there is localized, medium-to-high contamination. The bioremediation field is constantly changing. Previously thought to be untreatable contaminants are now easily broken down. However, bitumen, asphalt, and inorganic acid pollution still have not been addressed through bioremediation.

3.6.5 *Production of Biogas Through Anaerobic Digestion of Animal Waste*

Animal feces are a valuable source of nutrients and renewable energy. However, the inadequate management and disposal of this waste, such as storing it in lagoons or allowing it to degrade in open areas, pose significant risks to the environment. The decomposition of manure releases a range of atmospheric pollutants, including nitrous oxide, methane, hydrogen sulfide, ammonia, particulate matter, and volatile organic compounds. These pollutants have adverse effects on the environment and human health. Implementing proper solid waste management practices offers numerous benefits for both urban and rural communities. In rural India, for example, the use of bio gasifiers provides an alternative energy source for cooking, reducing reliance on traditional fuels.

3.6.5.1 Important Factors

The generation of biogas from animal dung is influenced by several factors, including the temperature and pH of the feedstock. It is widely recognized that a biogas plant operates most efficiently when the pH is neutral and the temperature is mesophilic, typically around 35 °C. Another crucial aspect is the carbon-to-nitrogen ratio in the feed material, which should ideally fall within the range of 20:1 to 30:1.

Maintaining a carbon-to-nitrogen ratio of approximately 25:1 in animal dung is considered optimal for maximizing biogas production. Furthermore, the solid concentration of the input material plays a vital role in facilitating gas generation and ensuring ease of mixing and handling. The hydraulic retention time (HRT) also significantly affects the capacity of the digester and consequently impacts the overall cost of the plant. Generally, a longer retention period translates to higher costs associated with the plant structure.

3.6.6 Production of Bioethanol from Agricultural Waste

Ensuring the sustainable use of energy has emerged as a pressing global priority in our efforts to reduce reliance on fossil fuels. Apart from their detrimental environmental impact, fossil fuel reserves are expected to diminish in the foreseeable future. In contrast, bio-based fuels offer a renewable energy source that can be produced without compromising the well-being of our planet. However, the initial generation of biofuels, which relied heavily on food grains, faced challenges regarding the food-versus-fuel dilemma. Nonetheless, significant strides have been made in overcoming these obstacles and finding alternatives to first-generation biofuels. Currently, biomass-derived bioethanol accounts for approximately 10–14% of the world's energy supply. This sustainable energy source not only contributes to mitigating global warming but also helps address the issue of fossil fuel depletion. By embracing bio-based fuels, we can pave the way for a greener and more sustainable energy future. The combustion of petroleum, diesel or a mix of the two has contributed greatly to the current worldwide issue of bioethanol in the fight to minimize pollution. Waste from fruits and vegetables has a high sugar content that can be used as a raw material to create sustainable energy using *Saccharomyces cerevisiae*. Although currently, starch and sugar are used to produce 80% of the bioethanol. Recently, lignocellulosic biomass has engrossed substantial attention (Khandaker et al. 2018a, b). It has been established that the first-generation bioethanol production method uses conventional feedstocks, such as feedstocks high in starch (corn, potato, and sugarcane). However, they face social and economic hindrances. The second-generation bioethanol method is gaining traction. The second-generation process uses lignocellulosic biomass (sugarcane bagasse, corn stover, and straws). Producing bioethanol as an alternative method is a viable approach to address the need for sustainable energy sources and reduce reliance on fossil fuels. One advantageous aspect of bioethanol is its compatibility with the existing high-octane engines, making it easier to integrate this biofuel into current transportation systems (Wheals et al. 1999; Grad 2006).

Bioethanol production utilizes plant materials with high sugar content as raw materials. Notable examples include sugarcane, pineapple, and potatoes, which yield substantial amounts of bioethanol due to their abundant sugar concentration (Khandaker et al. 2018a, b).

The conversion process involves fermenting the sugar content of starchy materials under controlled climatic conditions, leading to the production of ethanol and carbon dioxide (CO2) (Sun and Cheng 2002). This anaerobic fermentation process, known as the Embden–Meyerhoff pathway (EMP), relies on the enzymatic activity of bacteria and fungi. The same fermentation process is employed in the production of alcoholic beverages. The initial step involves the digestion of starchy materials into fermentable sugars, which are then fermented using specific yeast species. Yeast and heat are utilized to convert complex carbohydrates into simpler sugars. During fermentation, yeast cells metabolize a portion of the sugar, resulting in the production of glycerol, acetaldehydes, and lactic acid. While bioethanol production from edible food materials poses challenges, utilizing lignocellulosic materials such as corncobs, corn stalks, corn husks, sugarcane bagasse, and sugarcane bark can offer a sustainable alternative. These residual components contain abundant energy and are renewable worldwide (Zhang et al. 2010). Furthermore, utilizing agricultural waste for bioethanol production can optimize land use and avoid competition with food supplies. Overcoming hurdles associated with the crystalline structure and lignin coating of lignocellulosic materials has been achieved through pretreatments such as acid hydrolysis (Gould and Freer 1984). Additionally, challenges related to the concentration of reducing sugars in ethanol derived from agricultural waste have been addressed (Zhang et al. 2010).

To overcome the challenges associated with food dependency and promote sustainable biofuel production, the utilization of second-generation biofuels derived from waste biomass is a viable approach (refer to Table 2.2 and Table 3.2).

Table 3.2 Comparison bioethanol/fossil ethanol

Bioethanol	Fossil ethanol
Waste plant material	Fossils source
Renewable	Non-renewable
The least pollutants are resulted	Many hazardous pollutants
Cost-effective	Expensive

Table 3.3 Characteristics of ethanol (C_2H_5OH)

Parameter	Characteristics
Molecular mass	46.07 g/mol
Molecular formula	C2H5OH
Appearance	Colorless liquid
Pka	15.9
Density	0.789 kg/l
Octane number	99
Refractive index (nD)	1.36 (25 °C)
Boiling temperature	78.5 °C
Explosion limits	Lower 3.5% (v/v) upper 19%(v/v)
Lower heating value (at 20 °C)	21,090 KJ/kg
Vapor pressure at 38 °C	50 mm Hg
Ignition temperature	Lowest temperature of ignition
Viscosity	1.200 mPa.s (20 °C)
Specific heat	Kcal/kg 60 °C
Water solubility	Between −117 and 78 °C
Higher heating value (at 20 °C)	29,800 KJ/kg
Flash point	12.8 °C
Freezing point	−117 °C

Walker (2011)

Lignocellulosic biomass, in particular, plays a crucial role in the production of these advanced biofuels (as highlighted by Sarangi and Nayak 2021) (Table 3.3).

3.6.6.1 Relative Account of Gasoline and Ethanol

Gasoline, a vital fuel used in transportation, is produced as a result of fractional distillation of crude oil. It is a hydrophobic substance with a flash point of approximately −45 °F, although this can vary depending on its octane grade. The density of gasoline vapor ranges between 3 and 4, and its specific gravity is typically between 0.72 and 0.76, making it insoluble in water. Due to these characteristics, gasoline tends to float on the surface of water. The combustion point of gasoline falls within a temperature range of 280–456 °C. Its boiling point varies between 37 °C and 204 °C. Prolonged exposure to gasoline can have adverse effects on the respiratory system (Ameri et al. 2008) (Fig. 3.4).

In the modern era, fossil fuels such as oil, coal, and natural gas dominate as the primary energy sources. However, their availability is expected to diminish within the next few decades, prompting the need for a shift toward renewable and environmentally friendly energy alternatives. This transition is necessary to mitigate the anticipated environmental consequences such as global warming, acid rain, and urban smog. To achieve this, it is crucial to reduce carbon emissions by 80% (v/v) and embrace sustainable sources like solar power, wind energy, biofuels, and other

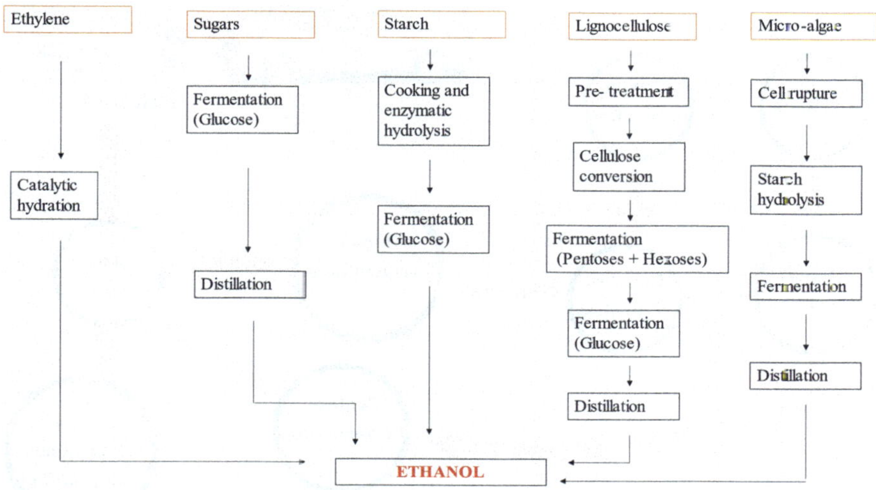

Fig. 3.4 Flowchart showing how ethanol is produced from five distinct source ingredients (Corncob sugarcane bagasse, cornhusk, sugarcane bark cornstalk)

renewable options. Ethanol is one of the most promising alternative biofuels. Despite having a 68% lower energy equivalent than petroleum fuel, ethanol burns more cleanly since it contains oxygen, making it a potential gasoline substitute made from renewable sources. Ethanol has traditionally been added to gasoline blends in concentrations between 10 and 85% (v/v). Ethanol has more recently been recognized as fuel for biofuel cells and direct ethanol fuel cells (DEFC). The two major feedstocks for ethanol production are sugarcane and corn (Fig. 3.1). Though just enough to match the demand at the moment. Although lignocellulosic biomass provides an alternate source, little is known about its accessibility. Vohra et al. (2014) addressed the present state of ethanol production (Fig. 3.5) from various feedstocks and the current technological state of such ethanol production.

Bioethanol is a fuel that contains 35% oxygen. Nitrogen oxides and particulate particles are reduced by ethanol. A polar solvent with a flash point of 55oF, pure ethanol is water soluble. Point. The fact why it weighs more than vaporized air is due to its density of 1.59. Ethanol has a specific gravity of 0.79. It hardly dissolves in water at all. When blended with gasoline, bioethanol is less dangerous than regular gasoline and is used as fuel for automobiles. It thereby reduces the requirement for crude oil. It increases the fuel's oxygen content, which improves gasoline combustion and reduces greenhouse gas emissions (Wang et al. 1999). E10 is the name for the ethanol concentration that results from blending 10% bioethanol with 90% gasoline and is usually referred to as "gasohol."

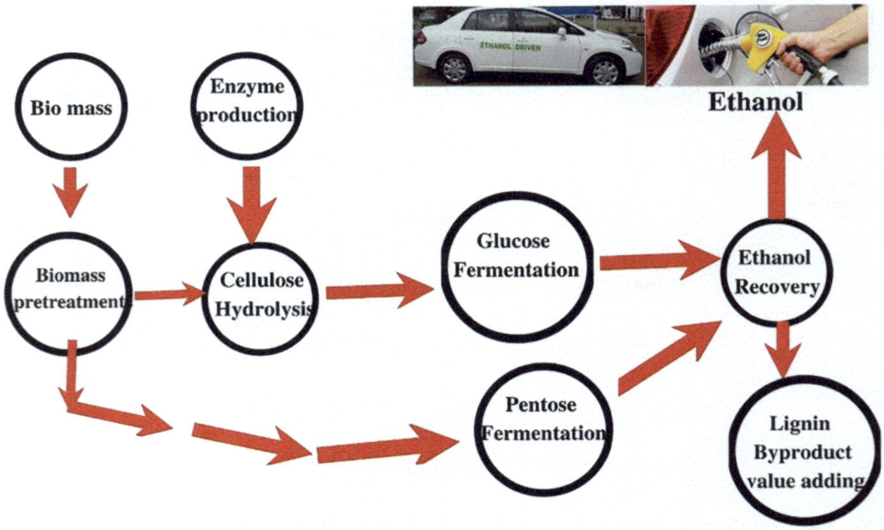

Fig. 3.5 Bioethanol production process

3.6.7 Applying Microorganisms to Cleanup Oil Slicks in the Water

Oil spills are now frequently seen in oceans all over the world. These incidents are a direct result of human mistakes while crude oil or refined petroleum products are being transported across international borders. Oil spill cleanup has gained significant attention from governments and environmental organizations all around the world. In addition to coastal facilities, gasoline freighters, and oil leaks can also occur spontaneously through mechanisms like petroleum seeps offshore production or extraction wells (Wang and Stout 2010).

3.6.8 Some Important Oil Spill Events

During an oil spill, significant quantities of liquid hydrocarbons are mistakenly released into the environment. The local environment is harmed by the broad, ongoing contamination that results from oil spills. Only tanker mishaps caused 5.72 million tonnes of crude oil to leak into the ocean between 1970 and 2015. While in 2015, environmental losses from tankers, pipelines, etc. were almost 7000 tonnes (ITOPF 2015). Nevertheless, the Exxon Valdez oil leak, which spewed 11 million gallons of crude oil, stands out as the most notable oil spill disaster in recent human history. It is not, however, the largest oil leak in recorded history. Iraqi forces intentionally dumped between 380 and 250 million gallons of oil into Persian Gulf during 1991 in Gulf War (Moss 2010).

3.6.9 Impact of Oil Spill on Microbial Communities

A coating of oil that floats on the water's surface shortly after an oil spill causes enormous populations of marine species (like marine mammals, sea algae, birds, and fish) to perish through poisoning and asphyxia. While scattered oil droplets affect the benthic community by sinking to the ocean's bottom. The fast alterations in the local microbial communities are yet another important effect of oil spills. This is as a result of a sizable amount of hydrocarbons that they depend on as a source of energy and carbon suddenly becoming unavailable. Oil spills in the natural environment are primarily cleaned up by the microbial breakdown of oil hydrocarbons. This has also been used to artificially boost remediation in circumstances of severe spills. Although yeast, fungus, and bacteria may all break down a complicated oil-hydrocarbon mixture, bacteria do it with the greatest efficiency. Therefore, the primary degraders of oil hydrocarbons are bacteria.

Such oil spill accidents cause changes in the structure and ecology of the local marine microbial population. In such cases, the dominating species in the community becomes a group of microorganisms that can use various hydrocarbon molecules as a source of energy, which subsequently aids in the biodegradation process (Table 2.1). Following the Exxon Valdez and Deepwater Horizon accidents in the Gulf of Mexico, numerous studies have been carried out to monitor the nature and rate of changes in the local microbial communities. These studies emphasize a range of factors influencing community interactions, genetic and metabolic diversity, as well as chemical and geological processes in addition to past contamination histories (Bordenave et al. 2007).

3.6.10 Hydrocarbon Degraders Types

Numerous species of bacteria and archaea have been demonstrated which are capable of decomposing hydrocarbons by chemical, culture-based, and metagenomics studies, as shown in Table 2.1. However, some bacterial taxa, including *Alcanivorax, Cycloclasticus, Oleispira, Oleiphilus*, and *Thalassolituus*, have been identified as being obligate hydrocarbon degraders. Although other organisms like *Marinobacter* and *Pseudomonas* have also demonstrated a variety of hydrocarbon degradation capabilities (Brooijmans et al. 2009; Das and Chandran 2011).

3.6.11 Oil Hydrocarbons Fall into Four Categories

1. Resins (sulfoxides quinolones, pyridines, and amides,)
2. Asphaltenes (ketones, fatty acids, phenols, and esters)
3. Aromatics
4. Saturates

Together, these nearly 17,000 chemicals make up the broad spectrum of oil hydrocarbons. Alkanes breakdown quickly during processes for hydrocarbon degradation. However, due to their high molecular weights and hydrophobicity, polycyclic aromatic hydrocarbons (PAHs) are the least effectively destroyed (Atlas and Bragg 2009).

3.6.12 The Importance of Biosurfactants in Oil Spill Bioremediation

Furthermore, the natural environment's ability to break down oil spills is greatly helped by the biosurfactants that bacteria produce. As a result, research on oil spill bioremediation agents has been conducted on it. These substances are a class of surface-active substances. Surface and interfacial tensions can be decreased by them. These also help bacteria by emulsifying hydrophobic hydrocarbon substrates to increase their stability and bioavailability. By two basic methods, microbial biosurfactants aid in the bioremediation of hydrophobic substances.

(a) The first is the emulsification process, which increases surface area.
(b) Second, making hydrophobic water-insoluble substrates more bioavailable (by forming micelles or lowering interfacial tension and boosting solubility).

Bacteria must get close to the oil substrate in order for the breakdown process to start. Hydrophobic hydrocarbons can be emulsified for this purpose, or high molecular weight polymers that facilitate surface bonding can be produced. As a result, it ensures that oxygenases linked to cell membranes may access the substrate (Mukherjee et al. 2006). As a result, hydrophobic oil substrates, whose low bioavailability has previously limited biodegradation processes, can now be digested by the microorganisms. However, the complex chemical makeup of the oil poses another obstacle to oil spill bioremediation, resulting in the oil's variable vulnerability to microbial attack. This is because varying amounts of saturates, aromatics, asphaltenes, and resins are often present in oil in varying proportions. While some of these types of hydrocarbon molecules, such as alkanes, are readily degradable by microbes, others, such as polycyclic aromatic compounds (PAHs), are exceedingly difficult to decompose (Das and Chandran 2011). Such resistant substances that are incapable of degrading may still be present in the environment and show no evidence of being eliminated.

Plastics are extremely robust and resistant to physical and chemical deterioration because of their intrinsic qualities. Furthermore, there is more pressure on scientists and policymakers to reduce pollution since plastic wastes are found in greater quantities in marine areas.

3.7 New Developments in the Management of Microbiological Waste and Their Difficulties

Due to the recognition of the harmful effects caused by soil erosion, sediment migration, chemical fertilizers, pesticides, and inadequate waste management practices, there has been an increased emphasis on the beneficial utilization of microorganisms in environmental management and biotechnology. Recent scientific developments have contributed to our understanding of how microbes can be effectively applied to address these environmental concerns. We must explore beyond conventional physical and chemical technology for solutions to the major environmental and societal issues that these detrimental events have caused around the world (Satyanarayana et al. 2012). Notably, improvements in environmentally friendly cleaning methods have been made that are more compatible with waste disposal methods, such as biological processes and biotechnological equipment. Biodegradation techniques including bioremediation, bio stimulation, bioaugmentation, phytoremediation, and other similar techniques are examples of such biotechnological tools.

The process of biodegradation is responsible for the conversion of organic material into nutrients that can be utilized by other species. Chahal (2015) highlights that a variety of organisms, including bacteria, fungi, insects, worms, and other saprophytic organisms, contribute to the breakdown and recycling of dead material. Their collective action facilitates the decay process and transforms organic matter into new forms. When the right environmental factors are present, there are enough nutrients, bacteria, and their by-products for bioremediation to employ contaminants, including hazardous substances, to degrade. To minimize, eliminate, or alter harmful chemicals that are present in sediments, soils, the air and water, bioremediation technology uses microorganisms (Adams et al. 2015). By altering the environment, existing microorganisms that can do bioremediation are stimulated. Enhancing biodegradation processes involves introducing various types of nutrients and electron acceptors, such as carbon (e.g., molasses), oxygen, nitrogen, or phosphorus. Typically, these substances are present in limited quantities, which can inhibit microbial activity. Bioaugmentation is a technique that involves introducing biodegradable microbial cells into an ecosystem to supplement the native populations. These introduced microbes, known as bio-remediators, assist in the breakdown of pollutants. The timeframe required for microbial cleanup can vary depending on factors such as high pollutant concentrations, localized contaminant areas, or the nature of the contamination.

In mycoremediation, fungi are utilized to disinfect a site. Fungal mycelium releases extracellular enzymes and acids that facilitate the breakdown of lignin and cellulose. This method serves as an alternative approach to bioremediation.

Bioreactor landfills, which employ advanced microbiological processes, are considered superior to traditional sanitary landfills and controlled dumps. These landfills rely on technology advancements to quickly convert and stabilize waste components. By maintaining ideal moisture levels through the constant supply of fluids, microbial digestion is effectively regulated. Recirculating landfill leachate serves as the liquid component, while sewage sludge may be used when leachate volume is insufficient. Three types of bioreactors are commonly used: anaerobic, aerobic, and hybrid systems combining both anaerobic and aerobic methods. In all three approaches, collected leachate and water are reintroduced into the landfill to maintain moisture, promoting faster decomposition by the responsible microorganisms.

Aerobic bioreactors make use of vertical or horizontal pipe systems to introduce air into landfills, creating an environment rich in oxygen that accelerates the decomposition process. This technique effectively reduces levels of volatile organic compounds, toxicity, and methane emissions. On the other hand, anaerobic bioreactors employ leachate circulation and produce methane at higher rates compared to conventional landfills. The concentrated methane produced can be harnessed for industrial purposes and reduces the required monitoring period for methane production.

Hybrid bioreactors employ a combination of aerobic and anaerobic cycles to expedite decomposition in the upper layers of waste while simultaneously generating methane in the lower layers. In comparison to conventional landfills, bioreactor landfills generally produce lower levels of volatile organic compounds, with the exception of hydrogen sulfide (H2S), which tends to be higher in bioreactor landfills. Anaerobic digestion, considered an advanced method of in-vessel waste treatment, is widely recognized as an innovative technology in the industry. It involves a multi-step biological process where anaerobic bacteria convert organic waste into biogas and fertilizer without the need for oxygen. The process consists of hydrolysis/liquefaction, acidogenesis, acetogenesis, and methanogenesis, with various metabolic interactions occurring between different types of bacteria.

Anaerobic digestion has broad applications and can effectively treat wastewater, including organic solid waste. It is a rapid process that not only enables the conversion of biogas into energy but also allows for the utilization of the waste as fertilizer. Additionally, the combination of bio-dung composting and vermiculture technology offers an efficient approach for large-scale recycling of organic waste. Vermiculture technology, which utilizes earthworms to convert organic waste into vermicompost, has demonstrated its effectiveness in managing organic waste with minimal complexity and economic viability. It finds wide applications in waste management and sustainable organic farming.

In a study, a combination of grass clippings, water hyacinth, and cattle dung was employed as organic waste for the bio-dung composting and vermicomposting processes. The results indicated successful treatment of the organic wastes over a period of 60 days using partial bio-dung composting and vermicomposting methods. This helps to give vital micronutrients and contains compounds that encourage growth, such as auxins and cytokinin. Dig a pit that is 0.5 meters square and 1.0 m deep for vermiculture. Fill the hole with straw, dried grass, or leaves. Introduce a worm culture (also available commercially). Every day, make sure to sprinkle some soil and dried leaves on top of the contents. To keep the pit wet, water it once or twice a week. The garbage will eventually degrade through the action of microorganisms in around 45 days. This process produces nutritious soil that is full of nutrients.

3.8 Environmental Protection in Microorganism Applications

Beneficial microorganisms (BMs) are now being applied in a wide range of ways, finding increasing utilization. They can be introduced into the environment through three different methods: as a single strain, a combination of strains, or a combination of strains along with specific synergistic agents. Their role in promoting the decomposition of organic matter helps create favorable conditions for aquatic life and maintains a dynamic ecological balance among various organisms from different kingdoms (refer to Table 3.4).

Consequently, BMs serve diverse functions within microbial ecology and contribute significantly to maintaining ecological equilibrium, both in sediments and water including:

- Modifying the population of algae in water bodies to stop the deterioration of water quality.
- It stops the growth of fish illnesses and the putrefaction of certain aquatic plants throughout the summer.
- Enhancing the immune systems of aquatic creatures to increase their resistance to disease.
- Reducing the negative effects of oxidation by generating antioxidant molecules and the corresponding antioxidant wave emission (which also reduces negative effects of oxidation).
- In addition, BMs have the ability to neutralize naturally occurring free radicals in materials and living things.

Table 3.4 Applications of EM in the treatment of waste water

Type of wastewater	Type of pollutants	Reactor/process	Reference
Samples taken from a sewer pipe	Heavy metals	Batch	Zhou et al. (2008)
Domestic sewage	Alkalinity, TDS, BOD, COD	Batch	Namsivayam et al. (2011)
Beet sugar wastewater	COD, BOD, TDS	Batch	Embaby et al. (2010).
Municipal wastewater	COD, TN	Sequencing batch reactor	Lee and Cho (2010)
Raw sewage	BOD, TSS, COD, and TDS	Batch	Monica et al. (2011)
Sago industry wastewater	TSS and COD	Anaerobic sludge hybrid up-flow blanket	Priya et al. (2015)
Synthetic wastewater	Heavy metals elements	Adsorption	Ting et al. (2013)
Raw sewage	COD, TP, and BOD	Sludge with modified contact stabilization	Rashed and Massoud (2015)
Aquaculture wastewater	NH3, TP	Batch	Lananan et al. (2014)
Rubber processing wastewater	COD, TSS, BOD, and turbidity	Anaerobic sequencing batch reactor	Anwar et al. (2013)
Gray water	TSS, COD, BOD grease, and oil	Jar test	Abdel-Shafy et al. (2014)
Primary settled municipal wastewater	NH_3, COD, BOD, TSS, and TP	Anaerobic sequencing batch reactor	Ahmed et al. (2013)
Synthetic wastewater	Rudimentary dyes	Adsorption	Bhagavathi Pushpa et al. (2016)
Internal wastewater	NO3, COD, TSS, TDS, and BOD	Batch	El Shafei et al. (2017)
Primary stable municipal wastewater	TAN and COD	Moving bed biofilm reactor	Safwat (2018)

3.9 Recommendations and Prospective Outcomes

The global impact of increasing garbage has disrupted the natural recycling equilibrium of ecosystems. Over the past century, the magnitude and quality of waste released into the environment have exacerbated the situation, posing a growing threat to environmental safety. Factors such as the introduction of xenobiotic substances, excessive exploitation of natural resources, and the alarming rate of waste generation have contributed to this alarming scenario.

To mitigate the adverse effects of waste, numerous waste management solutions have been developed. Microorganisms, being essential inhabitants of ecosystems, play a critical role in processes such as biodegradation, bioremediation, nutrient cycling, and detoxification. By employing carefully selected microorganisms in controlled environments, microbial technology offers a means to cleanup environmental waste. It is worth noting that the utilization of microorganisms is not limited to culture-dependent techniques, as culture-independent methods have also been used to study a wide range of discovered bacteria. Technological advancements have facilitated the exploration of microbial diversity for waste management applications.

Both traditional and contemporary methods such as bioremediation, anaerobic digestion, and composting have been successfully applied to solid waste management. For wastewater treatment, techniques such as fixed-film systems, activated sludge, biosorption technology, and microbial electrochemical technology have yielded positive results. Moreover, microbial technology has proven effective in removing harmful substances from wastewater, including synthetic dyes and oil spills. The application of microbial technology for waste management has been implemented worldwide, contributing to the removal of trash from the environment. Relevant government parastatals ought to coordinate their initiatives, like Integrated Solid Waste Management System (ISWM) program that instructs citizens on correct waste management practices and collection management. To collect and manage waste more effectively and efficiently, waste should be separated at the source in the following ways:

1. Techniques for managing microbiological waste should be created and applied for their value-added advantages as well as for environmental cleanup.
2. As a last step, garbage collection systems should be improved for a more sanitary and sustainable environment, particularly in and around populated regions of the communities.

References

Abdel-Shafy HI, Al-Sulaiman AM, Mansour MSM (2014) Greywater treatment via hybrid integrated systems for unrestricted reuse in Egypt. J Water Process Eng 1:101–107. https://doi.org/10.1016/j.jwpe.2014.04.001

Adams GO, Fufeyin PT, Okoro SE, Ehinomen I (2015) Bioremediation, biostimulation and bioaugmention: a review. Int J Environ Bioremediat Biodegrad 3(1):28–39. Available from URL: http://www.sciepub.com/journal/IJEBB

Ahmed S, Abdelhalim H, Rozaik E (2013) Treatment of primary settled wastewater using anaerobic sequencing batch reactor seeded with activated EM. Civ Environ Res 3:130–137

Ameri M, Ghobadian B, Baratian I (2008) Technical comparison of a CHP using various blends of gasohol in an IC engine. Renew Energy 33(7):1469–1474. https://doi.org/10.1016/j.renene.2007.09.015

Anwar ZR, Ariffin M, Hassan A, Mahmood I, Khamis AK (2013) Treatment of rubber processing wastewater by effective microorganisms using anaerobic sequencing batch reactor. J Agrobiotechnol 4:1–15

Atlas R, Bragg J (2009) Bioremediation of marine oil spills: when and when not—the Exxon Valdez experience. MicrobBiotechnol. [Internet]. Blackwell Publishing Ltd; Mar [cited 2016 Dec 1] 2(2):213–221. https://doi.org/10.1111/j.1751-7915.2008.00079.x

Banerjee A, Jhariya MK, Yadav DK, Raj A (2018) Micro-remediation of metals: a new frontier in bioremediation. In: Hussain C (ed) Handbook of environmental materials management. Springer. https://doi.org/10.1007/978-3-319-58538-3_10-1. ISBN:978-3-319-58538-3

Bhagavathi Pushpa T, Vijayaraghavan J, Vijayaraghavan K, Jegan J (2016) Utilization of effective microorganisms-based water hyacinth compost as bio sorbent for the removal of basic dyes. Desalin Water Treat 57(51):24368–24377. https://doi.org/10.1080/19443994.2016.1143405

Bordenave S, Goñi-Urriza MS, Caumette P, Duran R (2007) Effects of heavy fuel oil on the bacterial community structure of a pristine microbial mat. Appl Environ Microbiol 73(19):6089–6097. [Internet] American Society for Microbiology; [cited 2016 Nov 30] Available from: http://www.ncbi.nlm.nih.gov/pubmed/17704271

Brewer LJ (2001) Maturity and stability evaluation of composted yard debris. M.Sc. Thesis. Oregon State University, Corvallis, USA

Brim H, McFarlan SC, Fredrickson JK, Daly JM, Venkateswaran A, Kostandarithes MH (2000) Engineering Deinococcusradiodurans for metal remediation in radioactive mixed waste environments. Nat Biotechnol 18:85–90

Brooijmans RJW, Pastink MI, Siezen RJ (2009) Hydrocarbon-degrading bacteria: the oil-spill clean-up crew. Microb Biotechnol 2(6):587–594. [Internet]. Blackwell Publishing Ltd; ; [cited 2016 Dec 1]. https://doi.org/10.1111/j.1751-7915.2009.00151.x

Chahal K (2015) Bioremediation, bioaugmentation and phytoremediation. Available from URL: https://hubpages.com/education/BIOREMEDIATION-BIOAUGMENTATION-PHYTOREMEDIATION

Das N, Chandran P (2011) Microbial degradation of petroleum hydrocarbon contaminants: an overview. Biotechnol Res Int:1–13

El Shafei MM, Elmoteleb EMA, El MM (2017) Investigate the effect of effective microorganism (EM) on improving the quality of sewage water from Al-Gabal Al-Asfar area in Egypt. 1st Int Conf Towar a Better Qual Life 1–9 https://doi.org/10.2139/ssrn.3164096

Embaby AA, El-Shahawy Abd-Allah MA, Dawoud IA (2010) Application of effective microorganisms in treatment of wastewater of beet sugar factory at Bilqas, Dakahlia Governorate, Egypt. J Environ Sci 39:151–158

Ezeonu CS, Tagbo R, Anike EN, Oje OA, Onwurah INE (2012) Biotechnological tools for environmental sustainability: prospects and challenges for environments in Nigeria—a standard review. Biotechnol Res Int

Gerdes B, Brinkmeyer R, Dieckmann G, Helmke E (2005) Influence of crude oil on changes of bacterial communities in Arctic sea-ice. FEMS Microbiol Ecol 53(1):129–139. [Internet]. Blackwell Publishing Ltd Available from: http://femsec.oxfordjournals.org/cgi/doi/10.1016/j.femsec.2004.11.010

Gould JM, Freer SN (1984) High-efficiency ethanol production from lignocellulosic residues pretreated with alkaline H_2O_2. Biotechnol Bioeng 26:628–631

Grad P (2006) Biofuelling Brazil—an overview of the bioethanol success story in Brazil. Biofuels 7(3):56–59

Hassan MM, Alam MZ, Anwer MN (2013) Biodegradation of textile Azo dyes by bacteria isolated from dyeing industry effluent. Int Res J Biol Sci 2:27–31

Hawken P, Lovis A, Lovins LH (1999) Natural capitalism. The 25th Anniversary Report of the Council on Environmental Quality Council on Environmental Quality, 1996. In: Beamon BM (eds) Designing the Green Supply Chain. Logistics Information Management Little Brown & Co., Boston, 12(4): p. 332e342

ITOPF (2015) Oil tanker spill statistics available from: the international tanker owners pollution federation limited. ITOPF, London, United Kingdom, p 4

Ivanov V (2010) Microbiology of environmental engineering systems, in environmental biotechnology. Humana Press, Totowa, NJ, pp 19–79

Karigar CS, Rao SS (2011) Role of microbial enzymes in the bioremediation of pollutants. Enzyme Res. https://doi.org/10.4061/2011/805187

Kehinde FO, Isaac SA (2016) Effectiveness of augmented consortia of Bacillus coagulans, Citrobacter koseri and Serratia ficaria in the degradation of diesel polluted soil supplemented with pig dung. Afr J Microbiol Res 10:1637–1644

Khandaker MM, Qiamuddin K, Majrashi A, Dalorima T (2018b) Bio-ethanol production from fruit and vegetable waste by using saccharomyces cerevisiae. Bioethanol Technologies

Khandaker MM, Qiamuddin KB, Majrashi A, Dalorima T, Sajili MH, Hossain SABM (2018a) Bio-ethanol production from fruit and vegetable waste by using Saccharomyces cerevisiae. Biosci Res 15(3):1703–1711

Lananan F, Abdul Hamid SH, Din WNS, Ali N, Khatoon H, Jusoh A, Endut A (2014) Symbiotic bioremediation of aquaculture wastewater in reducing ammonia and phosphorus utilizing effective microorganism (EM-1) and microalgae (Chlorella sp.). Int Biodeterior Biodegrad 95:127–134. https://doi.org/10.1016/j.ibiod.2014.06.013

Lee J, Cho MH (2010) Removal of nitrogen in wastewater by polyvinyl alcohol (PVA)-immobilization of effective microorganisms. Korean J Chem Eng 27(1):193–197. https://doi.org/10.1007/s11814-009-0330-4

Maghraby DM, Hassan J (2018) Heavy metals bioaccumulation by the green alga Cladophora herpestica in Lake Mariut, Alexandria, Egypt. J Pollut 1:1

Martinez AG, Sihvonen M, Palazon BM, Sanchez AR, Mikola A, Vahala R (2018) Microbial ecology of full-scale wastewater treatment systems in the polar Arctic circle: archaea, bacteria and fungi. Sci Rep 8:2208. https://doi.org/10.1038/s41598-018-20633-5

Mohamed AT, El Hussein AA, El Sidding MA, Osman AG (2011) Degradation of oxyfluorfen herbicide by soil microorganisms: biodegradation of herbicides. Biotechnology 10:274–279

Monica S, Karthik L, Mythili S, Sathiavelu A (2011) Formulation of effective microbial consortia and its application for sewage treatment. J Microb Biochem Technol 3:51–55

Moss L (2010) The 13 largest oil spills in history [internet] Mother Nature Network Available from: http://www.mnn.com/earth-matters/wilderness-resources/stories/the-13-largest-oil-spills-in-history

Mukherjee S, Das P, Sen R (2006) Towards commercial production of microbial surfactants. Trends Biotechnol 24(11):509–515

Namsivayam SKR, Narendrakumar G, Kumar JA (2011) Evaluation of effective microorganism (EM) for treatment of domestic sewage. J Exp Sci 2:30–32

Priya M, Meenambal T, Balasubramanian N, Perumal B (2015) Comparative study of treatment of Sago wastewater using HUASB reactor in the presence and absence of effective microorganisms. Procedia Earth Planet Sci 11:483–490. https://doi.org/10.1016/j.proeps.2015.06.048

Rashed EM, Massoud M (2015) The effect of effective microorganisms (EM) on EBPR in modified contact stabilization system. HBRC J 11(3):384–392. https://doi.org/10.1016/j.hbrcj.2014.06.011

Safiyanu I, Isah AA, Abubakar US, Rita Singh M (2015) Review on comparative study on bioremediation for oil spills using microbes. Res J Pharm Biol Chem Sci 6:783–790

Safwat SM (2018) Performance of moving bed biofilm reactor using effective microorganisms. J Clean Prod 185:723–731. https://doi.org/10.1016/j.jclepro.2018.03.041

Sarangi PK, Nayak MM (2021) Agro-waste for second-generation biofuels. Liquid Biofuels: Fundamentals, Characterization, and Applications 697–709. https://doi.org/10.1002/9781119793038.ch20

Satyanarayana T, Bhavdish-Narain J, Prakash A (2012) Microorganisms in environmental management: microbes and environment. Springer Publishers, Netherlands

Simarro R, Gonzalez N, Bautista LF, Molina MC (2013) Assessment of the efficiency of in situ bioremediation techniques in a creosote polluted soil: change in bacterial community. J Hazard Mater 262:158–167

Sun Y, Cheng J (2002) Hydrolysis of lignocellulosic materials for ethanol production: a review. Bioresour Technol 83:1–11

Ting ASY, Rahman NHA, Isa MIHM, Tan WS (2013) Investigating metal removal potential by effective microorganisms (EM) in alginate-immobilized and free-cell forms. Bioresour Technol 147:636–639. https://doi.org/10.1016/j.biortech.2013.08.064

USEPA (2002) Latest Findings on National Air Quality (2001) Status and Trends, EPA 454-K-02-001. U.S. Environmental Protection Agency, Office of Air Quality Planning and Standards, Research Triangle Park, NC

Varma D, Meena RS, Kumar S, Kumar E (2017) Response of mungbean to NPK and lime under the conditions of Vindhyan Region of Uttar Pradesh. Leg Res 40(3):542–545

Vila J, Nieto JM, Mertens J, Springael D, Grifoll M (2015) Microbial community structure of a heavy fuel oil-degrading marine consortium: linking microbial dynamics with polycyclic aromatic hydrocarbon utilization. FEMS Microbiol Ecol 73:2). [Internet]. Blackwell Publishing Ltd; 2010 May 7 [cited 2016 Nov 30]. https://doi.org/10.1111/j.1574-6941.2010.00902.x

Vohra M, Manwar J, Manmode R, Padgilwar S, Patil S (2014) Bioethanol production: feedstock and current technologies. J Environ Chem Eng 2(1):573–584

Walker GM (2011) Fuel alcohol: current production and future challenges. J Inst Brew 117(1):3–22. https://doi.org/10.1002/j.2050-0416.2011.tb00438.x

Wang M, Saricks C, Santini D (1999) Effect of fuel ethanol use on fuel-cycle energy and greenhouse gas emission. Argonne National Laboratory, Argonne, IL

Wang Z, Stout S (2010) Oil spill environmental forensics: fingerprinting and source identification. Academic Press, San Diego, p 620

Wheals AE, Bassoc LC, Alves DMG, Amorimd HV (1999) Fuel ethanol after 25 years. Trends Biotechnol 17(12):482–487

Zhang MJ, Wang F, Su RX, Qi W, He ZM (2010) Ethanol production from high dry matter corncob using fed-batch simultaneous saccharification and fermentation after combined pre-treatment. Bioresour Technol 101:4959–4964

Zhou S, Wei C, Liao C, Wu H (2008) Damage to DNA of effective microorganisms by heavy metals: impact on wastewater treatment. J Environ Sci 20(12):1514–1518. https://doi.org/10.1016/S1001-0742(08)62558-9

Chapter 4
Microbial-Meditated Remediation of Crude Oil-Contaminated Soil

Shanthi Natarajan, Murugesan Subbiah, Vishnu Kiran Manam, Abdel Rahman Mohammad Said Al-Tawaha, Abdel Razzaq Al-Tawaha, and Marwa Adel Qotb

Abstract Crude oil production and its uses are considered very important for human needs. The major waste components in crude oil are petroleum hydrocarbons and other toxic substances, which affect the soil fertility, nature, and its physical and chemical properties to a considerable degree. Therefore, this complication is significant for the community and environment. By remediating the soil damaged by the oil contaminant, microbial-mediated remediation is extensively viewed as an eco-friendly and potent technology for the degradation of the contaminated substance in crude oil. Different types of bacteria, fungi, and algae are used in this bioremediation technology, and strategies are being developed to solve this problem. By improving the relationship between these microorganisms and crude oil-contaminated soil, new biotechnologies can be used to improve the organisms that degrade these toxic substances and develop their pathways. And by exploring its functional genes and creating new forms through molecular activity and microbial modification, we can develop a way to degrade this crude oil-contaminated soil. Future research should concentrate on ways to improve bioremediation technology efficiency in order to further alleviate environmental stress on land and aquatic ecosystems.

S. Natarajan (✉) · M. Subbiah
PG and Research Department of Botany, Pachaiyappa's College, Chennai, India

V. K. Manam
R&D Division, Indian Broiler Group, Indamara, Chhattisgarh, India

A. R. Mohammad Said Al-Tawaha
Department of Biological Sciences, Al Hussein Bin Talal University, Ma'an, Jordan

A. R. Al-Tawaha
Department of Crop Science, Faculty of Agriculture, Universiti Putra Malaysia, Serdang, Selangor, Malaysia

National Agricultural Research Center, Baqa, Jordan

M. Adel Qotb
Soil Fertility and Plant Nutrition Department, Soils, Water and Environmental Research Institute, ARC, Giza, Egypt

A. Karnwal, A. R. Mohammad Said Al-Tawaha (eds.), *Microbial Applications for Environmental Sustainability*, https://doi.org/10.1007/978-981-97-0676-1_4

Keywords Bioremediation · Microbes · Oil spill pollution · Sludge · Nanoremediation

4.1 Introduction

The discovery of crude oil is the most significant factor in the advancement of human life. Such oil is a very significant resource for energy and numerous chemical productions. Crude oil is the world largest nonrenewable energy sources accounting for 30% of total energy (Ezeji et al. 2007). It is an energy that is rapidly available and usable, and it improves the human economy. However, neglecting the extraction, transfer, processing storage, and consumption of this crude oil can cause huge impact on the environment (Daraei et al. 2013), it has been found 33.6 million barrel of water are contaminated globally each day. At the same time, this pollution increases age of the soil and reduces the resistant capacity of the soils, causing changes in soil properties and, this waste seeps from soil into ground water, posing serious and irreversible problem in the world (Marinescu et al. 2011). Also, the toxicity of the compound in this oil has been found to be carcinogenic to human and various animals.

4.2 Source of Crude Oil Contaminant

4.2.1 Wastewater from Oil Extraction

The volume of wastewater generated globally from crude oil production is approximately 80 million barrels per day, in comparison to the daily oil production of 250 million barrels (Veil 2015). The flow rate of wastewater fluctuates over the lifespan of an oil and gas well; initially, there may be little to no conventional oil and gas wastewater, but it tends to increase over time. In addition to dissolved oil and hydrocarbons, gases such as hydrogen sulfide (H_2S), carbon dioxide (CO_2), and acid carbonate, were present. Also, phenolic chemicals are part of the complex makeup of wastewater from oil and gas production (Tibbetts et al. 1992). The presence of oil, grease, and other polluting organic substances in wastewater discharge is a major contributor to the pollution of aquatic habitats. Because of the high water consumption during the refining process, a lot of wastewater is produced. The quantity of water needed is contingent upon factors such as the scale, intricacy of processes, and nature of the crude products involved. Wastewater generated by petroleum refineries is known to include a diverse range of hydrocarbons, making it a notable contributor to the contamination of aquatic environments. According to projections, it is anticipated that the worldwide demand for oil will see a surge in the forthcoming two decades, reaching a level of 107 million barrels per day (mbpd). Consequently, oil is expected to contribute to around 32% of the world's energy consumption by the year 2030. The growing need for oil necessitates the ongoing generation and discharge of effluents by the oil sector into aquatic environments. The effluents include contaminants that exhibit high toxicity and constitute a substantial environmental risk. A lot of different methods are used to clean up wastewater, including biological processes, adsorption, coagulation, chemical oxidation, and biological processes. More advanced methods include membrane filtration and microwave-assisted catalytic moist air oxidation. The appeal of advanced oxidation processes (AOPs), such as heterogeneous photocatalytic degradation, lies in their ability to effectively and completely mineralize organic waste, a feature appreciated by various individuals. This technique is predominantly attractive due to its cost-effectiveness and utilization of nontoxic catalysts. The increasing need for oil highlights the continuous production of waste materials by the oil sector, which consist of highly dangerous pollutants that pose a threat to both land-based and water-based ecosystems. A lot of research has been done on the types of chemicals that oil refineries use to make effluent, the effects of releasing this wastewater into waterways and ecosystems, and the creation of different treatment technologies to help handle wastewater and use it again in operational processes (Singh and Shikha 2019). The inadequate management of wastewater can result in the occurrence of several illnesses in both people and animals, namely those pertaining to the neurological, circulatory, and integumentary systems (Rahman et al. 2018). Hence, it is imperative to do a comprehensive assessment of wastewater prior to commencing treatment procedures.

4.2.2 *Crude Oil Spill Pollution*

On the contrary, oil spill contamination may be observed in several environments such as the ocean, seashores, and terrestrial regions. According to Chong et al. (2014), oil spills can occur at several stages of the oil production and trading process, including during production, transit for storage, and trade. What are the primary factors contributing to significant issues faced by marine ecosystems and marine organisms? Furthermore, the potential hazards posed by oil wastes in close proximity to coastal areas during oil drilling operations, oil transfer activities, oil refining facilities, and during the transportation of oil from one location to another, as well as during oil storage, pose significant threats not only to marine ecosystems but also to terrestrial organisms (Chang et al. 2014). From 1970 to 2017, the ocean had a significant influx of crude oil, with over five million units being spilled. In the year 2017 alone, an estimated 7000 tons of oil were lost in the saltwater (ITOPF 2018). Since the 1940s, the ocean has witnessed the occurrence of over 25 significant oil spills, resulting in the devastation of both ecosystems and the socioeconomic well-being of the afflicted regions. According to Nelson et al. (2018), the presence of this phenomenon has adverse effects on the tourism industry and recreational pursuits within the region. The presence of highly resilient oil slicks on the ocean's surface poses a significant risk to the preservation of marine biodiversity. The presence of this phenomenon poses a threat to the advancement and proliferation of autochthonous creatures inhabiting aquatic environments, hence impacting both coastal and marine ecosystems. Oil spills may be regarded as the unauthorized discharge of fluid petroleum hydrocarbons into the environment. These incidents have the ability to significantly influence many human activities (Sojinu and Ejeromedoghene 2019). Ordinioha and Brisibe (2013) study suggests that oil spills have the potential to significantly impact household food security, with a recommended reduction of 60%. Furthermore, the research revealed that the occurrence of oil spills might result in a reduction of up to 36% in the ascorbic acid concentration of vegetables and a fall of 40% in the crude protein content of cassava. The results of this study suggest a possible rise of 24% in the occurrence of childhood malnutrition (Ordinioha and Brisibe 2013). In order to preserve the natural ecosystem and aquatic organisms, it is imperative to use effective purification methods and prompt remediation strategies for the removal of oil slick layers present on both water and soil surfaces (Singh et al. 2020a, b). The elements copper (Cu), cadmium (Cd), cobalt (Co), nickel (Ni), lead (Pb), and zinc (Zn) are of interest in this context.

4.2.3 Exploitation of Crude Oil

The extraction of crude oil has resulted in notable detrimental consequences, encompassing soil degradation, pollution, the decline of aquatic organisms, and the destruction of biodiversity. The aforementioned outcomes highlight the insufficiency of international corporations involved in the extraction of crude oil, namely in the Niger Delta area (Imasogie and Odia 2009). The water-holding capacity of soils is negatively impacted by oil pollution, resulting in the repulsion of water and the obstruction of soil pores. Hewelke et al. (2018) and Wei et al. (2020a, b) found that this disturbance impedes the movement and distribution of soil water. Furthermore, it has been observed that oil-contaminated soils experience substantial changes in their physicochemical and biological characteristics (Dong et al. 2020). According to a research done in Ondo State, Nigeria, it has been proven that soil degradation has occurred as a consequence of oil and fuel-related activities (Olujimi et al. 2011).

The inadequate methods of disposing trash that are linked to the consumption of oil provide a significant risk to both human and animal populations. The discharge resulting from these activities has the potential to infiltrate aquatic ecosystems, hence posing a threat to the well-being of marine organisms (Adeola et al. 2022). The elements copper (Cu), cadmium (Cd), cobalt (Co), nickel (Ni), lead (Pb), and zinc (Zn) are the focal points of interest within this particular setting.

4.2.4 Automobile Exhaust Emission

Crude oil profoundly impacts the environment, primarily through the emissions of motor vehicle exhaust gases, which are a leading cause of atmospheric pollution. Carbon monoxide, nitrogen oxides, sulfur oxides, hydrocarbons, and solid particles are only some of the very dangerous compounds that may be found in these emissions at increasing quantities. These emissions also include higher levels of other harmful chemicals. Among these pollutants, nitrogen oxides and hydrocarbons are especially concerning in the photocatalytic process, posing significant health risks, particularly to the eyes and respiratory system (Dong-Sheng and Chang-Yuan 2022). The emission of exhaust gases is the principal factor responsible for surpassing acceptable levels of harmful chemicals and carcinogens in the ecosystems of major urban areas. Additionally, it results in the generation of smog, which can serve as a significant contributor to interior air pollution. According to Sassykova et al. (2019), two aspects affect automotive pollution: mass emission and exhaust gas composition.

Based on the 2014 environmental indicators circulated by the Department of Environmental Protection, there was a recorded reduction of 0.5% in the aggregate outdoor emissions of pollutants originating from motor vehicles, resulting in a total of 54,573,000 tons. This figure is in comparison to the emissions data from the preceding year, 2013.

4.2.5 Stacking of Oil and Sludge

This crude oil affects the atmosphere more when people use it for their needs. Oil sludge generated during crude oil production is a highly flammable and toxic substance that many countries have classified as a hazardous solid waste. This oil sludge is formed through transportation, mining, and storage in petroleum technology. Oil sludge is a complex water in oil (W/O) emulsion that usually contains 30–50% water, 30–50% oil, and 10–12% solids by mass (Ramaswamy et al. 2007). The accumulation of this oil sludge without removing it from a place long time affects the surrounding soil, atmosphere, and ground water to great extent (Qin et al. 2016; Chen et al. 2020). Therefore, this oil sludge poses a threat to the usual growth of plants, animals, and human health in the area.

The production of crude oil paved the way for economic growth and the simultaneous improvement of the lives of all people. But environmental damage has increased due to this energy consumption. The production of crude oil has indeed facilitated economic growth and improved the lives of many people. However, it has also led to increased environmental damage as a result of energy consumption. This underscores the dual challenge of balancing economic development with environmental protection (Dong-Sheng and Chang-Yuan 2022).

Nevertheless, despite the several approaches that have been investigated by researchers, microbial remediation is often regarded as the foremost way to address this issue. Bioremediation is a widely employed approach in the mitigation of hydrocarbon pollutants, involving the utilization of microorganisms to detoxify or eliminate these toxins through enzymatic activities (Margesin et al. 2003; Wolicka et al. 2009). Numerous bacterial genera possess the capacity to break down and/or consume hydrocarbons (Leahy and Colwell 1990). In the context of soil environments, the bacterial genera that play a significant role in the degradation of hydrocarbons are *Achromobacter, Acinetobacter, Alcaligenes, Arthrobacter, Bacillus, Burkholderia, Collimonas, Micrococcus, Mycobacterium, Nocardia, Nocardioides, Pseudomonas, Ralstonia, and Rhodococcus* (Chikere et al. 2009; Obayori and Salam 2010). Certain species have the potential to remediate oil pollution (Ron and Rosenberg 2014; Lea-Smith et al. 2015). Certain organisms possess a notable capacity for the breakdown of contaminants and are commonly employed in the remediation of environmental pollutants (Kuppusamy et al. 2017). The present growth and advancement of microbial remediation technology has also given rise to a fresh approach for addressing petroleum hydrocarbon contamination, a matter that has garnered considerable attention (Dombrowski et al. 2016; Dvořák et al. 2017).

Additionally, *Aspergillus flavus, Aspergillus niger, Cochliobolus lunatus,* and *Varicosporium elodeae* have been examined for their potential to digest crude oil pollutants (Kumar et al. 2022). Both types of species have seen biological degradation of monoaromatic hydrocarbons like benzene, toluene, ethylbenzene, and xylene to five-ring PAHs like benzo(a)pyrene. Many fungi have the potential to live and thrive in environments with hazardous metals due to their inherent biochemical and structural characteristics, as well as physiological and genetic adaptations. These adaptations may include morphological alterations and modifications in the surrounding environment that affect metal speciation, bioavailability, and toxicity (Sullivan and Gadd 2019). According to Adenipekun and Lawal (2012), *Phanerochaete chrysosporium, Agaricus bisporus, Trametes versicolor, and Pleurotus ostreatus* are suggested as viable options for the decontamination of polluted sites. Using bacteria, fungus, and microalgae to remediate hydrocarbons, especially aromatic chemicals like naphthalene, was originally described about 50 years ago (Cerniglia et al. 1980). Lei et al. (2002) found that green microalgae (Chlamydomonas, Chlorella, Scenedesmus, Selenastrum) and cyanophytes (Synechocystis) degraded pyrene from 34 to 100% after 7 days. Skeletonema costatum and Nitschia sp. degraded phenanthrene and fluoranthene, according to further studies. The green microalga Chlorella vulgaris may bioremediate crude oil-contaminated waters with 88–94% efficiency. In another study, pre-isolated oil-degrading bacteria and three plant species remedied contaminated silt-loam soil better than silt, loam, and sandy loam soils, reducing total petroleum hydrocarbons by 80% (Ghosh and Syed 2001). However, these microbes have different development needs and may degrade petroleum hydrocarbons differently depending on parameters including the hydrocarbon's chemical makeup, bioavailability, and environmental circumstances.

Crude oil production has been substantial innovative in the improvement of human civilization and economic growth. However, the waste generated by crude oil significantly impacts soil quality. This waste comprises hundreds of complex chemical compounds, and soils contaminated with petroleum waste pose environmental problems that pose serious risks to human health and wildlife. To address this complex and hazardous situation and protect the environment while restoring its quality, it is crucial to expand and integrate biobased technologies effectively and efficiently. Simple, integrated microbial-mediated methods are needed to address this complicated and possibly dangerous issue. Thus, addressing the main factors and mechanisms that improve microbial joint remediation degradation efficiency and designing a cycle with improved degradation performance, sustainability, and environmental friendliness is an urgent challenge.

4.3 Bacterial-Mediated Remediation

Chemical pollutants especially crude oil have been a major problem of elevated soil pollution worldwide nowadays (Hussain et al. 2018). The heterogeneous crude oil is a liquid that comprises hydrogen, carbon, nitrogen, sulfur, phosphorus, and oxygen with vanadium and nickel as trace elements. The soil pollution due to crude oil can be addressed by traditional remediation methods and bioremediation methods. The bioremediation methods of cleaning crude oil soil pollution are considered beneficial over the traditional methods (Xu and Zhou 2017). Bioremediation is defined as the mixture of certain rich nutrients, co-substrates that enhance the growth, and relevant species-specific microorganisms for the catalysis of the biodegradation process (Zhu et al. 2004).

Microbial biodegradation involves the consortium of microbes in degrading the crude oil-contaminated soil and each process has its uniqueness in degrading the chemical pollutant. The traditional methods of crude oil soil degradation include excavation and backfilling which have their demerits in the fragmented soils (Lance et al. 2023), whereas microbial degradation involves the breakdown of hydrogen and carbon into simpler forms present in the oil-contaminated soil and is of biological importance since there are no residues present after the degradation process. According to the 16s rRNA nucleotide analysis strains belonging to the genera *Gordonia, Acinetobacter, Rhodococcus, Pseudomonas, Cobetia, Marinobacter, Alcanivorax,* and *Microbacterium* were used to degrade the contaminated soil with crude oil. *Acinetobacter calcoaceticus* and *Alcanivorax dieselolei* showed higher efficacy in degrading crude oils from soil contamination as per the research study (Mehdi et al. 2012). The strains possessed a high emulsification and biosurfactant activity where the alkane hydroxylase gene was responsible for the activity (Fig. 4.1) (Mehdi et al. 2012). The hydrocarbons present in crude oil soil contamination possess a huge threat to humans, animal health, and plant habitation since they are mutagenic, carcinogenic, and immunotoxicants (Vasudevan and Rajaram 2001).

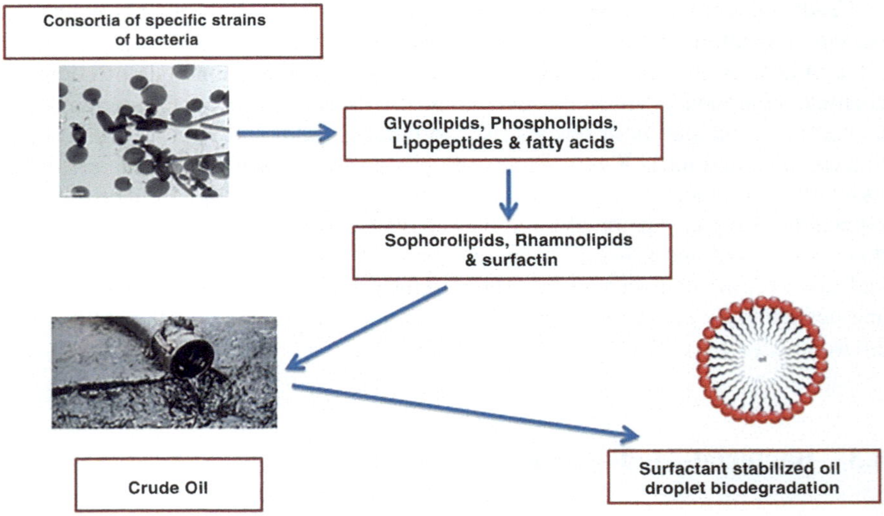

Fig. 4.1 Mechanism of microbial biodegradation of crude oil soil pollution

4.4 Microalgal Culture Based Remediation

The hydrocarbons in the soil contaminants due to oil spills and aquatic ecosystems pose a great threat to human life since they possess persistent carcinogenic compounds such as PAHs and PCBs (IARC 1983). The hydrocarbons from the oil spill contaminated soil and other aquatic environments can be disintegrated, adsorbed, and assimilated by organisms such as algae, fungi, and other microorganisms. The microalgae which are known to be wonder organisms bioremediate the hydrocarbons present in the oil spill by the way of two mechanisms, namely bio-assimilation and bio-sorption (Nidhin et al. 2020).

Microalgae are considered one of the most important organisms in the world of microbes in the bioremediation process of the ecosystem. The hydrocarbons from oil spills contain mostly nitrogen, sulfur, and oxygen which are resistant to microbial degradation. The enzymes produced by some microalgae are necessary and helpful in degrading the nitrogen and sulfur compounds present in hydrocarbons from the crude oil spill into less toxic compounds which are further biodegradable (Fig. 4.2) (Davies and Westlake 1979). The microalgae genus *Rhodococcus* is considered important and has a great role in the biodegradation of alkanes found in oil spills (Cappelletti et al. 2020). Similarly, the microalgae *Scenedesmus obliquus* GH_2 along with consortia with microbial population has been proven to degrade oil contamination (Tang et al. 2010). The production of O_2 by microalgae is utilized by microorganisms in degrading pollutants like aromatic hydrocarbons and phenolics present in the oil spill and other contaminates of the soil (Muñoz et al. 2003). Studies revealed the use of eight algal species namely *Chlorella autotrophica, Synechococcus*

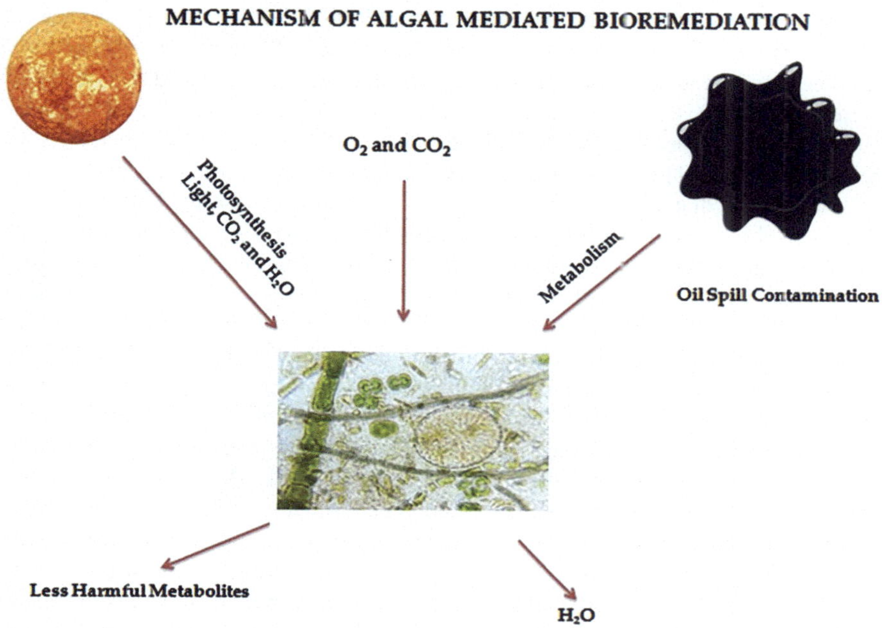

Fig. 4.2 Mechanism of algal biodegradation of crude oil pollution

elongatus, Aphanocapsa sp., *Coccochloris elabens, Volvox aureus, Oscillatoria* sp., *Scenedesmus obliquus,* and *Dunaliella tertiolecta* were efficient in biodegrading 35% of the oil contamination (Xiuqi et al. 2013). The next step in microalgae biodegradation of oil spills is the creation of the pathway for the efficient oxidation process by which the algal species will metabolize the hydrogen and carbons in oil and then cloning the complementary DNA for utilizing certain enzymes for better oxidation results, thus creating proper overexpression vector by transfection in algae (Ibraheem 2010). Studies suggest that green algae species such as *Chlorella vulgaris, Scenedesmus, and Selenastrum* are active in degrading aromatic hydrogen and carbons, such as pyrene, alpha-naphthalene, and phenanthrene by immobilizing the metals (Takáčová et al. 2014; De Llasera et al. 2016). The exact process of the microalgae-mediated biodegradation of oil spills is the proliferation of exopolysaccharides by the algae which mediates the uptake of the hydrocarbons onto their cell surface and converts them to less complex toxic forms (Khullar and Reddy 2019).

The biomass production and high growth efficiency of algae like green, brown, red, and cyanobacteria make them an efficient source for the biodegradation of hydrogen and carbons, especially in oil-contaminated soil and aquatic ecosystems (Kumar and Chandra 2020). The green algae species like *Chlorella vulgaris, Chlamydomonas, Scenedesmus, Selenastrum,* and *Synechocystis* revealed pyrene degradation in oil spill contamination ranging from 35 to 90% similarly the diatoms *Skeletonema costatum* and *Nitzschia* sp. showed a promising degradation toward phenanthrene and fluoranthene (Hong et al. 2008). The marine cyanobacteria *Microcoleus Chthonoplastes, A. quadruplicatum,* and *Phormidium corium* have also been studied in degrading phenanthrene (Ghasemi et al. 2011). The synergy of bacteria and microalgae-mediated bioremediation may pave a way for oil degradation which might involve the exudates production from microalgae, thereby supporting the bacterial growth for the oil biodegradation process (Mcgenity et al. 2012).

The algal-mediated nanoparticle synthesis can be used as a remediation delivery system in the biodegradation process especially the silver nanoparticles via seaweeds and algae which possess large surface areas where the eminent bioremediation molecules can be entrapped and reduce the oil biodegradation process (Vishnu and Murugesan 2020a, b). The advantage of silver nanoparticle algal-mediated synthesis is that it can be used to entrap bioremediation molecules as well as the exopolysaccharides present in the algal cell wall which entraps the contaminates of oil by converting the hydrocarbons and other contaminates into less toxic compounds.

The revolution in genetic engineering also has paved a pathway in the biodegradation of contaminates from the oil spills from the soil. The cyanobacteria and microalgae mixotrophy can damage organic contaminants and has also the advantage of examining the efficacy of the degradation process under certain field trials (Subash Chandra Bose et al. 2013). The genetic engineering in consortia of algae and microbes and algae-mediated nanoparticle synthesis has created a revolution as a pivotal solution in the adsorption and biodegradation of contaminates and hydrocarbons in the oil-spilled soil and aquatic ecosystems.

4.5 Fungi-Mediated Remediation

Fungi are among nature's most active decomposers of waste materials and are a vital part of the soil food web (Rhodes 2012). *Aspergillus* and other molds effectively break down sugar polymers like starches, hemicelluloses, celluloses, pectins, and others. Some of them can even break down substrates that are difficult to break down, like fats, oils, chitin, and keratin. These molds easily deteriorate materials like paper and textiles (cotton, jute, and linen), and the procedure is regularly regarded as biodeterioration. To successfully mycoremediate a pollutant, a simple screening approach is needed to choose the right fungal species (Matsubara et al. 2006).

Numerous fungi have been studied for their ability to degrade various organic pollutants, including azo dyes, in aerobic conditions. Among all fungi, white-rot fungi stand out because they produce lignocellulolytic enzymes such as laccase, manganese peroxidase, and lignin peroxidase, primarily used to break down the lignin found in lignocellulosic substrates. The degradation of dyes is a direct consequence of the ligninolytic system, making white-rot fungi highly valuable for bioremediation due to their capacity to break down a wide range of xenobiotic and textile dyes (Christian et al. 2005). It is well-documented that various fungi, including

T. versicolor (Erkurt et al. 2007), *Pleurotus ostreatus* (Faraco et al. 2009), *T. trogii,* (Levin et al. 2010), can effectively damage and decolorize different synthetic dyes.

The use of fungus in bioremediation has lately gained popularity (Riva 2006), since these organisms may convert harmful compounds into less hazardous forms. *Trametes versicolor* and *Pleurotus ostreatus* laccase, for instance, have been used to break down PCBs and phenol (Keum and Li 2004). Trametes versicolor, when given a mediator such as ABTS or HBT, is capable of oxidizing 3–5 ring PAHs. In addition, laccase from *Trametes pubescens* efficiently decomposed a combination of xenobiotic compounds, including PCP, 2-chlorophenol, 2,4-dichlorophenol, and 2,4,6-trichlorophenol (Gaitan et al. 2011). Zhao et al. (2010) verified the decomposition of DDT in soil. In this oxidative process, PAHs are broken down. Soil bacteria are more likely to be able to break down fungal metabolites because they are more water-soluble and chemically reactive than the parent PAHs (Collins et al. 1996). The fungal cell wall contains functional groups like carboxyl, amine, hydroxyl, and phosphate groups, which play a role in metal chelation (Couto et al. 2004). Fungi such as *A. niger* (Kapoor et al. 1999) and *Agaricus macrosporus* have been shown to effectively remove heavy metals like Pb, Cd, and Cu from contaminated waste. Rhizopus arrhizus binds metals depending on pH, Fe (III)-cyanide complex anions, and sorbent concentration (Aksu and Balibek 2007).

As per Bollag (1974), fungi break down pesticides by alkylation, dealkylation, amide or ester hydrolysis, dehydrogenation, hydroxylation, oxidation, reduction, and conjugate formation Sulfate is produced by soil fungus P. chrysosporium BU-1 decomposing endosulfan (Kullman and Matsumura 1996). Fungus have shown a notable capacity to decompose persistent and hydrophobic substances such as dioxins and furans, as evidenced by the use of a monooxygenase system by filamentous fungus for the degradation of dibenzofuran (Hammer and Schauer 1997). Various fungal strains have been shown to enzymatically break down dibenzofuran into intermediate molecules that possess hydrophilic characteristics. For instance, *Trichosporon mucoides* has been identified as a producer of six notable metabolites derived from dibenzofuran (Hammer et al. 1998). Furthermore, Nakamiya et al. (2005) have developed a novel mechanism for the breakdown of dioxin, using the cyclic ether-degrading fungus *Cordyceps sinensis*.

4.6 Nano-Microbial Remediation

Because of its small particulate substance, high surface-to-volume ratio, comfort of formation at contact locations, versatility, and other perquisites, nanotechnology has been occupied to discharge contaminants more and more newly (Khalkhal et al. 2020). The Environmental Protection Agency has identified nanoremediation as a technology with positive environmental effects. It is accepted as a workable approach for conventional site cleaning (Rajput et al. 2022).

Various methods utilizing nanomaterials (NMs) have been documented in the literature for the purpose of soil and water remediation. These methods include nano-phytoremediation as reported by Romeh and Saber (2020), nanobioremediation as described by Singh et al. (2020a, b), the utilization of nano-Fe_3O_4 as outlined by Tran et al. (2020), the application of nano-zerovalent iron as investigated by Gamallo et al. (2020), the use of nano-hydroxyapatite as explored by Liao and Yang (2020), the implementation of nano-zeolite as examined by Liu et al. (2020), the utilization of nano-zerovalent iron as studied by Lv et al. (2020), the application of ZnO-nanoparticles (NPs) as investigated by Ahmad et al. (2019), the use of nano-TiO_2 as explored by Sundararaghavan et al. (2020), the implementation of stabilized NPs as described by Sarkar et al. (2019), and the application of nano-silica as investigated by Wang et al. (2020).

4.6.1 Nanobioremediation of Heavy Metals

For the discharge of heavy metals, the handling of many nanoparticles, including biogenic nanoparticles, is regaining popularity (Rajput et al. 2021). AgNPs are among the well-known biogenic nanoparticles that *Morganella psychrotolerans* makes (Capeness et al. 2019). *Halomonas* sp., a Gram-negative microbial species, and FeO nanoparticles generated by coating polyvinylpyrrolidone (PVP) were successfully combined to enhance the bioremediation of Pb and Cd contamination. An adsorbent consisting of permanently magnetized Fe_3O_4 NPs was employed with *S. aureus* as well as a powder enclosed in phthalic acid to remove Cu, Ni, and Pb (as an n-Fe_3O_4–Phth–*S. aureus* complex).

By using heavy metal-resistant bacteria, like *Bacillus cereus* (PMBL-3) and *Lysinibacillus macroides* (PMBL-7), this study looks into how copper (Cu), cadmium (Cd), chromium (Cr), and lead (Pb) leave the body. Adding zinc oxide nanoparticles (ZnO NPs) at a concentration of 5 mg/L along with both bacterial strains removed 60% of Cr, 70% of Cu, and 85% of Pb in moderate conditions, according to the results. Moreover, when compared *with B. cereus* (80% and 60%), L. macroides exhibited removal rates of 55% for Cr and 50% for Pb at a neutral pH. From the information we have, we know that the *B. cereus* strain $XMCr^{-6}$ can lower Cr^{6+} by using enzymes to help the process. It is found that Cr6+ attached itself to the cell lines by reacting with functional clusters on the external of the bacterial cell membrane. Along with Se^{5+} and Se NPs, the ability of probiotics (specifically L. *casei* and L. *fermentum*) to effectively remove Cd from water was also studied. Rhizofiltration helps some plant species, like *Pedioplanis burchelli, Amaranthus spinosus*, and *Alternanthera pungens*, to grow in places where heavy metals are present in large amounts. These plants have exhibited effective methods to escape the negative effects of heavy metal exposure in their surroundings (Rizwan et al. 2019). According to Sun et al. (2016), excessive levels of heavy metals have detrimental effects on both human health and plant growth. In contrast to non-hyperaccumulating plant species, hyper-accumulating plant species have the ability to uptake significant quantities of metals from polluted soils. However, they do so by transporting and accumulating these metals in higher concentrations within their aboveground organ systems without causing noticeable harm to the plants themselves (Rascio and Navari-Izzo 2011). The technique of rhizofiltration has been employed for the purpose of heavy metal removal, and it has been observed that the surrounding environment exhibits coping mechanisms to prevent heavy metal contamination (Rizwan et al. 2021).

4.6.1.1 Soil Nanoremediation

Gillham and O'Hannesin (1994) proposed the practice of zerovalent iron (ZVI) as a penetrability barrier for the decontamination of water-halogenated contaminants, drawing upon their expertise in utilizing NPs for this purpose. This early study hypothesis was further supported by Roehl et al. (2005). Nanoparticles (NPs) have been found to be effective in the degradation of several types of pollutants, such as organic halocarbons, nitrates, heavy metals, herbicides, and dyes (Machado et al. 2015; Elliott et al. 2009). There is a limited amount of research that has utilized nanoparticle technology for the purpose of soil detoxification. Instead, investigations in this field have predominantly focused on its application in the purification of water or other aqueous solutions (Shi et al. 2011). The initial exploration of utilizing zerovalent iron (ZVI) as a permeability barrier was conducted by Gillham and O'Hannesin (1994), drawing upon their prior experience with nanoparticles in the remediation of water contaminated with halogenated substances (Roehl et al. 2005). Nanoparticles (NPs) have been validated to be operational in the deprivation of many pollutants, including organic halocarbons (Shih and Tai 2010), nitrates, heavy metals (Elliott et al. 2009), insecticides, and dyes (Machado et al. 2015). There has been limited research conducted on the application of nanoparticle technology for the purpose of remediating soil pollutants. Previous investigations in this area have mostly focused on the decontamination of water or aqueous solutions, as documented by Shi et al. (2011). Previous research has provided evidence indicating that polybrominated diphenyl ethers, a type of environmental contaminant with a propensity for accumulating in soil, can undergo degradation by the utilization of zerovalent iron nanoparticles confined within silica microspheres (Qiu et al. 2011; Xie et al. 2016). In their study, Yu et al. (2023) utilized tetrahydrofuran (THF) as a solvent to facilitate the degradation of decabromodiphenyl ether in an aqueous medium. The use of a lipid results in choline-coated silica nanoparticles (NPs) for the bioremediation of polycyclic aromatic hydrocarbons (PAHs). Previous studies have examined alternative nanomaterials (NMs) with the purpose of enhancing the remediation of mercury-contaminated soils with instance, Gong et al. (2012) investigated the efficacy of carboxymethylcellulose-stabilized iron sulfide NMs in this context. Trujillo-Reyes et al. (2014) have shown that zerovalent iron nanoparticles (ZVI NPs) have exceptional efficacy in the remediation of aqueous soil solutions polluted with ibuprofen. In their study, Olson et al. (2014) employed metal nanoparticles, namely iron (Fe) and magnesium (Mg), to effectively mitigate the presence of polychlorinated biphenyls (PCBs) in soil. The researchers achieved a substantial reduction of 56% in PCB levels, accompanied with a breakdown rate of 19 mg kg^{-1} $week^{-1}$.

Biochars have accomplished reputation in current years as capable additions to clean up polluted soil and water (Azadi and Raiesi 2021; Chen et al. 2022). According to studies, biochar may effectively remove contaminants from wastewater and lower the bioavailability of pollutants in soils (Li et al. 2020; Ren et al. 2022; Sun et al. 2022). In oxygen-limited conditions, biochar is produced by pyrolyzing biomass source materials at low to high temperatures (300–700 °C) (Li et al. 2022). Because it uses agricultural and forestry wastes, biochar is a cost-effective and ecologically friendly carbon sequestration addition (Lyu et al. 2020).

Nanobiochar refers to biochar that exhibits particle sizes ranging from 1 to 100 nm, as described by Zhang et al. (2022). Broadly speaking, there are two primary methods for acquiring nanobiochar. One of the substances is synthetically manufactured, while the other substance is the result of the natural degradation of biochar over time in the surrounding environment (Huang et al. 2021). According to Ramanayaka et al. (2020) and Pratap et al. (2022), nanobiochar exhibits a much higher specific surface area (SSA), a smaller hydrodynamic radius, an increased negative zeta potential, and a higher concentration of oxygen-containing functional groups compared to bulk biochar.

Nanobiochar has been studied and used in a variety of disciplines, including as pollution adsorbents, soil amendments, biosensors, and photocatalytic materials (Zhang et al. 2022). Furthermore, 51% of investigations indented on using nanobiochar to clean up contaminated soils and water. According to Zhang et al. (2021), nanobiochar can be an efficient preservative for discharging pollutants from polluted water. It can also be used to maintain pollutants in polluted soils (Sun et al. 2022).

Currently, systematic study on nanobiochar has not been conducted. Long-term investigation of nanobiochar effects on the ecosystem is also insufficient. Despite the fact that some papers have reviewed related studies on nanobiochar, these evaluations either concentrate on the substance characteristics or attempt to provide wide overviews of its use in a variety of industries. The objective of this study was to synthesize the findings from multiple studies regarding the following topics: (1) the effects of nanobiochar on the reduction of water and soil pollution; (2) the mechanisms involved in the formation of nanobiochar; and (3) the factors that influence the efficacy of nanobiochar in the remediation of environmental pollutants and associated systems. The primary objective of this work is to provide theoretical underpinning and practical guidance for the secure and efficient use of nanobiochar in the mitigation of environmental contamination.

It is still completely unknown how nanomaterials affect different ecosystems, as well as how they operate, live, and emit metal ions. Nanoremediation has a number of benefits, including the ability to be used on a wide scale, cheaper prices, and quicker clean-up times for polluted areas. However, in-depth study is necessary to determine the impacts of nanoremediation at the ecosystem level in order to stop unfavorable environmental side effects.

4.6.1.2 Mechanism of Microbial Remediation

The complications of soil pollution and land degradation are common and have both anthropogenic and natural sources (Malla et al. 2018). Heavy metals, insecticides, and polycyclic aromatic hydrocarbons (PAHs) are the most constant pollutants in soil. A natural and long-lasting solution for soil remediation after crude oil contamination is bioremediation, which uses plants and microorganisms. The concentrations of heavy metals in soil have risen recently as a result of heavy metal discharges in significant quantities from mining and industrial activities, which ultimately end up as deposits in the soil. An increase in soil heavy metal concentrations may also have been caused by widespread use of fertilizers and pesticides (Malidareh et al. 2014).

Two strategies are used to address heavy metal pollution in soil: (1) conventional chemical reclamation procedures established on curing and leaching and (2) ecological restoration systems based on adsorption and transfer (Lloyd 2002). The employment of hyperaccumulators to discharge heavy metals from contaminated soils is known as phytoremediation, which is the traditional method of ecological restoration (Marques et al. 2009).

However, microbial remediation has benefited demand and is extensively viewed as auspicious due to its bountiful benefits, along with the preservation of soil structure, the ability to almost completely remove pollutants and microbes from soils, and the ability to prevent secondary pollution (Mullen et al. 1989; Mattuschka and Straube 1993). A new area of study and development in bioremediation technology, microbial remediation offers fresh approaches for exchange with the concern of heavy metal pollution in soil and delivers as an effective model for restoring the health of the soil ecosystem. A natural, affordable, and environmentally acceptable technique of degrading pesticides and other xenobiotics on-site is known as microbial biosurfactant-based remediation. Biosurfactants make pesticides available, and bacteria exploit them as an origin of carbon, nitrogen, and phosphorus (Odukkathil and Vasudevan 2016; Bhatt et al. 2021; Lamilla et al. 2021).

4.6.1.3 Microbial Remediation of Heavy Metal-Contaminated Soil

According to Alloway (1995), the two primary mechanisms that raise the concentration of metal ions in soil are called adsorption and absorption. Microbes may employ any of these processes to accumulate heavy metals. Adsorption is different from absorption in that a liquid or solid dissolves or permeates a fluid (Jovancicevic et al. 1987). Adsorption is therefore a surface action, whereas absorption changes the entire volume of the substance. The precise immobilization mechanisms have not been fully understood because of the limits of current analytical techniques and the complexity of soil matrix (Wuana and Okieimen 2011). Numerous ions in the functional groups on the cell surface, including nitrogen, oxygen, sulfur, and phosphorus (Brady and Duncan 1994), are capable of forming complexes with metal ions as coordination atoms. Also, the microbial cell wall accommodates negatively charged phosphoric acid anions and carboxyl anionic groups, allowing the mass of heavy metals to either comply with or pass through the cell membrane (Sarret et al. 1998). Adsorption typically does not rely on energy metabolism. In general, heavy metal ions are quickly absorbed by microorganisms in high quantities. *Bacillus* can adsorb 60% of its Cu_2^+ capacity at pH 7.2 within the first minute and reach adsorption equilibrium within 10 min, according to research (He and Tebo 1998). While the clearance rates can be increased by 26.3–31.5% by adding substances like lemon oil or ethylene diamine tetra acetic acid (EDTA), absorption takes time and is often inefficient (Gan et al. 2012).

Brady and Duncan (1994) confirmed that while yeast is adsorbing Cu_2^+, it discharges approximately 70% of K^+ rapidly and 60% of Mg_2^+ gradually. Additionally, research has demonstrated that complexation can result in ion exchange (Brunetti et al. 2012). The elevations of discharged cations (Ca_2^+ and Mg_2^+) are always less than those of heavy metal ions, attributing that ion exchange is not the primary mechanism of microbial remediation, according to another study by Brady and Duncan (1994).

4.6.2 Bioleaching

Brunetti et al. (2012) suggest that the concept of "biomineralization" may include both bioleaching and bio-oxidation processes. Bioleaching is a method that facilitates the mobilization of inactive heavy metal ions from ores that are not soluble. Conversely, bio-oxidation often entails biological dissolution or complexation mechanisms (Rahman and Sathasivam 2015). Rahman et al. assert that the word "biomineralization" encompasses both of these processes. Chanmugathas and Bollag (1988) conducted a study in nutrient-rich settings and confirmed that microorganisms have the ability to efficiently use nutrients and energy in order to release organic acids, hence facilitating the leaching of Cd. Previous research has provided empirical evidence supporting the ability of some bacteria, such as Citrobacter, to produce free inorganic phosphate. This process leads to the formation of a solid metal phosphate layer, which effectively immobilizes a considerable quantity of toxic metals (Marchenko et al. 2015).

4.6.3 Plant–Microbial Remediation

The capability of plants to absorb or adsorb heavy metals may be stepped forward by a diversity of microorganisms, consisting of mycorrhizal fungi and other individuals in the rhizosphere (Bojórquez and Voltolina 2016). The uptake of Cd through mycorrhizal plants was revealed by Joner and Leyval (1997) to be 90, 127, and 131% higher than that of non-mycorrhizal plants at soil Cd_2^+ concentrations of 1,

10, and 100 mg/kg. For instance, mycelia produced by mycorrhizal fungi cultivate into the soil, gaining the surface area of plant roots (Trellu et al. 2016).

After the inoculation of mycorrhiza, Bissonnette et al. (2010) demonstrated a significant improvement in the plant's ability to absorb Cu^{2+}, Cd^{2+}, and Zn^{2+}. Endophytic mycorrhizal associations have been found to contribute to the development of heavy metal ion resistance in host plants. This synergy is primarily achieved through several key mechanisms, including acidification, the production of chelating agents, the facilitation of iron transporters, organic acid secretion, and the activation of metal phosphates. The mucus released by the fungal cell wall can form bonds with polyphosphate and organic acid ions within the fungal tissue, effectively immobilizing heavy metal ions in the soil when their concentrations reach alarming levels.

Arbuscular mycorrhizal fungi have been shown to bind Mn^{2+}, Zn^{2+}, and Cd^{2+} at levels equivalent to 1.6%, 2.8%, and 13.3% of their dry weights, respectively, as reported in the study by Chen et al. (1997). Moreover, Cervantes et al. (2001) reported that Furthermore, plant mycorrhizae have a defense mechanism that binds heavy metals to cell walls, preventing them from penetrating plant tissues.

4.7 Implementing Microbial Remediation

Microbial remediation has considerable potential to eliminate these pollutants from our environment. Prior, the major target was on microbial remediation using genetic engineering; however, the current study delivers the integrative access through GMOs to the multiomics. The introduction of certain strains of microbes or genetically modified organisms (GMOs) into a contaminated area necessitates comprehensive pilot testing and an assessment of potential alterations to the local environment. Furthermore, the utilization of multiomics methodologies plays a crucial function in comprehending and formulating the fundamental and pragmatic characteristics of microbial-mediated remediation strategies in environmentally polluted areas (Monu et al. 2020).

While microbial processes have the ability to use a wide range of substances for their metabolic activities, it has been shown that these processes may be costly in terms of their design, implementation, and long-term sustainability. Undoubtedly, the primary constraint of microbial remediation lies in the very precise inherent characteristics of their metabolic activities. Hence, in order for effective remediation to occur, it is essential that the environmental conditions be conducive and that appropriate quantities of nutrients or pollutants, serving as sources of energy, are available to support the synergistic growth of microbial populations.

Furthermore, the process of natural bioremediation requires a significant amount of time to effectively eliminate the toxins. The incorporation of appropriate foreign microorganisms capable of expedited repair has encountered ethical and management challenges. According to Sharma (2021), the introduction of alien microorganisms might potentially have significant and adverse consequences on the local ecology and the biological variety of the habitat. In addition, the use of microbial genetically modified organisms (GMOs) might give rise to uncertainty and perhaps result in the transfer of genes, so impacting the indigenous microbial populations of the environment (Wozniak et al. 2012). Genetically modified organisms (GMOs) have often been associated with crops, hence encountering increasing regulatory complexities in several countries.

The presence of certain organic contaminants in the environment has emerged as a significant concern in terms of pollution. Bioremediation has emerged as a viable approach for mitigating the release of harmful chemicals into the environment, and its significance has been recognized by researchers in recent decades. The scientific literature on the interactions between microorganisms and specific contaminants in recent decades has provided substantial evidence for the mitigation of environmental contamination. The use of traditional bioremediation procedures is constrained by certain limits. Consequently, it becomes imperative to explore novel bioremediation approaches that include biotechnological interventions in order to achieve optimal outcomes. It is anticipated that advancements in various techniques would enhance the efficacy of bioremediation methods and facilitate ecologically sustainable practices. The authors of the study conducted a comprehensive analysis of the potential outcomes and risks associated with various treatment strategies for the cleanup of contaminants (Mishra et al. 2021).

The book *Microbes and Microbial Biotechnology for Green Remediation* provides a comprehensive overview of sustainable microbial treatment approaches. The study presented focuses on the significantly important microbial species involved in remediation, the methods by which various microorganisms facilitate remediation, and strategies for the future development of bioremediation technology (Malik 2022). The introduction of pollutants into the environment due to rapid urbanization and human activities leads to changes and distress in the physicochemical systems, which also affect living creatures. Therefore, there exists a pressing global mandate for the mitigation of these pollutants and xenobiotics since they have the potential to adversely impact many forms of life (Miglani et al. 2022).

Over the course of time, efforts in microbial remediation have been accelerated in order to develop enhanced, environmentally sustainable, and highly biodegradable solutions for the comprehensive dispersal of these xenobiotic substances. The advancement in the fields of microbiology and biotechnology commenced with the emergence of microbial biotechnology as a distinct field of study. This development has significantly contributed to the progress in various domains including agriculture, environment, biopharmaceutics, and fermented foods (Biju and Krishnaswamy 2021). Microorganisms occupy a crucial and advantageous position as an environmentally friendly and cost-effective alternative to conventional treatment methods. The research provides economically viable and environmentally friendly options. According to Malik (2022), the likelihood of success in remediation efforts is much enhanced when a systematic approach is used. This entails thorough consideration of the situation at hand, allocation of suitable resources, and collaboration with key stakeholders.

4.8 Future Impact

Environmental pollution can be greatly reduced by microbial remediation. Available limited knowledge on the variables for the control of degradation process in the current environmental circumstances limits its widespread application. However, the development of biotechnology makes it possible to accelerate the breakdown of pollutants found in crude oil. Transcriptomics, proteomics, genomics, and metabolomics are examples of cutting-edge methods that are considered crucial to eliminate these contaminants in a sustainable way.

Genetic engineering-based microbial remediation was the main focus in the past; however, today's integral access through GMOs to the multiomics modes is being used to recognize and expose the elemental and functional components of these microorganism-mediated remediation measures within environmental contaminated sites. Future research should concentrate on a larger variety of contaminants that can be treated successfully. Nonbiodegradable pollutants should be included among the substances that need to be metabolized and broken down. More investigation is required to take advantage of the bioremediation capabilities of developing pollutants, boost bioadsorption, improve biodegrading enzymes, and optimize growth conditions in microalgae species, which should result in improving bioremediation technology efficiency to further lessen environmental stress on land and aquatic ecosystems. Carbon nanotubes (CNTs) are better nanomaterials for organic and inorganic cleanup because of their massive surface area and elevated adsorption capacity. Research should focus on these issues as a trial-and-error to advance microbial remediation technologies. Last but not least, the greatest challenge facing researchers in the future is to expand the scope of their understanding of the microbial populations that break down crude oil contamination.

4.9 Conclusion

Any nation that permits the use of natural resources while causing the least amount of harm to the environment must prioritize sustainable development in order to experience economic progress. It is crucial to get rid of soil contamination brought on by crude oil leaks since it negatively affects human health and plant growth. Given that crude oil contains a number of dangerous, toxic, and carcinogenic substances, the development of remediation technologies to clean up places that have become contaminated as a result of crude oil spills during transportation is crucial.

Crude oil removal from soil can be accomplished using a variety of techniques, including physical, chemical, thermal, and biological ones. Bioremediation is a very effective and self-propelled economic process when compared to physical and chemical cleanup procedures. Microbes may be employed in bioremediation techniques because they are effective, affordable, and environmentally acceptable. To successfully degrade crude oil-contaminated locations, environmental conditions, in particular tilling the soil to access oxygen for microbial metabolism, managing the soil moisture content, and soil pH, are also crucial. Growing microalgae is the best technique to remove pollutants from the soil environment. The cleaning techniques employed will not be established as general techniques but rather as alternatives for the bioremediation of soils affected by crude oil because each bioremediation technology used has its own unique properties. Based on this, the mixed design of bioremediation technologies—which permits the integration of two or more bioremediation technologies given that they have a higher performance and effectiveness in soil restoration, as in the case of bioaugmentation + biostimulation—that were used in other investigations—is proposed.

During the course of the research, new proposals for the remediation of contaminated soils were noticed. One of these proposals is nanoremediation, a strategy that is currently being developed and has a promising outlook for enhancing bioremediation. It incorporates various types of food, agro-industrial, and tree waste materials as significant sources of polyphenols in the successful removal of environmental pollutants. The final proposal is genetic engineering, which suggests the use of the technique to improve bioremediation. By accelerating the uptake and accumulation of pollutants in plants, speeding up the rate at which pollutants are broken down by microorganisms, and decreasing the remediation time and overall cost, as well as the fact that no disposal of the contaminated soil is necessary, the use of nanomaterials can enhance the effects of bioremediation. nZVI nanoparticles are frequently employed for environmental cleanup because of their high reactivity and powerful capacity to immobilize heavy metals such as Cd, Ni, and Pb.

Each of these technologies was thoroughly examined, along with the benefits, drawbacks, developments, and upcoming work related to them. But it is vital to understand that no one remediation technique is thought to be the ideal one for cleaning up oil-contaminated soil. Finally, integrating optimal environmental, edaphic, and climatic parameters of a typical contaminated site into the process from the outset is a difficulty in the advancement of bioremediation from laboratory research to large-scale field applications on crude oil-contaminated soils. This study will be helpful in finding a safe and practical answer for the issue of how to dispose of oil-contaminated soils by reusing the soil in construction applications.

References

Adenipekun CO, Lawal R (2012) Uses of mushrooms in bioremediation: a review. Biotechnol Mol Biol Rev 7(3):62–68

Adeola AO, Akingboye AS, Ore OT, Oluwajana OA, Adewole AH, Olawade DB, Ogunyele AC (2022) Crude oil exploration in Africa: socio-economic implications, environmental impacts, and mitigation strategies. Environ Syst Decis 42(1):26–50

Ahmad A, Ghufran R, Al-Hosni TK (2019) Bioavailability of zinc oxide nano particle with fly ash soil for the remediation of metals by Parthenium hysterophorus. J Environ Health Sci Eng 17:1195–1203

Aksu Z, Balibek E (2007) Chromium (VI) biosorption by dried Rhizopus arrhizus: effect of salt (NaCl) concentration on equilibrium and kinetic parameters. J Hazard Mater 145:210–220
Alloway BJ (1995) Heavy metals in soils, 2nd edn. Springer, Dordrecht
Azadi N, Raiesi F (2021) Biochar alleviates metal toxicity and improves microbial community functions in a soil co-contaminated with cadmium and lead. Biochar 3:485–498
Bhatt P, Verma A, Gangola S, Bhandari G, Chen S (2021) Microbial glycoconjugates in organic pollutant bioremediation: recent advances and applications. Microb Cell Fact 20:1–18
Biju LM, Krishnaswamy VG (2021) Bioremediation and phytoremediation: the remedies for xenobiotics. In: Handbook of research on microbial remediation and microbial biotechnology for sustainable soil. IGI Global, pp 38–64
Bissonnette L, St-Arnaud M, Labrecque M (2010) Phytoextraction of heavy metals by two Salicaceae clones in symbiosis with arbuscular mycorrhizal fungi during the second year of a field trial. Plant Soil 332:55–67
Bojórquez C, Voltolina D (2016) Removal of cadmium and lead by adapted strains of Pseudomonas aeruginosa and Enterobacter cloacae. Rev Int Contam Ambie 32:407–412
Bollag JM (1974) Microbial transformation of pesticides. Adv Appl Microbiol 18:75–130
Brady D, Duncan JR (1994) Cation loss during accumulation of heavy metal cations by Saccharomyces cerevisiae. Biotechnol Lett 16:543–548
Brunetti G, Farrag K, Soler-Rovira P, Ferrara MN, F and Senesi, N. (2012) The effect of compost and Bacillus licheniformis on the phytoextraction of Cr, Cu, Pb and Zn by three Brassicaceae species from contaminated soils in the Apulia region, Southern Italy. Geoderma 170:322–330
Capeness MJ, Echavarri-Bravo V, Horsfall LE (2019) Production of biogenic nanoparticles for the reduction of 4-Nitrophenol and oxidative laccase-like reactions. Front Microbiol 10:997
Cappelletti M, Presentato A, Piacenza E et al (2020) Biotechnology of Rhodococcus for the production of valuable compounds. Appl Microbiol Biotechnol 104:8567–8594
Cerniglia CE, Van Baalen C, Gibson DT (1980) Metabolism of naphthalene by the cyanobacterium Oscillatoria sp., strain JCM. J Gen Microbiol 116:485–494
Cervantes C, Campos-García J, Devars S, Gutiérrez-Corona F, Loza-Tavera H, Torres-Guzmán JC, Moreno-Sánchez R (2001) Interactions of chromium with microorganisms and plants. FEMS Microbiol Rev 25:335–347
Chang SE, Stone J, Demes K, Piscitelli M (2014) Consequences of oil spills: a review and framework for informing planning. Ecol Soc 19(2):26
Chanmugathas P, Bollag JM (1988) A column study of the biological mobilization and speciation of cadmium in soil. Arch Environ Contam Toxicol 17:229–237
Chen TB, Wong JWC, Zhou HY, Wong MH (1997) Assessment of trace metal distribution and contamination in surface soils of Hong Kong. Environ Pollut 96:61–68
Chen H-S, Zhang Q-M, Yang Z-J, Liu Y-S (2020) Research on treatment of oily sludge from the tank bottom by ball milling combined with ozone-catalyzed oxidation. ACS Omega 5(21):12259–12269
Chen H, Gao Y, Li J, Sun C, Sarkar B, Bhatnagar A, Bolan N, Yang X, Meng J, Liu Z, Hou H, Wong JWC, Hou D, Chen W, Wang H (2022) Insights into simultaneous adsorption and oxidation of antimonite [Sb(III)] by crawfish shell-derived biochar: spectroscopic investigation and theoretical calculations. Biochar 4:37
Chikere CB, Okpokwasili GC, Chikere BO (2009) Bacterial diversity in a tropical crude oil-polluted soil undergoing bioremediation. Afr J Biotechnol 8:2535–2540
Chong AYL, Ooi KB, Bao H, Lin B (2014) Can e-business adoption be influenced by knowledge management? An empirical analysis of Malaysian SMEs. J Knowl Manag 18(1):121–136
Christian V, Shrivastava R, Shukla D, Modi HA, Vyas BRM (2005) Degradation of xenobiotic compounds by lignin-degrading white-rot fungi: enzymology and mechanisms involved. Indian J Exp Biol 45:301–312
Collins PJ, Kotterman M, Field JA, Dobson A (1996) Oxidation of anthracene and benzo [a] pyrene by laccases from Trametes versicolor. Appl Environ Microbiol 62(12):4563–4567

Couto SR, Sanroman MA, Hoefer D, Gubitz GM (2004) Stainless steel: a novel career for the immobilization of the white rot fungus Trametes hirsute for decolorization of textile dyes. Bioresour Technol 95:67–72
Daraei H, Mittal A, Noorisepehr M, Daraei F (2013) Kinetic and equilibrium studies of adsorptive removal of phenol onto eggshell waste. Environ Sci Pollut Res 20:4603–4611
Davies JS, Westlake DWS (1979) Crude oil utilization by fungi. Can J Microbiol 25:146–156
De Llasera MPG, Olmos-Espejel JJ, Díaz-Flores G, Montaño-Montiel A (2016) Biodegradation of benzo(a)pyrene by two freshwater microalgae *Selenastrum capricornutum* and *Scenedesmus acutus*: a comparative study useful for bioremediation. Environ Sci Pollut Res 23:3365–3375
Dombrowski N, Donaho JA, Gutierrez T, Seitz KW, Teske AP, Baker BJ (2016) Reconstructing metabolic pathways of hydrocarbon-degrading bacteria from the Deepwater Horizon oil spill. Nat Microbiol 1:16057
Dong ZL, Wang BS, Li J (2020) Effects of petroleum hydrocarbon contamination on soil bacterial diversity in the permafrost region of the Qinghai-Tibetan Plateau. Soil Sedim Contam 29(3):322–339
Dong-Sheng Q, Chang-Yuan G (2022) Control measures for automobile exhaust emissions in PM2.5 governance. Discrete dynamics in nature and society. pp 1–14
Dvořák P, Nikel PI, Damborský J, de Lorenzo V (2017) Bioremediation 3.0: engineering pollutant-removing bacteria in the times of systemic biology. Biotechnol Adv 35:845–866
Elliott DW, Lien HL, Zhang WX (2009) Degradation of lindane by zero-valent iron nanoparticles. J Environ Eng 135:317–324
Erkurt EA, Ünyayar A, Kumbur H (2007) Decolorization of synthetic dyes by white rot fungi, involving laccase enzyme in the process. Process Biochem 42(10):1429–1435
Ezeji E, Anyadoh SO, Ibekwe VI (2007) Clean up of crude oil contaminated soil Terrestrial and Aquatic. Environ Toxicol 1(2):54–59
Faraco V, Pezzella C, Miele A, Giardina P, Sannia G (2009) Bio-remediation of colored industrial wastewaters by the white-rot fungi Phanerochaete chrysosporium and Pleurotus ostreatus and their enzymes. Biodegradation 20(2):209–220
Gaitan IJ, Medina SC, González JC, Rodríguez A, Espejo ÁJ, Osma JF, Sarria V, Alméciga-Díaz CJ, Sánchez OF (2011) Evaluation of toxicity and degradation of a chlorophenol mixture by the laccase produced by Trametes pubescens. Bioresour Technol 102(3):3632–3635
Gamallo M, Fernández L, Feijoo G, Moreira M (2020) Nano-based technologies for environmental soil remediation. In: Nanomaterials for sustainable energy and environmental remediation. Elsevier, Amsterdam, pp 307–331
Gan WJ, Yue HE, Zhang XF, Shan YH, Zheng LP, Lin YS (2012) Speciation analysis of heavy metals in soils polluted by electroplating and effect of washing to the removal of the pollutants. J Ecol Rural Environ 28:82–87
Ghasemi Y, Rasoul-Amini S, Fotooh-Abadi E (2011) The biotransformation, biodegradation, and bioremediation of organic compounds by microalgae. J Phycol 47:969–980
Ghosh S, Syed H (2001) Influence of soil characteristics on bioremediation of petroleum-contaminated soil. Geological Society of America Annual Meeting, Nov. 5–8, Boston, Massachusetts, USA
Gillham RW, O'Hannesin SF (1994) Enhanced degradation of halogenated aliphatics by zero-valent iron. Groundwater 32:958–967
Gong Y, Liu Y, Xiong Z, Kaback D, Zhao D (2012) Immobilization of mercury in field soil and sediment using carboxymethyl cellulose stabilized iron sulfide nanoparticles. Nanotechnology 23:294007
Hammer E, Schauer F (1997) Fungal hydroxylation of dibenzofuran. Mycol Res 101:433–436
Hammer E, Krowas D, Schafer A, Specht M, Francke W, Schauer F (1998) Isolation and characterization of dibenzofuran-degrading yeast: identification of oxidation and ring cleavage products. Appl Environ Microbiol 64:2215–2219
He LM, Tebo BM (1998) Surface charge properties of and Cu(II) Adsorption by Spores of the Marine Bacillus sp. Strain SG-1. Appl Environ Microbiol 64:1123–1129

Hewelke E, Szatyłowicz J, Hewelke P, Gnatowski T, Aghalarov R (2018) The impact of diesel oil pollution on the hydrophobicity and CO_2 efflux of forest soils. Water Air Soil Pollut 229(2):51

Hong YW, Yuan DX, Lin QM, Yang TL (2008) Accumulation and biodegradation of phenanthrene and fluoranthene by the algae enriched from a mangrove aquatic ecosystem. Mar Pollut Bull 56:1400–1405

Huang X, Zhu S, Zhang H, Huang Y, Wang X, Wang Y, Chen D (2021) Biochar nanoparticles induced distinct biological effects on freshwater algae via oxidative stress, membrane damage, and nutrient depletion. ACS Sustain Chem Eng 9:10761–10770

Hussain I, Aleti G, Naidu R, Puschenreiter M, Mahmood Q (2018) Microbe and plant assisted-remediation of organic xenobiotics and its enhancement by genetically modified organisms and recombinant technology: a review. Sci Total Environ 628–629:1582–1599

IARC (International Agency for Research on Cancer) (1983) In: IARC (ed) IARC monographs on the evaluation of the carcinogenic risk of chemicals to humans: polynuclear aromatic compounds part I. IARC Press, Lyon

Ibraheem M (2010) Biodegradability of hydrocarbons by cyanobacteria. J Phycol 46:818–824

Imasogie S, Odia LO (2009) Oil exploitation and conflict in The Niger-Delta region of Nigeria April 2009. J Hum Ecol (Delhi, India) 26(1):25–23

Joner EJ, Leyval C (1997) Uptake of 109Cd by roots and hyphae of a Glomus mosseae/Trifolium subterraneum mycorrhiza from soil amended with high and low concentrations of cadmium. New Phytol 135:353–360

Jovancicevic V, Bockris JO, Carbajal JL, Zelenay P, Mizuno T (1987) Abstract: adsorption and absorption of chloride ions on passive iron systems. ChemInform 18:2219–2226

Kapoor A, Viraraghavan T, Cullimore DR (1999) Removal of heavy metals using the fungus Aspergillus niger. Bioresour Technol 70:95–104

Keum YS, Li QX (2004) Fungal laccase-catalyzed degradation of hydroxy polychlorinated biphenyls. Chemosphere 56(1):23–30

Khalkhal K, Asgari Lajayer B, Ghorbanpour M (2020) An overview on the effect of soil physicochemical properties on the immobilization of biogenic nanoparticles. In: Biogenic nanoparticles and their use in agro-ecosystems. Springer, Singapore, pp 133–160

Khullar S, Reddy MS (2019) Cadmium and arsenic responses in the ectomycorrhizal fungus Laccaria bicolor: glutathione metabolism and its role in metal (loid) homeostasis. Environ Microbiol Rep 11:53–61

Kullman SW, Matsumura F (1996) Metabolic pathways utilized by Phanerochaete chrysosporium for degradation of the cyclodiene pesticide endosulfan. Appl Environ Microbiol 62:593–600

Kumar A, Chandra R (2020) Ligninolytic enzymes and their mechanisms for degradation of lignocellulosic waste in the environment. Heliyon 6:e03170

Kumar G, Varsha T, Natesan M (2022) Chapter 18—Bacterial- and fungal-mediated biodegradation of petroleum hydrocarbons in soil. In: Development in wastewater treatment research and processes microbial degradation of xenobiotics through bacterial and fungal approach. p 407

Kuppusamy S, Thavamani P, Venkateswarlu K, Lee YB, Naidu R, Megharaj M (2017) Remediation approaches for polycyclic aromatic hydrocarbons (PAHs) contaminated soils: technological constraints, emerging trends and future directions. Chemosphere 168:944–968

Lamilla C, Schalchli H, Briceño G, Leiva B, Donoso-Piñol P, Barrientos L (2021) A pesticide biopurification system: a source of biosurfactant producing bacteria with environmental biotechnology applications. Agronomy 11:624

Lance WF, Michele A, Coklin N, Chris S (2023) Bioremediation and phytoremediation of oil contaminated salt marsh in South Louisiana. The Benthic Zone Newsletter Jan 4, 2023

Leahy JG, Colwell RR (1990) Microbial degradation of hydrocarbons in the environment. Microbiol Rev 54:305–315

Lea-Smith DJ, Biller SJ, Davey MP, Cotton CA, Sepulveda BMP, Turchyn AV (2015) Contribution of cyanobacterial alkane production to the ocean hydrocarbon cycle. Proc Natl Acad Sci USA 112:13591–13596

Lei AP, Wong YS, Tam NFY (2002) Removal of pyrene by different microalgal species. Water Sci Technol 46:195–201
Levin L, Melignani E, Ramos AM (2010) Effect of nitrogen sources and vitamins on ligninolytic enzyme production by some white-rot fungi. Dye decolorization by selected culture filtrates. Bioresour Technol 101(12):4554–4563
Li R, Zhang Y, Deng H, Zhang Z, Wang JJ, Shaheen SM, Xiao R, Rinklebe J, Xi B, He X, Du J (2020) Removing tetracycline and Hg(II) with ball-milled magnetic nanobiochar and its potential on polluted irrigation water reclamation. J Hazard Mater 384:121095
Li DP, Ma J, Xu HC, Xu XY, Qiu H, Cao XD, Zhao L (2022) Recycling waste nickel-laden biochar to pseudo-capacitive material by hydrothermal treatment: roles of nickel-carbon interaction. Carbon Res 1:16
Liao Y, Yang J (2020) Remediation of vanadium contaminated soil by nano-hydroxyapatite. J Soils Sediments 20:1534–1544
Liu Y, Xu K, Cheng J (2020) Different nanomaterials for soil remediation affect avoidance response and toxicity response in earthworm (Eisenia fetida). Bull Environ Contam Toxicol 104:477–483
Lloyd JR (2002) Bioremediation of metals; the application of micro-organisms that make and break minerals. Microbiol Today 29:67–69
Lv Y, Huang S, Huang G, Liu Y, Yang G, Lin C, Xiao G, Wang Y, Liu M (2020) Remediation of organic arsenic contaminants with heterogeneous Fenton process mediated by SiO_2-coated nano zero-valent iron. Environ Sci Pollut Res 27:12017–12029
Lyu H, Xia S, Tang J, Zhang Y, Gao B, Shen B (2020) Thiol-modified biochar synthesized by a facile ball-milling method for enhanced sorption of inorganic Hg2+ and organic CH_3Hg^+. J Hazard Mater 384:121357
Machado S, Pacheco J, Nouws H, Albergaria JT, Delerue-Matos C (2015) Characterization of green zero-valent iron nanoparticles produced with tree leaf extracts. Sci Total Environ 533:76–81
Malidareh, H. B., Mahvi, A. H., Yunesian, M., Alimohammadi, M. and, Nazmara, S. (2014). Effect of fertilizer application on paddy soil heavy metals concentration and groundwater in North of Iran. Middle-East J Sci Res, 20, 1721–1727
Malik JA (2022) Microbes and microbial biotechnology for green remediation. Elsevier, pp vii–xxiii. https://doi.org/10.1016/B978-0-323-90452-0.00052-9. ISBN 9780323904520
Malla MA, Dubey A, Yadav S, Kumar A, Hashem A, Abd-Allah EF (2018) Understanding and designing the strategies for the microbe-mediated remediation of environmental contaminants using omics approaches. Front Microbiol 9:1132
Marchenko AM, Pshinko GN, Demchenko VY, Goncharuk VV (2015) Leaching heavy metal from deposits of heavy metals with bacteria oxidizing elemental sulphur. J Water Chem Technol 37:311–316
Margesin R, Labbe D, Schinner F, Greer CW, Whyte LG (2003) Characterization of hydrocarbon degrading microbial populations in contaminated and pristine Alpine soils. Appl Environ Microbiol 69:3085–3092
Marinescu M, Toti M, Tanase V, Plopeanu G, Calciu I, Marinescu M (2011) The effects of crude oil pollution on physical and chemical characteristics of soil. Res J Agric Sci 43(3):125–129
Marques APGC, Rangel AOSS, Castro PML (2009) Remediation of heavy metal contaminated soils: phytoremediation as a potentially promising clean-up technology. Crit Rev Environ Sci Technol 39:622–654
Matsubara M, Lynch JM, De Leij FAAM (2006) A simple screening procedure for selecting fungi with potential for use in the bioremediation of contaminated land. Enzym Microb Technol 39(7):1365–1372
Mattuschka B, Straube G (1993) Biosorption of metals by a waste biomass. J Chem Technol Biotechnol 58:57–63
Mcgenity TJ, Folwell BD, Mckew BA, Sanni GO (2012) Marine crude-oil biodegradation: a central role for interspecies interactions. Aquat Biosyst 8:10

Mehdi HS, Giti E, Simone C (2012) Isolation and characterization of crude-oil-degrading bacteria from the Persian Gulf and the Caspian Sea. Mar Pollut Bull 64(1):7–12
Miglani R, Parveen N, Kumar A, Ansari MA, Khanna S, Rawat G, Panda AK, Bisht SS, Upadhyay J, Ansari MN (2022) Degradation of xenobiotic pollutants: an environmentally sustainable approach. Metabolites 12(9):818
Mishra B, Varjani S, Kumar G, Awasthi MK, Awasthi SK, Sindhu R, Binod P, Rene ER, Zhang Z (2021) Microbial approaches for remediation of pollutants: innovations, future outlook, and challenges. Energy Environ 32(6):1029–1058
Monu J, Manish Y, Nitin KS, Suman Yadav IS, Swati D, Arti T (2020) Microbial remediation progress and future prospects. In: Bioremediation of pollutants. Elsevier, pp 187–214. ISBN 9780128190258
Mullen MD, Wolf DC, Ferris FG, Beveridge TJ, Flemming CA, Bailey GW (1989) Bacterial sorption of heavy metals. Appl Environ Microbiol 55:3143–3149
Muñoz R, Guieysse B, Mattiasson B (2003) Phenanthrene biodegradation by an algal-bacterial consortium in two-phase partitioning bioreactors. Appl Microbiol Biotechnol 61:261–267
Nakamiya K, Hashimoto S, Ito H, Edmonds JS, Yasuhara A, Morita M (2005) Degradation of dioxins by cyclic ether degrading fungus, Cordyceps sinensis. FEMS Microbiol Lett 248:17–12
Nelson JR, Grubesic TH, Sim L, Rose K (2018) A geospatial evaluation of oil spill impact potential on coastal tourism in the Gulf of Mexico. Comput Environ Urban Syst 68:26–36
Nidhin S, Aswathy U, Shantkriti S (2020) Algal bioremediation of heavy metals removal of toxic pollutants through microbiological and tertiary treatment. Elsevier, pp 279–307. ISBN 9780128210147
Obayori SO, Salam LB (2010) Degradation of polycyclic aromatic hydrocarbons: role of plasmids. Sci Res Ess 5:4093–4106
Odukkathil G, Vasudevan N (2016) Residues of endosulfan in surface and subsurface agricultural soil and its bioremediation. J Environ Manag 165:72–80
Olson MR, Blotevogel J, Borch T, Petersen M, Royer R, Sale T (2014) Long-term potential of in situ chemical reduction for treatment of polychlorinated biphenyls in soils. Chemosphere 114:144–149
Olujimi JAB, Emmanuel AA, Sogbon O (2011) Environmental implications of oil exploration and exploitation in the coastal region of Ondo State, Nigeria: a regional planning appraisal. J Geo Regional Plan 4(3):110–121
Ordinioha B, Brisibe S (2013) The human health implications of crude oil spills in The Niger delta, Nigeria: an interpretation of published studies. Niger Med J 54(1):10–16
Pratap T, Chaubey AK, Patel M, Mlsna TE, Pittman CU Jr, Mohan D (2022) Nanobiochar for aqueous contaminant removal. In: Mohan D, Pittman CU, Mlsna TE (eds) Sustainable biochar for water and wastewater treatment. Elsevier, Amsterdam, pp 667–704
Qin H, Ma J, Wang Q, Li J, Chi M, Zhang L, Liu H (2016) Treatment of oil shale sludge using solvent extraction and thermo-chemistry. Chin J Environ Eng 10:851–857
Qiu X, Fang Z, Liang B, Gu F, Xu Z (2011) Degradation of decabromodiphenyl ether by nano zero-valent iron immobilized in mesoporous silica microspheres. J Hazard Mater 193:70–81
Rahman MS, Sathasivam KV (2015) Heavy metal adsorption onto Kappaphycus sp. from aqueous solutions: the use of error functions for validation of isotherm and kinetics models. Biomed Res Int 2015:126298
Rahman WU, Khan MD, Khan GH (2018) Anaerobic biodegradation of benzene-laden wastewater under mesophilic environment and simultaneous recovery of methane-rich biogas. J Environ Chem Eng 6(2):2957–2964
Rajput VD, Minkina T, Kimber RL, Singh VK, Shende S, Behal A, Sushkova MS, Lloyd JR (2021) Insights into the biosynthesis of nanoparticles by the genus Shewanella. Appl Environ Microbiol 87:e01390–e01321
Rajput VD, Minkina T, Upadhyay SK, Kumari A, Ranjan A, Mandzhieva S, Sushkova S, Singh RK, Verma KK (2022) Nanotechnology in the restoration of polluted soil. Nanomaterials 12:769

Ramanayaka S, Vithanage M, Alessi DS, Liu WJ, Jayasundera ACA, Ok YS (2020) Nanobiochar: production, properties, and multifunctional applications. Environ Sci Nano 7:3279–3302
Ramaswamy D, Kar DD, De S (2007) A study on recovery of oil from sludge containing oil using froth flotation. J Environ Manag 85:150–154
Rascio N, Navari-Izzo F (2011) Heavy metal hyperaccumulating plants: how and why do they do it? And what makes them so interesting? Plant Sci 180:169–181
Ren SY, Xu X, Hu KS, Tian WJ, Duan XG, Yi JB, Wang SB (2022) Structure-oriented conversions of plastics to carbon nanomaterials. Carbon Res 1:15
Rhodes CJ (2012) Feeding and healing the world: through regenerative agriculture and permaculture. Sci Prog 95(4):345–446
Riva S (2006) Laccases: blue enzymes for green chemistry. Trends Biotechnol 24(5):219–226
Rizwan M, Ali S, Rehman MZ, Adrees M, Arshad M, Qayyum MF, Ali L, Hussain A, Chatha SAS, Imran M (2019) Alleviation of cadmium accumulation in maize (*Zea mays* L.) by foliar spray of zinc oxide nanoparticles and biochar to contaminated soil. Environ Pollut 248:358–367
Rizwan M, Ali S, Rehman MZ, Rinklebe J, Tsang DCW, Tack FMG, Abbasi GH, Hussain A, Igalavithana AD, Lee BC, Ok YS (2021) Effects of selenium on the uptake of toxic trace elements by crop plants: a review. Crit Rev Environ Sci Technol 51(21):2531–2566
Roehl KE, Meggyes T, Simon F, Stewart D (2005) Long-term performance of permeable reactive barriers. Gulf Professional Publishing, Houston
Romeh AA, Saber RAI (2020) Green nano-phytoremediation and solubility improving agents for the remediation of chlorfenapyr contaminated soil and water. J Environ Manag 260:110104
Ron EZ, Rosenberg E (2014) Enhanced bioremediation of oil spills in the sea. Curr Opin Biotechnol 27:191–194
Sarkar A, Sengupta S, Sen S (2019) Nanoparticles for soil remediation. In: Nanoscience and biotechnology for environmental applications. Springer, Cham, pp 249–262
Sarret G, Manceau A, Spadini L, Roux JC, Hazemann JL, Soldo Y, Eybert-Bérard, and Menthonnex, J. (1998) Structural determination of Zn and Pb binding sites in Penicillium chrysogenum cell walls by EXAFS spectroscopy. Environ Sci Technol 32:1648–1655
Sassykova YA, Aubakirov S, Sendilvelan ZHK, Tashmukhambetova MF, Faizullaeva K, Bhaskar AA, Batyrbayeva RG, Ryskaliyeva BB, Tyussyupova AA, Zhakupova MA, Sarybayev S (2019) The main components of vehicle exhaust gases and their effective catalytic neutralization. Orient J Chem 35(1):110–127
Sharma P (2021) Efficiency of bacteria and bacterial assisted phytoremediation of heavy metals: an update. Bioresour Technol 328
Shi LN, Zhang X, Chen Zl (2011) Removal of chromium (VI) from wastewater using bentonite-supported nanoscale zero-valent iron. Water Res 45:886–892
Shih Y-h, Tai Yt (2010) Reaction of decabrominated diphenyl ether by zerovalent iron nanoparticles. Chemosphere 78:1200–1206
Singh H, Bhardwaj N, Arya SK, Khatri M (2020a) Environmental impacts of oil spills and their remediation by magnetic nanomaterials. Environ Nanotechnol Monit Manag 14:100305
Singh R, Behera M, Kumar S (2020b) Nano-bioremediation: an innovative remediation technology for treatment and management of contaminated sites. In: Bioremediation of industrial waste for environmental safety. Springer, Singapore, pp 165–182
Singh S, Shikha (2019) Treatment and recycling of wastewater from oil refinery/petroleum industry. In: Singh R, Singh R (eds) Advances in biological treatment of industrial waste water and their recycling for a sustainable future. Applied environmental science and engineering for a sustainable future. Springer, Singapore
Sojinu SO, Ejeromedoghene O (2019) Environmental challenges associated with processing of heavy crude oils, processing of heavy crude oils—challenges and opportunities Ramasamy Marappa Gounder
Subash Chandra Bose SR, Ramakrishnan B, Megharaj M, Venkateswarlu K, Naidu R (2013) Mixotrophic cyanobacteria and microalgae as distinctive biological agents for organic pollutant degradation. Environ Int 51:59–72

Sullivan TS, Gadd GM (2019) Metal bioavailability and the soil microbiome. Adv Agron 155:79–120

Sun TY, Bornhöft NA, Hungerbühler K, Nowack B (2016) Dynamic probabilistic modeling of environmental emissions of engineered nanomaterials. Environ Sci Technol 50:4701–4711

Sun Y, Lyu H, Cheng Z, Wang Y, Tang J (2022) Insight into the mechanisms of ball-milled biochar addition on soil tetracycline degradation enhancement: physicochemical properties and microbial community structure. Chemosphere 291:132691

Sundararaghavan A, Mukherjee A, Suraishkumar GK (2020) Investigating the potential use of an oleaginous bacterium, Rhodococcus opacus PD630, for nano-TiO_2 remediation. Environ Sci Pollut Res 27:27394–27406

Takáčová A, Smolinská M, Ryba J, Mackuľak T, Jokrllová J, Hronec P, Čík G (2014) Biodegradation of Benzo[a]Pyrene through the use of algae. Cent Eur J Chem 12:1133–1143

Tang X, He LY, Tao XQ, Dang Z, Guo CL, Lu GN, Yi XY (2010) Construction of an artificial microalgal-bacterial consortium that efficiently degrades crude oil. J Hazard Mater 181:1158–1162

Tibbetts PJC, Buchanan IT, Gawel LG, Large R (1992) A comprehensive determination of produced water composition. In: Ray JP, Englehart FR (eds) Produced water. Plenum Press, New York

Tran TD, Dao NT, Sasaki R, Tu MB, Dang GHM, Nguyen HG, Dang HM, Vo CH, Inigaki Y, van Nguyen N (2020) Accelerated remediation of organochlorine pesticide-contaminated soils with phyto-Fenton approach: a field study. Environ Geochem Health 42:3597–3608

Trellu C, Mousset E, Pechaud Y, Huguenot D, van Hullebusch ED, Esposito G, Oturan MA (2016) Removal of hydrophobic organic pollutants from soil washing/flushing solutions: a critical review. J Hazard Mater 306:149–174

Trujillo-Reyes J, Peralta-Videa J, Gardea-Torresdey J (2014) Supported and unsupported nanomaterials for water and soil remediation: are they a useful solution for worldwide pollution? J Hazard Mater 280:487–503

Vasudevan N, Rajaram P (2001) Bioremediation of oil sludge contaminated soil. Environ Int 26:409–411

Veil J (2015) U.S. produced water volumes and management practices in 2012. Ground water protection council, Oklahoma city, OK, USA. Produced water report: regulations, current practices, and research needs. 2019

Vishnu KM, Murugesan S (2020a) Biosynthesis and characterization of silver nanoparticles from marine macroscopic red seaweed Halymenia porphyroides Boergesen (Crypton). J Nanosci Technol 6(2):886–890

Vishnu KM, Murugesan S (2020b) Biosynthesis and characterization of silver nanoparticles from marine macroscopic brown seaweed Colpomenia sinuosa (Mertens ex Roth) Derbes and Solier. J Adv Chem Sci 6(1):663–666

Wang Y, Liu Y, Zhan W, Zheng K, Lian M, Zhang C, Ruan X, Li T (2020) Long-term stabilization of Cd in agricultural soil using mercapto-functionalized nano-silica (MPTS/nano-silica): a three-year field study. Ecotoxicol Environ Saf 197:110600

Wei Z, Wang JJ, Gaston LA, Li J, Fultz LM, De Laune RD, Dodla SK (2020a) Remediation of crude oil-contaminated coastal marsh soil: integrated effect of biochar, rhamnolipid biosurfactant and nitrogen application. J Hazard Mater 396:122595

Wei Z, Wang JJ, Meng Y, Li J, Gaston LA, Fultz LM, DeLaune RD (2020b) Potential use of biochar and rhamnolipid biosurfactant for remediation of crude oil-contaminated coastal wetland soil: ecotoxicity assessment. Chemosphere 253:126617

Wolicka D, Suszek A, Borkowski A, Bielecka A (2009) Application of aerobic microorganisms in bioremediation in situ of soil contaminated by petroleum products. Bioresour Technol 100:3221–3227

Wozniak CA, McClung G, Gagliardi J, Segal M, Matthews K (2012) Regulation of genetically engineered microorganisms under FIFRA, FFDCA and TSCA. In: Regulation of agricultural biotechnology: the United States and Canada. Springer, Dordrecht, pp 57–94

Wuana RA, Okieimen FE (2011) Heavy metals in contaminated soils: a review of sources, chemistry, risks and best available strategies for remediation. ISRN Ecol 2011:402647

Xie Y, Cheng W, Tsang PE, Fang Z (2016) Remediation and phytotoxicity of decabromodiphenyl ether contaminated soil by zero valent iron nanoparticles immobilized in mesoporous silica microspheres. J Environ Manag 166:478–483

Xiuqi C, Ying X, Janelle L (2013) The effects of micro-algae characteristics on the bioremediation rate of Deepwater Horizon crude oil. J Emerg Investig 17:1–7

Xu Y, Zhou NY (2017) Microbial remediation of aromatics-contaminated soil. Front Environ Sci Eng 11:1

Yu J, Li R, Zhang X, Du Y, Ma R, Zhao X, Zuo S, Dong K, Wang R, Zhang Y, Gu Y, Sun J (2023) Bioremediation of petroleum hydrocarbon contaminated soil by microorganisms immobilized on sludge modified by non-ionic surfactant. Environ Sci Pollut Res 30:28010–28022

Zhang Y, Zhao G, Xuan Y, Gan L, Pan M (2021) Enhanced photocatalytic performance for phenol degradation using ZnO modified with nano-biochar derived from cellulose nanocrystals. Cellulose 28:991–1009

Zhang X, Wells M, Niazi N, Bolan N, Shaheeng S, Hou D, Gao B, Wang H, Rinklebe J, Wang Z (2022) Nanobiochar-rhizosphere interactions: implications for the remediation of heavy-metal contaminated soils. Environ Pollut 299:118810

Zhao YC, Yi XY, Zhang M, Liu L, Ma WJ (2010) Fundamental study of degradation of dichlorodiphenyltrichloroethane in soil by laccase from white rot fungi. Int J Environ Sci Technol 7(2):359–366

Zhu X, Venosa AD, Makram TS, Lee K (2004) Guidelines for the bioremediation of oil-contaminated salt marshes. EPA/600/R-04/074. US Environmental Protection Agency, Cincinnati, OH

Chapter 5
Microbial Degradation of Pesticides in the Environment

Adesh Kumar, Etalesh Goutam, Rainy Thakur, Geetika, and Shweta Meshram

Abstract It has long been understood that excessive and illogical pesticide usage is bad for the environment, as it affects soil fertility and causes biomagnification of toxicity in living tissues. Such hazardous pesticidal loads can currently be reduced by means of physical, chemical, biological, and enzymatic methods. Among these, microbial enzymatic methods have proven to be more economically viable and thermodynamically reliable than other measures. In a convenient environment, bacteria utilize pesticides to dispense sulfur, carbon, and act as electron donors. Microbes including bacteria and fungi are employed to efficiently eliminate or detoxify chlorinated toxins, polychlorinated diphenyl, polycyclic aromatic hydrocarbons, and organophosphorus. In this reference, *Sphingomonas, Aerobacter*, *Flavobacterium*, *Micrococcus*, *Bacillus*, *Pseudomonas*, *Burkholderia*, *Acinetobacter*, and *Alcaligenes* are considered a few of the major bacterial genera. Among, *Streptomyces* and *Actinomycetes* have been proven to successfully detoxify pesticides among the others. *Aspergillus niger*, *Oxysporum*, *Penicillium*, and *Fusarium* are some of the fungi that possess potential to break down toxic substances. Previous studies have revealed the role of several microbial enzymes including peroxidase, hydrolases, oxygenase, dehydrogenase, organophosphorus acid anhydrolase, phosphotriesterase, dehalogenase, and ligninase, for efficiently removing the persistent organic pollutants existing in the form of pesticides. In this chapter, we stated about the role of a variety of microbes in the bioremediation pesticides and their importance in agricultural sustainability.

Keywords Pesticides · Bioremediation · Microbes · Enzymatic methods · Soil fertility · Chlorinated toxins · Bacterial genera · Detoxification · Environmental impact · Agricultural sustainability

---

A. Kumar · R. Thakur · Geetika · S. Meshram (✉)
Department of Plant Pathology, Lovely Professional University, Phagwara, Punjab, India
e-mail: shweta.26662@lpu.co.in

E. Goutam
Department of Horticulture, Lovely Professional University, Phagwara, Punjab, India

A. Karnwal, A. R. Mohammad Said Al-Tawaha (eds.), *Microbial Applications for Environmental Sustainability*, https://doi.org/10.1007/978-981-97-0676-1_5

5.1 Introduction

Pesticides are inorganic substances that are probably meant for boosting agricultural production and minimizing losses brought on by crop pests and control insect vectors to prevent the dissemination of diseases. Additionally, enhanced herbicidal and pesticidal implementation has indeed been made in agriculture for the sake of food preservation (Parte et al. 2013). In the Indian context, pests cause crop losses of more than 6000 crores rupees each year, from which 26% caused by diseases, 20% by insects, 33% by weeds, 10% by birds and rodents, and rest 11% are attributable to other factors (Rao et al. 2010). Despite the consumption of more than two million tonnes of pesticides, around one-third of the world's agricultural production is sacrificed annually due to pests. Being required for the management of diseases, insect pests, and weeds, pesticides become one of the key components after it came into regular practice after the green revolution period. Irrespective of their need in crop production, only 1–2% of the chemical targets the pest, 10–20% residue on plant canopy, and 80–90% pollutes different ecological components, i.e., soil, water, air, flora, and fauna around (Ye et al. 2018). The introduction of pesticides increased agricultural production, but their indiscriminate application resulted in excessive hazardous contamination (Joint Food and Agriculture Organization and World Health Organization 2017). There have been instances where excessive pesticides application led to drifting or leaching through ground level water and contaminated surface and sub-surface water bodies (Kruger et al. 2019; Chekroun et al. 2014; Dsikowitzky and Schwarzbauer 2014; Hui et al. 2021). Phytoplankton and zooplankton, as well as terrestrial and aquatic plants and animals, are harmed directly or indirectly by this contamination, which is driven primarily by runoff from agricultural areas (Salinas and Hernández 2011). After the success of dichloro-diphenyl-trichloro-ethane (DDT) as an insecticide, many more chemicals were introduced and used for long time. These chemicals were banned time to time but for several chemicals long have residual period and still polluting the ecological bodies (Ye et al. 2018). Pyrethroids, carbamates, organochlorine, and organophosphorus are the only four primary types of chemicals used in insecticides, which serve the most for controlling of insect pests of plants and products at pre- and post-harvest level. In these, organochlorine pesticides have a long residual period, are biologically very stable, lipophilic, and end up causing slower natural degradation. In contrast, pyrethroids, carbamates, and organophosphorus have short residual periods of biodegradation and yet still persist and pollute ecological bodies to some extent (Rani and Dhania 2014). Since it is the chemical nature of pesticides that dictates their level of natural breakdown or persistence, they could be divided into two types: chlorinated and non-chlorinated pesticides (Aresta et al. 2015). The organochlorine pesticides that have been discussed are the chlorinated pesticides that strongly persist in nature (Belal et al. 2020), trigger bioaccumulation in biotic substances, and contaminate foodstuff (Briz et al. 2011). Pesticides and their association with ecosystem are represented in Fig. 5.1.

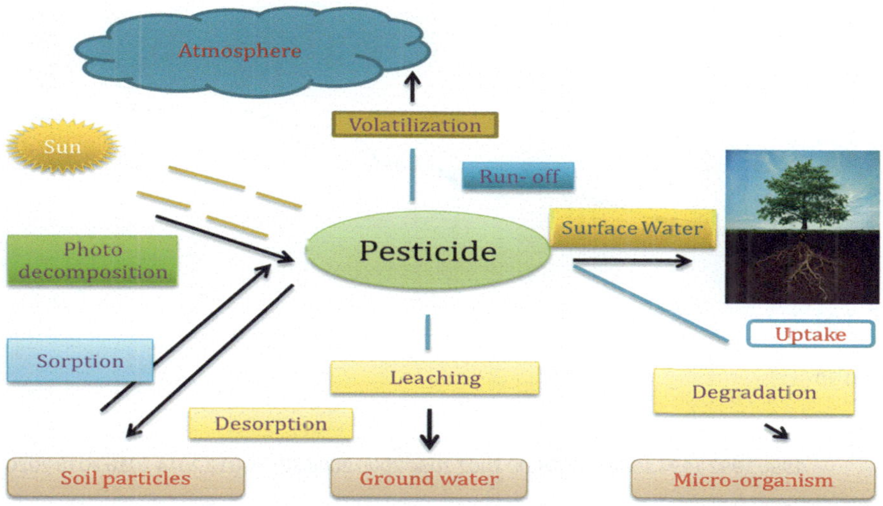

Fig. 5.1 Pesticides and their association with ecosystem

5.2 Introduction of Pesticides in Agriculture

The basic motive for the adoption of pesticides into the environment was to eliminate or control pests, animals, and unpleasant plant species (Yamada 2017). The agricultural consumption of pesticides worldwide from 1990 to 2020 is represented in Fig. 5.2. They are often employed during a variety of cultivation stages, such as seeding, growing, producing, early storing, distributing, and processing. They are additionally used to guard farmed animals against endoparasites including tapeworm, worms, nematodes, and ectoparasites like lice and bedbugs (Joint Food and Agriculture Organization and World Health Organization 2017). Commonly, pesticides consist of chemicals that are used to control plant growth, fruit thinning, restrict sprouting, and defoliation purposes. In addition, these chemicals may also be employed for maintaining the effective protection of plants from predatory organisms that could degrade or impair their quality (Fothergill and Abdelghani 2013). The concern about the consequence of population growth on food security has led to a further escalation in the utilization of pesticides in agriculture (Liu et al. 2015). Broad-spectrum insecticides, such as organophosphorus pesticides, are employed on a variety of crops, including ornamental plants, cereals, vegetables, and fruits. Although they were intended to kill or repel pests, they have also been shown to be lethal to people and other non-pest animals (Fulekar and Geetha 2008). General characteristics of some common group of pesticides are explained in Table 5.1, and classification of different pesticide groups on the basis of their chemical nature is stated in Table 5.2.

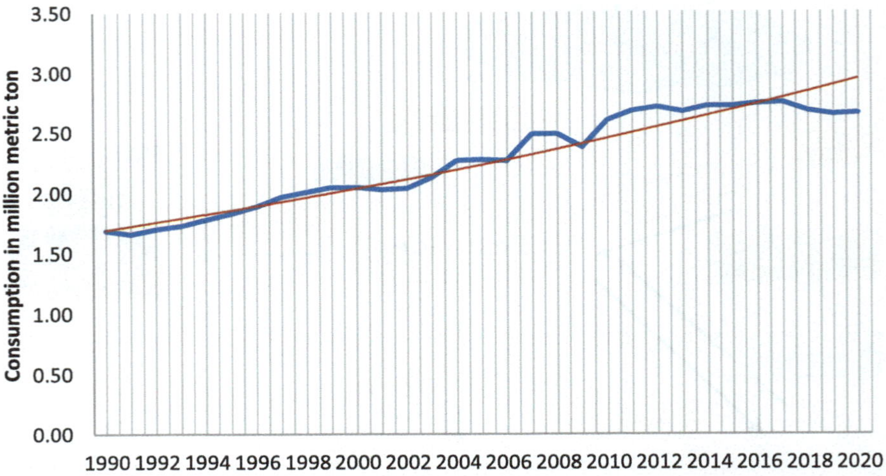

Fig. 5.2 Agricultural consumption of pesticides worldwide from 1990 to 2020 (FAO 2023)

Table 5.1 General characteristics of some common group of pesticides

Pesticide group	Major characteristics
Organophosphates	Water-soluble as well as soluble in organic solvents. Less persistent than chlorinated hydrocarbons, they enter groundwater and some of them have CNS effects. They are taken up by plants, moved to the leaves, and then fed on by insects that feed on leaves or on plant stems. Malathion, methyl parathion, and diazinon are a few examples
Organochlorines	Long-lasting and harmful to a range of animals, they are soluble in lipids, accumulate in animal fatty tissue, and travel up the food chain. DDT, aldrin, lindane, chlordane, and mirex are among examples
Pyrethroids	Have a negative impact on the nervous system, are less persistent than other pesticides, and are the safest to use; some are employed as insecticides in homes. For example, pyrethrin
Carbamates	Derivatives of carbamates acids are very poisonous to vertebrates but only kill a small range of insects. Very little persistence. Carbaryl and sevin are two examples

Table 5.2 Classification of pesticides according to their chemical nature

Pesticide group	Chemical nature
Organophosphate	Phosphorus present in the centre of molecule. More stable and less toxic than organochlorines and can be heterocyclic cyclic and aliphatic
Carbamates	Based on the chemical composition of the plant alkaloid, *Physostigma venenosum*
Organochlorine	Carbon atoms may be chlorine, hydrogen, and occasionally oxygen and non-polar and lipophilic in nature
Pyrethroids	Chemical nature is similar to the synthetic pyrethrin (alkaloids obtained from petals of *Chrysanthemum cinerariifolium*)

Pesticide group	Chemical nature
Thiocarbamates	Contain an S-group in its composition
Organotin	Presence of tin as a central atom in the molecule
Di-nitrophenols	Recognized by the two nitro groups (NO_2) bonded to a phenol ring
Urea derivatives	Those which include the urea bound the aromatic compounds
Organo-sulfur	Sulfur in the centre of molecule and are highly toxic to mites or insects
Diverse composition	Talimides, carboxyamide, triazines, trichloroacetic and trichloropicolinic acids derivatives, naphthoquinones, and guanidines

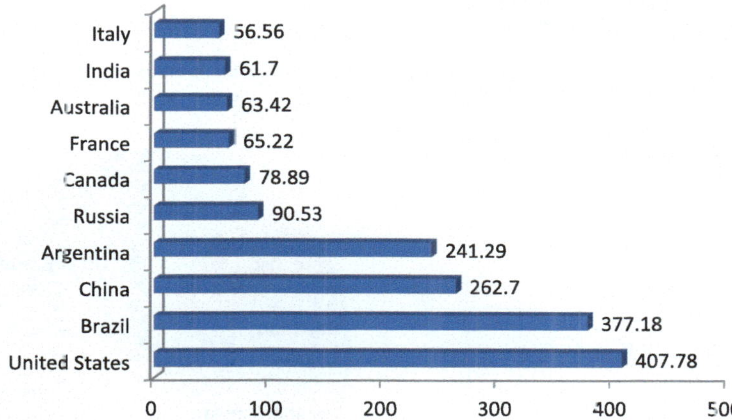

Fig. 5.3 Leading countries in agricultural consumption of pesticides worldwide in year, 2020 (*in 1000 metric tons*)

5.3 Harmful Effects of Pesticides

Pesticides mainly and intentionally were kept in the domain of public health which could improve quality and lifespan. For instance, the reduction in mortality rate was possible after implementation of pesticide-oriented programmes that managed the insects responsible for fatal diseases such as malaria, typhus, dengue, yellow fever, etc. The same way use of pesticides allows a cost-effective pest and disease management as an essential component of conventional agriculture after green revolution (Ortiz-Hernández et al. 2011). Leading countries in agricultural consumption of pesticides worldwide in year, 2020, are shown in Fig. 5.3. Although pesticides are used to prevent, eliminate, repel, or counteract pests, their elevated use poses a variety of health and environmental issues, particularly with regard to humans. These issues include poisoning of farmers and field workers, cardiopulmonary, neurological, and skin disorders, fatal deformities, miscarriages, and a decline in sperm

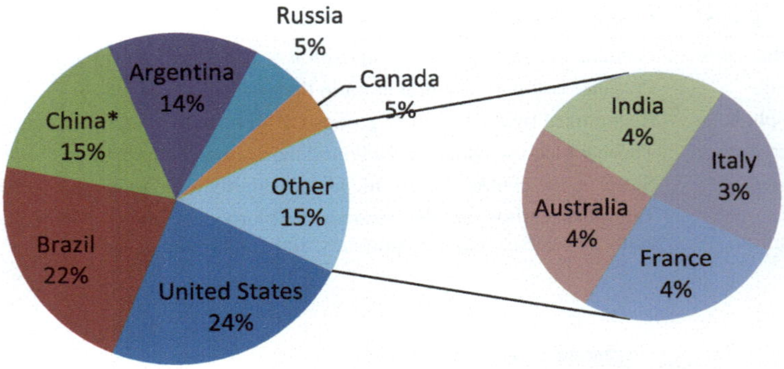

Fig. 5.4 Proportionate of leading countries in agricultural consumption of pesticides worldwide in 2020 (*in 1000 metric tons*) (FAO 2023)

Fig. 5.5 Top pesticide consuming states of India

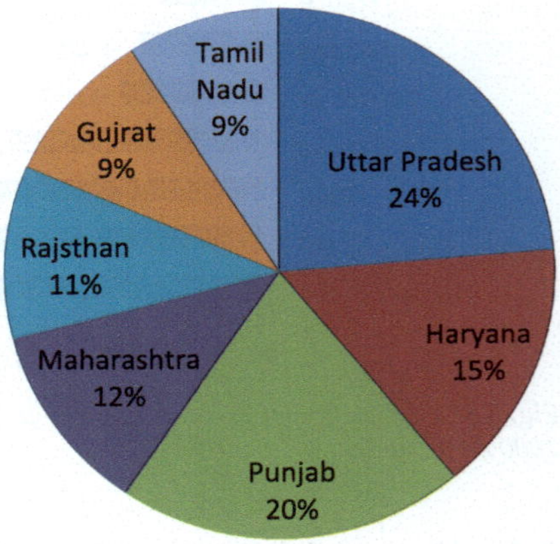

count (Abou-Donia 2003; Malizia et al. 2010). The proportionate of leading countries in agricultural consumption of pesticides worldwide in year, 2020 and top pesticide consuming states of India are depicted in Figs. 5.4 and 5.5, respectively.

5.3.1 *Harm to Environment, Soil, and Soil Microbiota*

Excessive use of pesticides can even contaminate water, despite the fact that they can only be damaging to soil. This contaminated water can further travel and merge into freshwater bodies, either with the rains or when it is used to irrigate the soil (Andreotti and Silverman 2012). Upon times obsolete pesticides have accumulated in developed and developing nations or economy, for instance, about 100,000 tonnes in Africa, 200,000 tonnes in Middle East Asia, and a comparable amount in East Europe and the former Soviet Union (Ortiz-Hernández et al. 2011). Worldwide

Table 5.3 Worldwide higher pesticide (insecticide, herbicides, fungicides, and bactericides) consuming countries with agricultural land area in year, 2010

Country	Insecticides	Herbicides	Fungicides and bactericides	Total pesticide consumption	Total land area	Total agricultural area
El Salvador	750.97	11,007.37	85.41	11,843.75	20,720	15,293.4
India	20,618.82	6704.98	13,055.44	40,379.24	2,973,190	1,797,590
Germany	1243.79	16,675.00	9666.70	27,585.49	348,570	167,000
United Kingdom	889.46	7488.83	5521.28	13,223.93	241,930	172,240
Hungary	1965.00	4479.00	2285.50	8729.5	90,530	53,430
Chile	7071.00	7234.00	3727.00	18,032	743,530	157,430
Peru	4352.16	6160.08	4045.60	14,557.84	1,280,000	214,700
Colombia	8717.60	17,587.15	22,313.72	48,618.47	1,109,500	425,030
Bolivia	9843.50	17,263.60	4459.66	31,566.76	1,083,300	369,650
Costa Rica	2148.52	4416.26	7438.56	14,003.34	51,060	18,800
Ecuador	7689.61	14,394.76	9118.73	31,203.1	248,360	749,770
Turkey	8215.57	745.59	17,545.58	60,792.4	769,630	390,120
Dominican Republic	609.43	3849.66	1323.83	5782.92	48,320	24,470
Bangladesh	2681.69	658.54	9883.70	13,223.93	130,170	31,650
Japan	2053.30	12,166.70	22,337.00	36,557	364,500	45,930
Guatemala	2026.92	8362.84	4508.48	14,898.24	107,160	43,950
Ukraine	8531.50	40,910.60	11,350.30	13,899.57	579,320	412,670
Honduras	314.00	2850.90	1877.60	5052.5	111,890	32,200
Portugal	370.97	2042.29	9475.36	7249.2	91,470	36,770
Netherlands	1634.90	2642.30	3613.10	7890.3	33,730	19,087
Cameroon	3467.28	3470.20	2457.46	9394.94	472,710	96,000
Italy	10,834.00	9934.00	42,537.00	63,305	294,140	143,228
Romania	1327.70	3688.90	2232.60	26,506.74	230,050	141,560
Norway	5.33	8321.09	1479.40	9805.82	365,268	10,060
Myanmar	6175.68	661.00	2141.33	8978.01	653,260	125,260

(Consumption in = tonnes per year and area in sq. km.) (Verma et al. 2014)

higher pesticide (insecticide, herbicides, fungicides, and bactericides) consuming countries with agricultural land area in year, 2010 is described in Table 5.3. Pesticides can lower the quality of both the soil and food's nutritious worth by polluting it (Aulakh et al. 2007). Pesticides get into soils where they become poisonous to worms, fungus, bacteria, and protozoa in addition to arthropods. Microorganisms are the key components since, they dominate both structure and function of ecosystems are targeted proportionally most. Pesticides have had a great negative impact on biodiversity, helpful natural enemies, and pest populations, disturbed equilibrium. Extensive application of pesticides has led to the evolution of pesticide resistance in addition to decimating natural enemy populations (Pimentel et al. 1992; Pimentel and Burgess 2014).

5.3.2 *Harm to Humans, Animals, and Plants Health*

Pesticides include varieties of compounds that are effective in eliminating fatal organisms that cause damage to crops, but this cannot happen that they have no impact on human health (Edwards 2013). The vegetables and fruits that we consume daily have now been exposed to numerous sorts of pesticides, which seem to poison when enters in our bodies and gradually deteriorate it (Aulakh et al. 2006). Due to soil erosion and water runoff, pesticides are deposited in aquatic habitats. During spraying, pesticides can potentially disperse and contaminate aquatic systems. Certain soluble insecticides often leak into lakes and streams. Pesticides cause harm to and obliterate wild birds and mammals, making them great "indicator species". Pesticides can kill wildlife directly or cause secondary poisonings by eating contaminated food. They can also lower survival, growth, and reproductive rates when exposed to sub-lethal doses, and they can limit habitat by removing food supplies and safe havens (Boudh and Singh 2019; de O Gomes et al. 2020). Due to soil erosion and water runoff, pesticides are deposited in aquatic habitats. Also, during spraying pesticides can potentially disperse and contaminate aquatic systems. Certain soluble insecticides often leach through lakes and streams. Pesticides cause harm to and obliterate wild birds and mammals that play a key role in the environment as indicator species. Pesticides kill wildlife directly or indirectly via secondary poisoning after eating contaminated food. They can also impair survival, growth, and reproductive rates when their sub-lethal doses are exposed (Özkara et al. 2016).

5.4 Microbial Biodegradation of Pesticides

Biotic and abiotic means cause biodegradation of pesticides in the environment. Naturally in the environment, pesticides get breakdown into small particles after the biodegradation by microorganisms (Atlas 1988). The rate and mode of biodegradation of pesticides vary; some pesticides such as dieldrin and DDT take a very long time and get accumulated into food, soil, and water (Aberdeen 1993; Kannan et al. 1994), while other pesticides like simazine and atrazine biodegrade slowly so the chances of leaching to soil and mixing in groundwater are more (Kookana and Aylmore 1994). Organophosphates get biodegraded more readily and become first choice by growers and adapted by manufactures. There also a category of pesticides does exist that are so readily biodegradable and result in less effective insect pest management (Felsot 1989). Pesticides can be broken down by microbes from a variety of sources. Microorganisms that have been recognized as pesticide degraders have typically been isolated from a range of pesticide-contaminated areas (Alexander 1977). When these chemicals are applied to agricultural crops, they primarily end up in the soil. Additionally, the effluent from the pesticide industry, sewage sludge, activated sludge, wastewater, natural waters, sediments, and areas near the manufacture of pesticides are also rich sources of pesticide degraders (Aronstein et al. 1991). Pesticides are subjected to microorganism-oriented transformation processes (biodegradation, biomineralization, bioremediation, biotransformation, bioaccumulation, and co-metabolism) (Verma et al. 2014). Once they

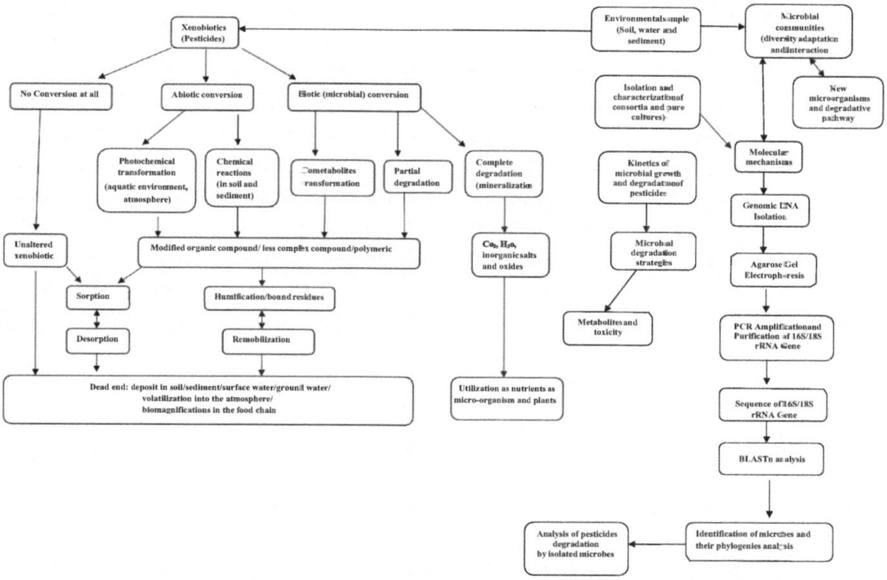

Fig. 5.6 Microbial degradation of pesticide affected by various environmental factors and characterization of pesticide-degrading microbes (Verma et al. 2014)

are introduced to the environment, they get into an enzyme led process. Degradation of pesticides by microorganisms is either after direct metabolism or after indirect metabolism in which microbes disturb chemical or physical environment (Bollag and Liu 1990). Microbial degradation of pesticide affected by various environmental factors and characterization of pesticide degrading microbes are depicted in Fig. 5.6.

One of the promising methods is biological method to remove pollutants from the different ecological environments. It has become an alternate strategy, which is a biological and environment friendly approach for the treatment of pesticide residue (Finley et al. 2010). The microbial biodegradation is done with the help of bacteria, fungi, actinomycetes, and viruses that do not produce toxic intermediates. These soil inhabitants decompose various xenobiotic compounds and break them into mineral salts that are now absorbed and used by plants. The microorganisms that live in the environment and use pesticides as a source of energy appear to be the greatest contenders for bioremediation of pesticides available in the soil or water when compared to chemical treatments (Uqab et al. 2016).

5.4.1 Bacteria-Mediated Biodegradation of Pesticides

Bacteria, due to their continuous exposure to environmental stress posed by pesticide residues, develop a system which is genetically determined against toxicants (Parsek et al. 1995). Large number of bacterial strains, viz. *Pseudomonas*, *Flavobacterium*, and *Alcaligenes* (Aislabie and Lloyd-Jones 1995), *Acetobacter*

(Shakoori et al. 2000), *Cryptococcus* (Evy et al. 2012), *Arthrobacter*, *Bacillus*, *Nocardiopsis*, *Ralstonia*, and *Rhodococcus* (Verma et al. 2014), *Klebsiella*, *Diaphorobacter*, *Agrobacterium*, and *Ochrobactrum* (Ataikiru et al. 2020) metabolize pesticides. Bacterial metabolism of aromatic compounds has shown that a number of enzymes, including hydrolase, monooxygenase, isomerase, oxygenase, and dehydrogenase are involved in the transformation of harmful pesticide chemicals into a variety of intermediate metabolites, including protocatechuate and catechols (Reiner 1972; Gibson and Subramanian 1984; Reineke 1984; Fewson 1988; Knackmuss and Reineke 1988; Commandeur and Parsons 1990). These dihydroxylated intermediates directly enter one of the two potential pathways, either a meta cleavage-type pathway or an ortho cleavage-type pathway, that provide major metabolic pathways like the tricarboxylic acid cycle (Harayama et al. 1989; van der Meer et al. 1992). The bacterial-mediated biodegradation of pesticides is being stated in Table 5.4.

Table 5.4 Biodegradation of pesticides by different bacterial species

Pesticide	Organism	Nature of study	References
Acibenzolar-S-methyl	*B. pumilus* SE34	The effects of certain strains of plant growth-promoting rhizobacteria (PGPR) on the bacterial growth and the biodegradation of soil-applied pesticides	Myresiotis et al. (2012)
Atrazine	*Arthrobacter nicotinovorans*	Characterization of the atrazine-degrading bacterium *Arthrobacter nicotinovorans* HIM	Aislabie et al. (2005)
Aldrin	*Pseudomonas fluorescens*	Aldrin was removed using activated carbon and free-living *Pseudomonas fluorescens* cultures	Erick et al. (2006)
Chlorpyrifos	*Bacillus pumilus* C2A1	*Bacillus pumilus* strain C2A1 biodegrades chlorpyrifos and its by product 3,5,6-trichloro-2-pyridinol through hydrolysis	Anwar et al. (2009)
	Pseudomonas nitroreducens PS-2	Using ryegrass, the rhizosphere can be cleaned of chlorpyrifos from mycorrhizosphere soil	Korade and Fulekar (2009)
DDT	*Eubacterium limosum*	*Eubacterium limosum*, a human intestine bacterium, dechlorinates DDT in a reductive manner under anaerobic circumstances	Yim et al. (2008)
	Sphingobacterium sp. D6	The best favouring conditions for degradation were noticed at 30 °C temperature with neutral pH and the successful detoxification of toxin accomplished in various steps starting from dichlorination, passing between steps and lastly completed with phenyl ring-cleavage reaction	Fang et al. (2010)

Pesticide	Organism	Nature of study	References
Diazinon	*Pseudomonas aeruginosa* S1, *Bacillus amyloliquefaciens* S2, *Bacillus pseudomycoides* S4, *and Bacillus licheniformis* S5	Decomposition of diazinon by soil microorganisms and detection of degradation by GC-MS	Thabit and El-Naggar (2013)
Dichlorvos	*Flavobacterium* sp.	A phyllosphere *Flavobacterium* strain that performs in situ improved bioremediation of dichlorvos	Ning et al. (2012)
	Acinetobacter sp.	This organo-phosphatic insecticide is biodegraded efficiently	Agarry et al. (2013)
Endosulfan	*Bacillus* sp.	It showed an efficient degradation of endosulfan with 88% rate	Krishnasamy et al. (2019)
Lindane	*Streptomyces* sp.	A better vigour index and 68% removal of lindane were observed when inoculated in the contaminated soil	Benimeli et al. (2008)

5.4.2 Fungal-Mediated Biodegradation of Pesticides

The use of fungi to remove pollutants from contaminated places is another viable alternative treatment method to cremation. Fungi are among the potential bioremediation alternatives since they naturally break down waste materials and secrete a number of extracellular enzymes that can break down lignin and cellulose, the two crucial components of plant fibre. For myco-remediation to be successful, it is essential to accurately identify and choose the fungal species that will target a certain contaminant (Gupta et al. 2017). Many fungi like *Auricularia auricula*, *Agrocybe semiorbicularis*, *Dichomitus squalens*, *Coriolus versicolor*, *Hypholoma fasciculare*, *Flammulina velutipes*, *Pleurotus ostreatus*, *Avatha discolor*, and *Stereum hirsutum* have proved their potential to biodegrade pesticides like dicarboximide, phenylamide, chlorinated, triazine, organophosphorus compounds, and phenyl urea. *Phanerochaete chrysosporium* was reported to destroy a wide range of chemicals. In soils with a variety of climatic conditions, fungi can develop and spread by dispersing spores into the air. They can also help to preserve the balance of an ecosystem. Due to their diverse habitats and ability to release a wide spectrum of enzymes, fungi are interesting candidates for bioremediation at various sites (Deshmukh et al. 2016). By producing various ligninolytic enzymes including laccases and peroxidases, white-rot fungi like *Pleurotus* spp., *Phanerochaete chrysosporium*, *Bjerkandera adusta*, and *Trametes versicolor* have been shown to have the potential for bioremediation in numerous investigations (dos Santos Bazanella et al. 2013). The ligninolytic enzymes from white-rot fungi have been employed to convert a variety of organic pollutants, including pesticides, from contaminated wastewaters by promoting microbial activity with the aid of a biopurification system (BPS) (Rodríguez-Rodríguez et al. 2013). *Aspergillus niger* and *Penicillium chrysosporium* demonstrated significant removal of petroleum hydrocarbons from soil contaminated with petrol and diesel at short incubation times, as seen by improved total organic carbon removal (Maruthi et al. 2013). The elimination of chlorpyriphos and its metabolite 3,5,6-trichloro-2-pyridinol (TCP) from polluted soils by the fungus strain *A. niger* JAS1 was examined by Silambarasan and Abraham (2013). Various mechanisms adopted by fungi for the bioremediation of toxic, recalcitrant compounds are depicted in Fig. 5.7.

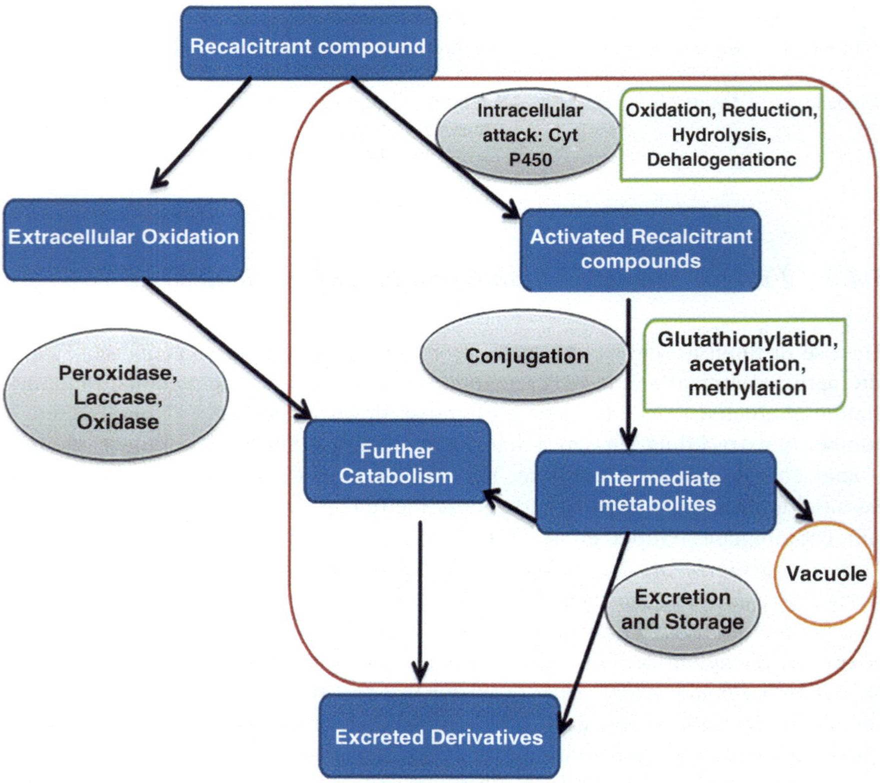

Fig. 5.7 Mechanisms adopted by fungi for bioremediation of toxic, recalcitrant compounds (Deshmukh et al. 2016)

5.4.3 Xenobiotics

Xenobiotics are recognized as chemicals (pollutants) present within an organism which do not release or synthesize naturally and are foreign to a whole biological system. These chemicals enforce toxicological intimidations in reference to pollutants and related side effects when exposed to biota. Chemicals which mainly considered pollutants are pesticides, synthetic fertilizers, polychlorinated biphenyls dyes, polycyclic aromatic hydrocarbons, and dioxins. Higher concentrations of these pollutants always create a biological imbalance in soil that leads to pollution of surface as well as groundwater. Several white-rot fungi including *Lentinula edodes*, *Bjerkandera adusta*, *Pleurotus tuber regium*, *Agaricus bisporus*, *Pleurotus ostreatus*, *Irpex lacteus*, *Pleurotus pulmonarius*, and *Trametes versicolor* can also degenerate persistent level of xenobiotic compounds (Singh 2006; Gupta et al. 2017). Microbial degradation pathway of xenobiotic compound is shown in Fig. 5.8.

Below, different processes which are involved in bioremediation of several pollutants utilizing different organisms are stated in Table 5.5.

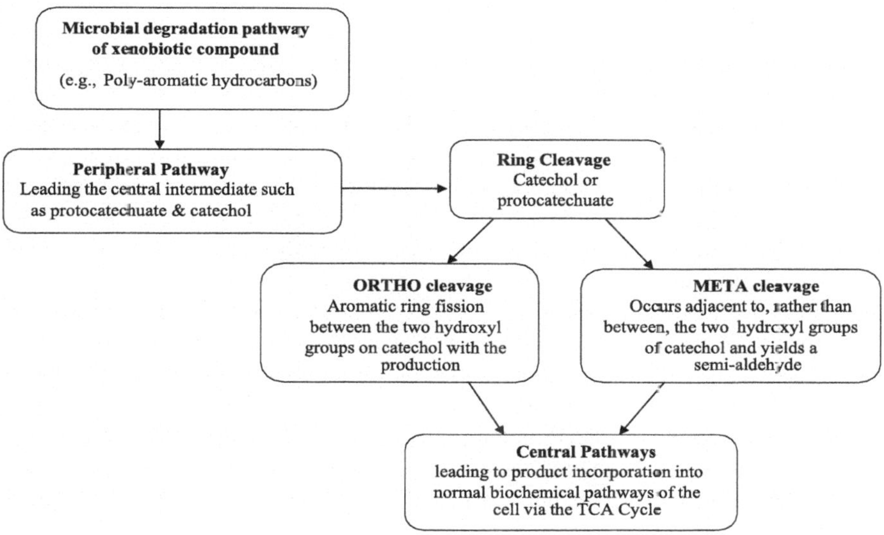

Fig. 5.8 Microbial degradation pathway of xenobiotic compound

Table 5.5 Processes involved in bioremediation of pollutants utilizing different organisms

Process	Source	Pollutant/s	Microbes/plants
Phytoremediation	Soil	Cd, Pb	*Eichhornia crassipes* and *Vetiveria zizanioides* with addition of zeolite
Biodegradation	Garden, beach, and mud	Saw	*Trichoderma* sp., *Aspergillus* sp.
Degradation	Cellulosic substances like saw dust, corncobs, and sugarcane	Blue dye 2B	*Bacillus* sp.
Phytoremediation, adsorption	Sewage irrigated soil wastewater	Heavy metals lead, iron, manganese, copper, zinc, etc.	*Dunaliella* algae a species of genus *Flagellate*
Adsorption	Soil	Pd, Pb, Cr, Cu, Mn, and Zn	*Urtica dioica* L., *Taraxacum officinale* L., *Plantago major* L., and *Trifolium pratense* L.

5.5 Conclusion

Considerable annual food loss can be prevented with an effective pest management system. Pesticides had a substantial impact on the farmer's economy. Their role in managing the fatal diseases and improving the health of an ecosystem is very crucial. However, indiscriminate use of these synthetics had a negative effect on the living world since only a small portion of the applied pesticides reached the targeted organisms and the remaining just entered into the soil and water bodies and polluting these ecological systems where they affect its health brutally. These pesticidal effects encountering an ecosystem can potentially be remedied to a great extent using bioremediation by means of a variety of microbes found around the system. The most crucial method for removing pollutants from a polluted environment is biopesticide enzymatic degradation, and studies have shown that this method had a significant potential for bioremediation. Therefore, bioremediation is a very promising method for dealing with pesticide pollution and may undoubtedly resolve the issue of pesticide contamination. Hence, for seeking a safer life, adoption of such eco-friendly and environmentally sustainable technologies would be of great significance.

Acknowledgements The authors are highly grateful to the Lovely Professional University, Punjab, for the infrastructural support. We are obliged to Dean, School of Agriculture, Lovely Professional University, Punjab, for his valuable suggestions and guidance.

References

Aberdeen I (1993) Litigation over dieldrin in Victoria. Agric Sci 6:42–45

Abou-Donia MB (2003) Organophosphorus ester-induced chronic neurotoxicity. Arch Environ Health 58:484–497. https://doi.org/10.3200/AEOH.58.8.484-497

Agarry SE, Olu-arotiowa OA, Aremu MO, Jimoda LA (2013) Biodegradation of *Dichlorvos* (organophosphate pesticide) in soil by bacterial isolates. J Nat Sci Res 3:12–16

Aislabie J, Lloyd-Jones G (1995) A review of bacterial-degradation of pesticides. Soil Res 33:925–942. https://doi.org/10.1071/SR9950925

Aislabie J, Bej AK, Ryburn J, Lloyd N, Wilkins A (2005) Characterization of *Arthrobacter nicotinovorans* HIM, an atrazine-degrading bacterium, from agricultural soil New Zealand. FEMS Microbiol Ecol 52:279–286. https://doi.org/10.1016/j.femsec.2004.11.012

Alexander M (1977) Introduction to soil microbiology, 2nd edn. Wiley, New York

Andreotti G, Silverman DT (2012) Occupational risk factors and pancreatic cancer: a review of recent findings. In special issue: epidemiology and etiological mechanisms in pancreatic cancer. Mol Carcinog 51:98–108. https://doi.org/10.1002/mc.20779

Anwar S, Liaquat F, Khan QM, Khalid ZM, Iqbal S (2009) Biodegradation of chlorpyrifos and its hydrolysis product 3,5,6-trichloro-2-pyridinol by *Bacillus pumilus* strain C2A1. J Hazard Mater 168:400–405. https://doi.org/10.1016/j.jhazmat.2009.02.059

Aresta A, Marzano CN, Lopane C, Corriero G, Longo C, Zambonin C, Stabili L (2015) Analytical investigations on the lindane bioremediation capability of the demosponge *Hymeniacidon perlevis*. Mar Pollut Bull 90:143–149. https://doi.org/10.1016/j.marpolbul.2014.11.003

Aronstein BN, Calvillo YM, Alexander M (1991) Effect of surfactants at low concentrations on the desorption and biodegradation of sorbet aromatic compounds in soil. Environ Sci Technol 25:1728–1731. https://doi.org/10.1021/es00022a008

Ataikiru TL, Okerentugba PO, Okpokwasili GC (2020) Identification of carbofuran and paraquat degrading microorganisms from soil. South Asian J Res Microbiol 7:40–52. https://doi.org/10.9734/sajrm/2020/v7i130164

Atlas RM (1988) Microbiology-fundamentals and applications, 2nd edn. Macmillan Publishing Co., New York, pp 352–353

Aulakh RS, Gill JPS, Bedi JS, Sharma JK, Joia BS, Ockerman HW (2006) Organochlorine pesticide residues in poultry feed, chicken muscle and eggs at a poultry farm in Punjab, India. J Sci Food Agric 86:741–746. https://doi.org/10.1002/jsfa.2407

Aulakh RS, Bedi JS, Gill JPS, Joia BS, Pooni PA, Sharma GK (2007) Occurrence of DDT and HCH insecticide residues in human biopsy adipose tissues in Punjab, India. Bull Environ Contam Toxicol 78:330–334. https://doi.org/10.1007/s00128-007-9187-6

Belal ESB, Sidkey NM, Gad WA, El-Gendy SS (2020) Biodegradation of organochlorine pesticides by means of *Pectobacterium wasabiae*. Int J Oceanogr Aquacult 4:1–14. https://doi.org/10.23880/ijoac-16000183

Benimeli CS, Fuentes MS, Abate CM, Amoroso MJ (2008) Bioremediation of lindane-contaminated soil by *Streptomyces* sp. M7 and its effects on *Zea mays* growth. Int Biodeterior Biodegrad 61:233–239. https://doi.org/10.1016/j.ibiod.2007.09.001

Bollag JM, Liu SY (1990) Biological transformation processes of pesticides. In: Pesticides in the soil environment: processes, impacts and modelling, vol 2, pp 169–211. https://doi.org/10.2136/sssabookser2.c6

Boudh S, Singh JS (2019) Pesticide contamination: environmental problems and remediation strategies. In: Emerging and eco-friendly approaches for waste management. Springer, Singapore, pp 245–269. https://doi.org/10.1007/978-981-10-8669-4_12

Briz V, Molina-Molina J-M, Sánchez-Redondo S, Fernandez MF, Grimalt JO, Olea N, Rodrıguez-Farre E, Sunol C (2011) Differential estrogenic effects of the persistent organochlorine pesticides dieldrin, endosulfan, and lindane in primary neuronal cultures. Toxicol Sci 120:413–427. https://doi.org/10.1093/toxsci/kfr019

Chekroun KB, Sánchez E, Baghour M (2014) The role of algae in bioremediation of organic pollutants. Int Res J Pub Environ Health 1:19–32

Commandeur LCM, Parsons JR (1990) Degradation of halogenated aromatic compounds. Biodegradation 1:207–220

de O Gomes H, Menezes JMC, da Costa JGM, Coutinho HDM, Teixeira RNP, do Nascimento RF (2020) A socio-environmental perspective on pesticide use and food production. Ecotoxicol Environ Saf 197:110627. https://doi.org/10.1016/j.ecoenv.2020.110627

Deshmukh R, Anshuman A, Khardenavis AA, Purohit HJ (2016) Diverse metabolic capacities of fungi for bioremediation. Indian J Microbiol 56:247–264. https://doi.org/10.1007/s12088-016-0584-6

dos Santos Bazanella GC, Araujo AV, Castoldi R, Maciel GM, Inacio FD, de Souza CGM, Bracht A, Peralta RM (2013) Ligninolytic enzymes from white-rot fungi and application in the removal of synthetic dyes. In: Polizeli TM, Rai M, De Lourdes M (eds) Fungal enzymes. CRC Press, Boca Raton, pp 258–279

Dsikowitzky L, Schwarzbauer J (2014) Industrial organic contaminants: identification, toxicity and fate in the environment. Environ Chem Lett 12:371–386. https://doi.org/10.1007/s10311-014-0467-1

Edwards CA (2013) Environmental pollution by pesticides In: Environmental science research series, vol 3. Plenum Pub Co Ltd Davis House, 8 Scrubs Lane, Harlesden, London, pp 1–459. ISBN: 13:978-1-4615-8944-0

Erick RB, Juan A-O, Paulino P, Torres LG (2006) Removal of aldrin, dieldrin, heptachlor, and heptachlor epoxide using activated carbon and/or pseudomonas fluorescens free cell cultures. J Environ Sci Health 41:553–569

Evy AAM, Lakshmi V, Nilanjana D (2012) Biodegradation of atrazine by *Cryptococcus laurentii* isolated from contaminated agricultural soil. J Microbiol Biotechnol Res 2:450–457

Fang H, Dong B, Yan H, Tang F, Yunlong Y (2010) Characterization of a bacterial strain capable of degrading DDT congeners and its use in bioremediation of contaminated soil. J Hazard Mater 184:281–289. https://doi.org/10.1016/j.jhazmat.2010.08.034

FAO (2023). https://www.fao.org/faostat/en/#data/RP/visualize

Felsot AS (1989) Enhanced biodegradation of insecticides in soil: implications for agroecosystems. Annu Rev Entomol 34:453–476. https://doi.org/10.1146/annurev.en.34.010189.002321

Fewson CA (1988) Microbial metabolism of mandelate: a microcosm of diversity. FEMS Microbiol Rev 4:85–110. https://doi.org/10.1111/j.1574-6968.1988.tb02737.x

Finley SD, Broadbelt LJ, Hatzimanikatis V (2010) In silico feasibility of novel biodegradation pathways for 1,2,4-trichlorobenzene. BMC Syst Biol 4:4–14

Fothergill A, Abdelghani A (2013) A review of pesticide residue levels and their related health exposure risks. In: Food and environment II. WIT transactions on ecology and the environment, vol 170, pp 195–205. https://doi.org/10.2495/FENV130181

Fulekar MH, Geetha M (2008) Bioremediation of chlorpyrifos by *Pseudomonas aeruginosa* using scale up technique. J Appl Biosci 12:657–660

Gibson DT, Subramanian V (1984) Microbial degradation of aromatic hydrocarbons. In: Gibson DT (ed) Microbial degradation of organic compounds. Marcel Dekker, New York, pp 181–252

Gupta S, Wali A, Gupta M, Annepu SK (2017) Fungi: an effective tool for bioremediation. In: Plant-microbe interactions in agro-ecological perspectives. Springer, Singapore, pp 593–606. https://doi.org/10.1007/978-981-10-6593-4_24

Harayama S, Rekik M, Wubbolts M, Rose K, Leppi RA, Timmis KN (1989) Characterization of five genes in the upper-pathway operon of TOL plasmid pWWO from *Pseudomonas putida* and identification of the gene products. J Bacteriol 171:5048–5055. https://doi.org/10.1128/jb.171.9.5048-5055.1989

Hui T, Xiujuan L, Qifa S, Qiang L, Zhuang K, Yan G (2021) Evaluation of drinking water quality using the water quality index (WQI), the synthetic pollution index (SPI) and geospatial tools in Lianhuashan district, China. Pol J Environ Stud 30:141–153. https://doi.org/10.15244/pjoes/120765

Joint Food and Agriculture Organization & World Health Organization (2017) Pesticide residues in food 2017. Food and Agriculture Organization of the United Nations (FAO), Rome, pp xv + 718

Kannan K, Tanabe S, Williams RJ, Tatsukawa R (1994) Persistent organochlorine residues in foodstuffs from Australia, Papua New Guinea and the Solomon Islands: contamination levels and dietary exposure. Sci Total Environ 153:29–49. https://doi.org/10.1016/0048-9697(94)90099-X

Knackmuss HJ, Reineke W (1988) Microbial degradation of halogenated aromatics. Ann Rev Microbiol 42:263–287

Kookana RS, Aylmore LAG (1994) Estimating the pollution potential of pesticides to ground water. Aust J Soil Res 32:1141–1155. https://doi.org/10.1071/SR9941141

Korade DL, Fulekar MH (2009) Rhizosphere remediation of chlorpyrifos in mycorrhizospheric soil using ryegrass. J Hazard Mater 172:1344–1350

Krishnasamy L, Sundaram CS, Sivakumar J (2019) Biodegradation of pesticides from the isolated microbial flora of crop field contaminated soil. Res J Life Sci Bioinform Pharm Chem Sci 5:150–163. https://doi.org/10.26479/2019.0502.12

Kruger GR, Klein RN, Ogg CL, Vieira BC (2019) Spray drift of pesticides. G1773. Nebraska Extension, pp 1–6. https://extensionpublications.unl.edu/assets/html/g1773/build/g1773.htm

Liu Y, Pan X, Li J (2015) A 1961-2010 record of fertilizer use, pesticide, application and cereal yields: a review. Agron Sustain Dev 35:83–93. https://doi.org/10.1007/s13593-014-0259-9

Malizia D, Giuliano A, Ortaggi G, Masotti A (2010) Environmental monitoring of heavy metal contamination in soil with common plants used as potentially reliable analytical instruments. In: Application of multivariate analysis and chemometry to cultural heritage and environment, 3rd edn, Taormina, Sicily Island, Italy, Europe, p 26

Maruthi YA, Hossain K, Thakre S (2013) *Aspergillus flavus*: a potential bioremediator for oil contaminated soils. Eur J Sustain Dev 2:57. https://doi.org/10.14207/ejsd.2013.v2n1p57

Myresiotis CK, Vryzas Z, Papadopoulou-Mourkidou E (2012) Biodegradation of soil-applied pesticides by selected strains of plant growth-promoting rhizobacteria (PGPR) and their effects on bacterial growth. Biodegradation 23:297–310

Ning J, Gang G, Bai Z, Qing H, Hongyan Q, Anzhou M, Zhuan X, Guoqiang Z (2012) In situ enhanced bioremediation of dichlorvos by a phyllosphere *Flavobacterium* strain. Front Environ Sci Eng 6:231–237

Ortiz-Hernández ML, Sánchez-Salinas E, Olvera-Velona A, Folch-Mallol JL (2011) Pesticides in the environment: impacts and their biodegradation as a strategy for residues treatment. In: Stoytcheva M (ed) Pesticides—formulations, effects, fate. InTech, pp 551–574. ISBN: 978-953-307-532-7. http://www.intechopen.com/books/pesticides-formulations-effects-fate/pesticides-in-theenvironment-impacts-and-its-biodegradation-as-a-strategy-for-residues-treatmentpp

Özkara A, Akyıl D, Konuk M (2016) Pesticides, environmental pollution, and health. In: environmental health risk-hazardous factors to living species. Intech Open, pp 3–27. https://doi.org/10.5772/63094

Parsek MR, McFall SM, Chakrabarty AM (1995) Microbial degradation of toxic, environmental pollutants: ecological and evolutionary considerations. Int Biodeterior Biodegrad 35:175–188. https://doi.org/10.1016/0964-8305(95)00032-Z

Parte S, Rokade K, Mali G, Kudale S (2013) Biodegradation of sulfonated aromatic amine by *Pseudomonas demolyticum* NCIM 2112. J Chem Pharm Res 5:335–339

Pimentel D, Burgess M (2014) Environmental and economic benefits of reducing pesticide use. In: Integrated pest management. Springer, Dordrecht, pp 47–71

Pimentel D, Acquay H, Biltonen M, Rice P, Silva M, Nelson J, Lipner V, Giordano S, Horowitz A, D'Amore M (1992) Environmental and economic costs of pesticide use. Bioscience 42:750–760. https://doi.org/10.2307/1311994
Rani K, Dhania G (2014) Bioremediation and biodegradation of pesticide from contaminated soil and water—a novel approach. Int J Curr Microbiol Appl Sci 3:23–33
Rao MA, Scelza R, Scotti R, Gianfreda L (2010) Role of enzymes in the remediation of polluted environments. J Soil Sci Plant Nutr 10:333–353. https://doi.org/10.4067/S0718-95162010000100008
Reineke W (1984) Microbial degradation of halogenated aromatic compounds. Microbiol Series 13:319–360
Reiner AM (1972) Metabolism of aromatic compounds in bacteria: purification and properties of the catecholforming enzyme, 3, 5-cyclohexadiene-1, 2-diol-1-carboxylic acid (NAD+) oxidoreductase (decarboxylating). J Biol Chem 247(16):4960–4965
Rodríguez-Rodríguez CE, Castro-Gutiérrez V, Chin-Pampillo JS, Ruiz-Hidalgo K (2013) On-farm biopurification systems: role of white rot fungi in depuration of pesticide-containing wastewaters. FEMS Microbiol Lett 345:1–12. https://doi.org/10.1111/1574-6968.12161
Salinas ES, Hernández LO (2011) Riesgos y estrategias enel uso de plaguicidas. Inventio 7:21–28
Shakoori AR, Makhdoom M, Haq RU (2000) Hexavalent chromium reduction by a dichromate-resistant gram-positive bacterium isolated from effluents of tanneries. Appl Microbiol Biotechnol 53:348–351. https://doi.org/10.1007/s002530050033
Silambarasan S, Abraham J (2013) Ecofriendly method for bioremediation of chlorpyrifos from agricultural soil by novel fungus *Aspergillus terreus* JAS1. Water Air Soil Pollut 224:1369. https://doi.org/10.1007/s11270-012-1369-0
Singh H (2006) Mycoremediation: fungal bioremediation. Wiley, Hoboken
Thabit TMAM, El-Naggar MAH (2013) Diazinon decomposition by soil bacteria and identification of degradation products by GC-MS. Soil Environ 32:96–102
Uqab B, Mudasir S, Nazir R (2016) Review on bioremediation of pesticides. J Bioremed Biodegr 7:1–5. https://doi.org/10.4172/2155-6199.1000343
van der Meer JR, de Vos WM, Harayama S, Zehnder AJB (1992) Molecular mechanisms of genetics adaptation to xenobiotic compounds. Microbiol Rev 56:677–694. https://doi.org/10.1128/mr.56.4.677-694.1992
Verma JP, Jaiswal DK, Sagar R (2014) Pesticide relevance and their microbial degradation: a-state-of-art. Rev Environ Sci Biotechnol 13:429–466. https://doi.org/10.1007/s11157-014-9341-7
Yamada Y (2017) Importance of codex maximum residue limits for pesticides for the health of consumers and international trade. Food Safety Assess Pestic Resid 269–282. https://doi.org/10.1142/9781786341693_0007
Ye X, Dong F, Lei X (2018) Microbial resources and ecology—microbial degradation of pesticides. Nat Reso Cons Res 1:22–28
Yim YJ, Seo J, Kang SI, Ahn JH, Hur HG (2008) Reductive dechlorination of methoxychlor and DDT by human intestinal bacterium *Eubacterium limosum* under anaerobic conditions. Arch Environ Contam Toxicol 54:4

Chapter 6
Plastisphere: Marine Microbial Assemblages for Biodegradation of Microplastics

Atif Khurshid Wani, Nafiaah Naqash, Nahid Akhtar, Tahir ul Gani Mir, Bilal Ahmad Mir, and Rattandeep Singh

Abstract The emergence of plastic waste from industries is a serious ecological concern. In recent years, plastic has been identified in various ecosystems, including human tissues, posing a threat to the biotic components of the Earth and giving rise to potential adverse consequences. The plastic waste is fragmented into smaller components like nanoplastics (NPs) and microplastics (MPs) by different physical and chemical processes. The surface of small plastic polymers is occupied by several microbial communities such as bacteria, fungi, and diatoms, giving rise to the plastisphere. The hydrophobic surface of MPs can provide suitable environment for aquatic microbial communities to colonize and form biofilms. The microbial composition and colonization on the surface of MPs are influenced by the environment and physiochemical properties of the plastisphere. Several techniques are used for the characterization and analysis of the plastisphere to frame a suitable strategy for plastic degradation. In this chapter, we provide a detailed account of the plastic pollution in different environments, factors affecting plastisphere, characterization of plastisphere, and microbial and enzyme-mediated degradation of MPs.

Keywords Plastic · Pollution: plastisphere, biofilms · Biodegradation

A. K. Wani · N. Naqash · N. Akhtar · B. A. Mir · R. Singh (✉)
School of Bioengineering and Biosciences, Lovely Professional University, Phagwara, Punjab, India
e-mail: rattan.19383@lpu.co.in

T. u. G. Mir
School of Bioengineering and Biosciences, Lovely Professional University, Phagwara, Punjab, India

Chemistry and Toxicology Division, State Forensic Science Laboratory, Jammu and Kashmir, India

A. Karnwal, A. R. Mohammad Said Al-Tawaha (eds.), *Microbial Applications for Environmental Sustainability*, https://doi.org/10.1007/978-981-97-0676-1_6

6.1 Introduction

Plastic goods are exceptionally durable, lightweight, economical, and adaptable in nature, which allows them to replace many conventional materials such as paper, metal, glass, and wood. Polymeric molecules with extended chains, consisting of carbon, silicon, hydrogen, oxygen, and chloride atoms, are harnessed in the production of plastics (Rojas-Parrales et al. 2018). Presently, the most prevalent polymers employed in various applications include polyethylene (PE), polypropylene (PP), polyvinyl chloride (PVC), polystyrene (PS), and polyethylene terephthalate (PET) (Andler et al. 2022). Overall, plastics have become a key commodity in our society, akin to electricity or the Internet. However, in addition to all of the benefits, these low-cost materials have fostered a "use and toss" consumer culture, resulting in several environmental challenges (Naqash et al. 2020). Despite the increasing recognition, a staggering quantity exceeding 300 million tonnes of plastic waste is produced annually, with the majority being left unrecycled following disposal. Plastic debris has now been included among the roster of global perils, alongside climate change, ocean acidification, and ozone depletion (Gallowaya and Lewisa 2016). Plastic pollution is reported in almost every environment on the planet, regardless of its origin. Water systems including rivers, lakes, and oceans are acting as "plastic collectors" from adjoining terrestrial systems/watersheds (Yang et al. 2020). Plastic marine debris has accumulated in the world's oceans as a result of improper plastic disposal. Their primary land-based entry points into the marine environment are accidental loss or improper handling of plastics, dumping plastic waste, and poorly managed landfills (Phelan et al. 2020).

Microplastic (MP) pollution, or plastic particles smaller than 5 mm, is particularly noteworthy. These tiny plastic particles emerge from various sources (Kasmuri et al. 2022). Primary MPs are formed deliberately and utilized in personal care items like toothpaste, cosmetics, etc., providing a direct supply of microplastic. Secondary MPs are formed when plastic debris is broken down by several abiotic factors including erosion, weathering, sunlight, and aquatic immersion (Yang et al. 2020). The extensive utilization of polymer-based items and their enduring resistance to physical and biological deterioration over extended periods is evident in the ubiquitous presence of MP across the globe, found in both surface water and sediment within marine ecosystems (Zalasiewicz et al. 2016). Given the substantial influx of plastic waste into marine ecosystems, the prevalence of microplastics in these environments comes as no surprise. In fact, specific regions with elevated concentrations of MPs, known as MP hotspots, have been identified in ten distinct estuaries located in North West England (Onyena et al. 2022). Microplastics were found in more than 90% of samples collected from 25 beaches along Hong Kong's coastline (Fok and Cheung 2015). Furthermore, numerous "microplastic hotspots" with a maximum microplastic concentration of about 517,000 particles m^{-2} were found in North West England (Hurley et al. 2018). Additionally, highly noticeable microplastic pollution has been observed in the coastal waters of Africa (Alimi et al. 2021; Oni et al. 2020).

Plastic particles can adsorb organic material and nutrients from their environment while also acting as inert surfaces in nutrient-depleted water bodies. This ability makes MPs essential substrates for the development of microbial biofilms on these synthetic particles (Shen et al. 2019). As a result, MP biofilms may be considered a new microbiological niche in the surroundings, particularly in pelagic waters (Arias-Andres et al. 2018). Earlier, the diversity and function of "plastisphere" microbial communities and further the future of plastics in the marine environment have been discussed. The significant differences, similarities, and influencing variables of MP-associated biofilm communities in freshwater and marine environments have also been outlined in past studies (Harrison et al. 2018). In this chapter, initially we compile the diversity and composition of the MP-associated microbial communities and their molecular and microscopic characterization. Secondly, we discuss the different factors that are affecting the development of MP-associated biofilms. Thirdly, we have summarized the MP degradation in marine environment, highlighting the necessity of more research on biodegradable plastic. Furthermore, the potential role of enzyme driven degradation linked to microplastics has been discussed in detail along with addressing the future challenges. The primary goal of this chapter is to discuss an overview of how microbes form on MPs, as well as the community structure, functions, and environmental functions of these biofilms in marine systems. We emphasize the role of MPs in pelagic systems as a recent, developing microbial niche with uncertain impact on environmental and biological changes, as well as biogeochemical cycling in water resources.

6.2 The Plastisphere: Marine Microbial Colonization

The increasing prevalence of MP contamination in marine systems throughout the world has clearly declared it a severe environmental issue. Numerous bacterial and fungal strains have been shown to be capable of degrading various forms of plastic (Krueger et al. 2015). MP biofilm populations, however, differ dramatically from bacterial communities on natural surfaces such as wood pellets, organic matter, and glass beads (Oberbeckmann et al. 2018; Ogonowski et al. 2018). Even previous innovative research investigations on marine MP-related biofilms indicated that these communities differed greatly from those in the surrounding water (Oberbeckmann and Labrenz 2020). It is anticipated that sessile organisms lead distinct lifestyles compared to free-living microorganisms, leading to variations in microbial community compositions between natural particles and free-living microorganisms. Consequently, unless the existence of a genuine microplastic-associated plastisphere, distinct from those associated with natural particles like wood, cellulose, or glass, is verified, the role of biofilms on microplastics will remain uncertain (Rieck et al. 2015). Certain environmental and biogeographical parameters, like salinity and nutrient content, have a significant influence on the production of MP biofilms in marine environments (Oberbeckmann and Labrenz 2020). The microplastic surfaces themselves have an impact on colonization processes, although a minor one. It has also been argued if some overrepresented and perhaps hydrocarbon clastic marine microplastic biofilms may use plastics as a source of energy because of their ability to decompose extremely complex biopolymers such as lignin and petroleum-based goods (Ogonowski et al. 2018). Oberbeckmann et al. (2016) discovered that, rather than the surface properties of the plastics, the microbial communities that colonized polyethylene terephthalate (PET) after weeks of exposure in the North Sea were predominantly driven by spatial and seasonal parameters (Oberbeckmann et al. 2016). Another study revealed that nutrients such as total phosphorus and total nitrogen, as well as salinity, influenced the average development rate of biofilms on plastic trash, with salinity being the most critical factor impacting bacterial colony diversity (Li et al. 2019). Furthermore, it was discovered that plastic colour influences the community structure and functional diversity of the plastisphere. However, several research found no effect of the sample site, environment, or polymer type on the microbial communities in plastispheres (Wen et al. 2020). Ecologists have been working hard to understand the processes that define microbial communities in plastisphere biofilms as they develop in situ. Understanding the relative role of deterministic and stochastic mechanisms in microplastic bacterial assembly dynamics in natural water settings is critical. Despite the fact that both stochastic and deterministic factors influence community assembly in a variety of environments such as the human body (Venkataraman et al. 2015), soils, and oceans (Hamdan et al. 2012), the mechanisms of plastisphere microbial community assembly remain unknown. The proportional contributions of stochastic and deterministic processes, especially the functions of basic ecological processes such as dispersion, selection, drift, and diversification that influence bacterial distribution patterns across plastisphere niches, are mainly understood.

6.3 Microplastic-Associated Biofilms

Microbes, notably bacteria, diatoms, and algae, may adhere to MPs in the aquatic environment and quickly colonize the surface or even the insides of MPs to create biofilms (Harrison et al. 2014). MPs' hydrophobic surfaces may provide an appropriate habitat for aquatic microbial populations to grow and build biofilms (Sun et al. 2020). Furthermore, MPs can serve as a separate and novel home for aquatic microbial communities since the taxonomic makeup of microbial communities on MP surfaces differs from that of microbial communities in surrounding water, whether in free living or particle attached forms (Yang et al. 2020; Deng et al. 2021). Furthermore, due to differences in roughness and surface morphology among different types of MP surfaces, the types of microorganisms present in the MP-related biofilm may vary with the type of MP (Miao et al. 2021; Qiang et al. 2021). The formation of microbial coatings on MPs might alter their density and convey them from the surface to deeper areas of water bodies (Cózar et al. 2014; Harrison et al. 2014). Furthermore, MP-associated biofilms have the potential to influence the weathering process and the physiological aspects of MPs (Deng et al. 2021).

Furthermore, MP-associated biofilms may influence MP aggregation and, as a result, their subsequent deposition or suspension (Lagarde et al. 2016). MP-associated biofilms can also serve as vectors for pathogenic bacteria in fish, animals, and humans, as well as aid in the transfer of biotoxins generated by certain toxic algae (Harrison et al. 2018). In addition to pathogenic germs, bacteria linked with antibiotic resistance and polycyclic aromatic hydrocarbon degrading bacteria have been identified in MP-related biofilms (Sun et al. 2020) Microplastic-associated biofilms have the potential to influence the environmental effect, dispersion, and degradation of microplastic contaminants (Harrison et al. 2018). UV irradiation, salinity, light availability, pressure, and temperature can all influence the production of microplastic-associated biofilms (Harrison et al. 2018). Furthermore, whether the habitat is freshwater or marine, the microbial community makeup of the microplastic-associated biofilms might vary (McCormick et al. 2014).

6.4 Microscopic and Molecular Characterization

Although the surface layer of plastic provides an optimal situation for microbial colonization, the succession of microbes requires extensive exploration. Utilizing scanning electron microscopy (SEM) alone is insufficient to analyse the makeup and structure of biofilms developed on plastisphere. However, SEM is still a beneficial firsthand experience observing bacteria, other protists, and diatoms for visualizing the structure and spatial patterns of microbial communities on plastic (Zettler et al. 2013). SEM observations have found the microbes hiding in grooves and pits, which indicated that the plastic's surface had deteriorated. Although SEM provides a thorough examination of the life that exists on the surface of plastic waste, the taxonomic resolution that can be obtained from this method is constrained. With the exception of morphologically distinct protists like diatoms and a few filamentous cyanobacteria, it is very difficult, if not impossible, to distinguish between different groups of microbes at the species level using SEM alone. SEM is not only expensive and labour-intensive but it also faces difficulties with data interpretation and quantification at the moment due to a lack of automation. Epifluorescence microscopy, in conjunction with the use of phylogenetic probes and through fluorescence in situ hybridization (FISH), has the ability to improve our understanding of how microbes interact on the plastic surface by allowing us to examine microbial diversity on the surface with significant taxonomic clarity (DeLong et al. 1989). To study plastisphere communities, FISH procedure with combinative labelling and spectral imaging was formed for investigating a probe set with four newly designed probes and three pre-existing phylogenetic probes. This technique for imaging using light, fluorescence, confocal, or electron microscopy was effective in keeping track of biofilm development. The technique further enabled not only the observation of surface coverage but also the observation of biofilm structures and microbial associations (Schlundt et al. 2020). In general, FISH offers several advantages, such as the capacity to provide precise abundance data rather than relative abundance information for the targeted cells and the ability to reveal both the spatial arrangement of the microbial community and taxonomic insights regarding the composition of the community on the plastic surface. Since the structure and function of biological systems are closely interconnected, investigating the spatial distribution of microbes can offer insights into their physiology and interactions. However, an inherent challenge associated with FISH techniques arises from the fact that phylogenetic probes specifically target and bind to ribosomal RNA (rRNA) within the cell (Valm et al. 2012). The fact that phylogenetic probes target and hybridize with ribosomal RNA (rRNA) in the cell presents a potential problem with FISH techniques. The fluorescent signal may be weak in communities that are not actively expanding, resulting in overlooked taxa. To enhance the fluorescent signal, multiple techniques such as catalysed reporter deposition FISH (CARD-FISH), nested FISH approach, or a next-generation in situ hybridization chain reaction (HCR) using multiple probes are obligated. However, using signal augmentation techniques can complicate a procedure. Another restriction of FISH strategies is that they require the purchase of fluorescently labelled probes, which necessitates molecular knowledge of the communities to be probed. For example, in one study, clone libraries were used in accordance with Sanger sequencing. They used genetic analysis and T-RFLP prior CARD-FISH to evaluate the phylogenetic associations of the microorganisms (Eich et al. 2015). The use of up to three different fluorophores simultaneously using the FISH technique was also constrained. The abundance of taxa which can be identified in an individual sample was consequently decreased. But a novel FISH technique called CLASI-FISH might be able to solve this issue by employing multiple fluorophores in parallel (six or more) (Valm et al. 2012). The amount of taxa which could be labelled in an individual sample is enhanced by labelling a given taxonomic group with a combination of different fluorophores (2–3) during the same period. Depending on the number of fluorophores used, this technique can significantly increase the number of distinguishable microbes, possibly reaching hundreds in a single FISH experiment. Microscopy, spectroscopy, statistical analysis, and digital fluorescence microscopy images can all be used to differentiate the fluorophores. This approach is currently being tested and applied to plastic debris after being effectively implemented to human oral microbes, opening up an exciting new field of study.

Sanger sequencing of rRNA genes, other biomarkers, as well as physiological characterization and ecological experimentation, are all facilitated by the capability to identify microorganisms in the lab. Because more than 99% of the microbes in any surroundings are not suitable for growing using traditional cultivation techniques, the majority of the microbial community is still unexplored (Amann et al. 1995). The development of culture-independent strategies enabled us to investigate the majority of the environment's uncultivable microbes. As far as molecular techniques are concerned, clone libraries, denaturing gradient gel electrophoresis (DGGE), terminal restriction fragment length polymorphisms (T-RFLP), and high-throughput sequencing have previously been used to identify microbes on plastic surface in aquatic systems. All of these methods rely on the polymerase chain reaction (PCR), which has historically been unable to provide info on absolute species

Table 6.1 List of techniques utilized for analysing microbial communities inhabiting plastic surfaces

Techniques	Plastic size	Plastic type	Habitat	References
SEM	Microplastic	PP, PC, ABS, PA, LDPE	Marine offshore	Feng et al. (2020)
SEM	Microplastic	LDPE	Microcosm	Harrison et al. (2014)
CARD-FISH	Microplastic	LDPE	Microcosm	Harrison et al. (2014)
Clone libraries	Macroplastic	PVC	Marine water	Dang et al. (2008)
	Microplastic	LDPE	Microcosm experiment	Harrison et al. (2014)
DGGE	Macro and microplastic	PET, PE, PP, PS	Marine coastal waters and offshore	Oberbeckmann et al. (2014)
	Environmental samples	LDPE, PET, PS	Seawater offshore	Delacuvellerie et al. (2019)
T-RFLP	Microplastic	LDPE	Microcosm	Harrison et al. (2014)
Amplicon sequencing	Microplastic	PET, HDPE, PVC, LDPE, PP, PS	Marine offshore	Dudek et al. (2020)
	Microplastic	PE, PP	Marine offshore	Zettler et al. (2013)
Shotgun metagenomics	Macro- and microplastic	Unidentified	Marine offshore	Bryant et al. (2016)
	Macroplastic	PET	Coastal waters	Kerfahi et al. (2022)

abundances due to PCR bias. Table 6.1 summarizes the main techniques used to analyse the microbial diversity on plastics in aquatic habitats. The 16S rRNA gene could be sequenced either entirely or partially in accordance with Sanger sequencing when it comes to clone libraries. Bacteria can typically be categorized to the species level, thanks to the remarkable taxonomic/phylogenetic assignments made possible by complete 16S rRNA gene sequence data. However, because of the time and labour involved in creating clone libraries, most studies only process below 1000 clones in a specified library (Konopka 2009). Furthermore, the genetic fingerprinting techniques DGGE and T-RFLP utilize the PCR products that have been amplified from environmental DNA (Muyzer and Smalla 1998).

Next-generation sequencing has enhanced read depth for 16S rRNA marker genes, overcoming many of the detriments of first-generation sequencing technologies. The sequencing reactions are enormously parallelized by the amplicon sequence analysis, which amplifies and sequences genetic markers like the 16S rRNA gene for multiple ecological samples at once (Sogin et al. 2006). Amplicon sequencing has until now been restricted to 500 bp marker genomic regions, rendering it difficult to categorize unknown reads using phylogenetic inference (Li et al. 2015). The established molecular procedures for characterizing bacterial species on plastic that have been previously discussed rely on sequencing the entire or a portion of the 16S rRNA gene. These methods can only identify taxa in a taxonomic sense. Shotgun metagenomics, however, has the capacity to both extract metabolic diversity and ascertain the taxonomic structure of all microbes present on plastic This molecular strategy aims to accomplish two aspects: initially, it enables for the independent analysis of annotated genomic DNA fragments, and further, it allows for the assembly of fragments into genomic bins for comparative genomic analysis. For two reasons, we assume that a thorough knowledge of the entire microbial community is necessary. For starters, plastic can act as a vector for pathogenic microorganisms (Barnes 2002). Pathogenic Vibrio and harmful algal blooming species have already been discovered on plastic surface (Masó et al. 2003; Kirstein et al. 2016). Therefore, understanding the microbiota is essential for transport to other regions. Second, it is important to look into community members other than bacteria, like fungi, in terms of the deterioration of plastic. Plastics can be degraded by both bacterial and fungal microorganisms in terrestrial environments (Koutny et al. 2006). Fungal organisms will likely be included in the plastic microbial community investigations, which will likely yield unique knowledge into the ecosystems and biological function of the plastisphere.

6.5 Factors Affecting Plastisphere

The microbial development on MPs believed to be a complex procedure impacted by a lot of considerations (Carson et al. 2013). This is significant information because the structure of microbiota on such biofilms differs from other situations and is essential to the functioning of marine habitats (Rummel et al. 2017). Understanding the elements having an impact on biofilm development and the makeup of microorganisms on MP particles is crucial. Using this knowledge, MP pollution can be decreased by identifying potential biodegradable MP species (Lobelle and Cunliffe 2011). Two aspects influence the expansion and makeup of microbial communities on MP particles (Carlén et al. 2001). The substrate's physicochemical characteristics, which include the hydrophobicity, plastic type, specific surface area, and roughness, are the first determinant. The environment, which also includes nutrients, temperature, salinity, and location, is the second factor (Stoodley et al. 2002).

6.5.1 Physiochemical Factors

Various types of MPs influence the development of biofilms and the colonization of microbial communities, according to numerous studies (Miao et al. 2019). The majority of research studies have used MPs made from polyethylene, polypropylene, polyethylene-terephthalate, polystyrene, and polyvinyl chloride, due to their frequent presence in water environments. Compared to biofilm communities on the surfaces of PP and PE, those on the surface of PS have more bacteria (Pinnell and Turner 2020). Studies revealed that various MP structures (like plastic form, item size, and roughness) or other MP characteristics are what causes different biofilm development (like solid particles, synthetic items, with more polymer quantity and 5 mm in size) (Frère et al. 2018). On the surface of PET, numerous Mycetozoa, Bacteroides, and Cyanobacteria have been observed in addition to algae like Bacillariophyta and brown algae (Pinto et al. 2019; Eich et al. 2015). But the procedure underlying the preferential colonization of common microorganisms on a particular MP surface remains poorly understood.

The microbial development rate and the density of microbes on MPs are positively correlated with roughness, as the rougher surface is considered better for the development and functioning of microbes on MPs (Miao et al. 2021; Hirai et al. 2015). PVC had more operational taxonomic groups than PP because of its rough surfaces and greater surface area, which made bacterial interaction relatively easy (Miao et al. 2019). Plastic surfaces are capable of absorbing both inorganic and organic material from the water environment, creating a layer that affects the adhesion of biofilms to MP surfaces (Rummel et al. 2017). The fact that the layer's formation is related to how rough the surface of MPs is evident that the layer's chemical composition had an impact on the plastisphere. During the last stages of biofilm development, hydrophobicity is firmly parallel with micro-community, adhesion, and microbial communities on MP particles (Pompilio et al. 2008). Even so, compared to other physicochemical characteristics of MPs, a small number of studies found that the plastic's hydrophobicity had a minor impact on the adherence and firmness of bacteria (Heistad et al. 2009; Hook et al. 2012). Due to unknown factors like cell size, MP fragment size, and shape, the correlation of hydrophilicity or hydrophobicity of MP boundary and level of cell adherence is still unclear (Foulon et al. 2016; MacHado et al. 2010).

6.5.2 Environmental Aspects

The variety of bacteria living on the surface correlates unfavourably with nutritional status, whereas the mean expansion rate of MP bacterial growth correlates with it positively (Oberbeckmann et al. 2018). The relationship between nutritional levels based on TN and TP and plastisphere formation has received the most consideration. Scientists have discovered that TN and TP quantity are mandatory for the development of biofilm and serve to the growth of particular microbial communities (Miao et al. 2021). The biosorption and decomposing activities of microbes on MP exterior surface vary with nutritional state. Green algae and autotrophic diatoms are the major species on the MP biofilm communities in lakes with average and little concentration of nutrients, while autotrophic species are taken up by predators like ciliates (heterotrophic), which shift MPs to a nutritious habitat (Miao et al. 2020).

Due to MPs' prolonged exposure to the surroundings and alteration in seasons, the environmental temperature where MPs are present alters, which obviously impacts the variety of microbes on biofilm (Carson et al. 2013). Both the enzyme production of cells and the physiological behaviour of microorganisms are significantly impacted by temperature (Xu et al. 2019). Summer temperatures caused the biofilm on PET surfaces to be thicker, and the comparative richness of *Cytophagales* and *Ignavibacteriales* varied seasonally (Pinnell and Turner 2020). Further, salinity and light conditions also have an impact on the formation of biofilms, in addition to seasonal temperature variations (Oberbeckmann et al. 2016). Environment salinity has a negative relationship with biofilm growth rate and a positive relationship with bacterial community diversity (Oberbeckmann et al. 2018). Salinity affects the MP surface biofilm's entire bacterial community as well as the distribution of particular bacteria (Guo et al. 2017). The overall process of biofilm formation on the MPs' surfaces is adaptable (Caruso 2020). To investigate the dynamic procedure of biofilm development in its initial, central, and late phases, hypothetical analyses have been utilized (Jiang et al. 2018). Based on the recent observations, the actual number of microbial population on the surface of MPs has grown consistently with exposure period (Tu et al. 2020).

6.6 Microplastic in Marine Environment

The concern of marine ecosystems being contaminated by MPs has been on the. MPs are pervasive and abundant in the marine environment, with higher concentrations occurring along coastlines. Numerous marine organisms have shown the ability to ingest MPs, which opens the door to the transmission of hydrophobic waterborne contaminants or chemical additives to the marine biota (Andrady 2011). The amount of plastic that contributes towards municipal waste accounts for 10% of all waste produced globally, which has been intensified by the widespread usage of disposable (use and throw) plastics (Ncube et al. 2021). Despite the fact that some plastic garbage is recycled, the bulk is disposed of in landfills, where it may take generations to break down and degrade (Moore 2008). Plastics that are entering the maritime environment via careless dumping have been of particular concern (Gregory 2009). Despite the fact that plastic is globally recognized as a pollutant and that laws are in place to reduce the quantity of plastic trash entering the marine, it has been estimated that up to 10% of plastics manufactured end up in the seas and thereby contaminating marine ecosystem. The irresponsible dumping of waste materials that enter our seas and oceans either directly or indirectly leads to marine litter (Lopez Lozano and Mouat 2009; Ryan et al. 2009). Approximately 80% of the plastics in marine litter come from terrestrial sources (Andrady 2011). Primary MPs used in cosmetics and air-blasting dumped disposal plastics, and plastic leachates from waste sites are some of the common sources of plastic found in marine litter since these types of plastic have a significant potential to reach the marine environment via wastewater systems, rivers and canals, drainage systems (Derraik 2002;

Thompson 2006; Moore 2008). Numerous studies have demonstrated that freshwater systems with the unidirectional flow are also responsible for the transfer of plastic waste into the marine ecosystem (Moore et al. 2002; Browne et al. 2010). Moore (2008) calculated the proportion of plastic particles that were present in water samples taken from two rivers in Los Angeles (California, USA) (Moore 2008). The extrapolated statistics showed that over the course of 3 days, only these two rivers will discharge almost two billion microplastic particles into the marine ecosystem.

Several worldwide studies have been conducted in recent years to assess the prevalence of floating MPs and microbeads in various water bodies across the world (Eriksen et al. 2013b; Cózar et al. 2014; Reisser et al. 2015). According to a study by Li et al. (2016), the North Atlantic subtropical gyres are the most affected by floating MPs and microbeads (Li et al. 2016). According to Lebreton et al. (2012), the subtropical gyres have become the hotspots for MPs as a consequence of the plastic waste that came from both terrestrial and marine sources (Lebreton et al. 2012). The South Pacific subtropical gyre was found to possess an approximate average of 26,898 particles per square kilometre, varying in size from 0.355 to over 4.750 mm (Eriksen et al. 2013a). Similarly, the North Pacific subtropical gyre region has been documented as accumulating approximately 22,290 tonnes of floating plastic waste (Law et al. 2010). According to Eriksen et al., the South Pacific subtropical gyre has a high concentration of microplastic debris because of natural processes including wind currents around the coasts of Indonesia and Ecuador (Eriksen et al. 2013a). The Arctic Ocean has been found to contain between 38 and 234 particles/m^3 of MPs (Obbard et al. 2014), which is particularly alarming given that the amount was two times more than that previously recorded for the Pacific Gyre (Goldstein et al. 2012). MPs are persistent in the marine environment, and because of their tiny size, they are permeable to a wide variety of marine organisms, including fish, corals, sea urchins, zooplanktons, lobsters, worms, and many more (Browne et al. 2008). After being consumed by these marine species, the non-degradable MPs bioaccumulate up the food chain (Gregory 1996, 2009) until they reach higher tropic levels (Carpenter and Smith Jr 1972). Microplastic waste has been found in fish, crustaceans, turtles, and seabirds all around the world (Derraik 2002; Cole et al. 2011) and thus obstructing the digestive system, inhibiting stomach enzyme release, imbalanced hormone levels, delayed ovulation, and infertility (Azzarello and Van Vleet 1987; McCauley and Bjorndal 1999; Derraik 2002; Wright et al. 2013). Compared to marine habitats, MP pollution in terrestrial and freshwater ecosystems has received less attention (Ncube et al. 2021). About 10% of microplastic ends up in the ocean, and 7–8 million pieces of plastic accidentally wind up there from terrestrial sources (Lacerda et al. 2019). The Mediterranean Sea, which has an average depth of 1500 m, is known as a hotspot for plastic pollution since its MP concentration is four times higher than that of the North Pacific Ocean. This is explained by the Mediterranean Sea's distinctive semi-enclosed topology and the nearby nations that produce large amounts of plastic waste (Sharma et al. 2021).

6.7 Microbial and Enzymatic Degradation of Microplastic

The degradation of MPs is essential for maintaining the ecological balance. Mechanical and chemical degradation processes are widely used. The rate of degradation depends mostly on polymer structure, and composition, besides the temperature, humidity, and the site of deposition. The exposure of MPs to ultra-violet rays showed variation in rate of degradation (Arpia et al. 2021). The chemical degradation is mediated by hydrolysis, oxidation, and photooxidation. This leads to the release of gases into the environment which can disturb the ecosystem (Da Costa et al. 2018). Biodegradation of MPs is a preferred strategy for minimizing the plastic contamination in the environment (Rojas-Parrales et al. 2018; Wani et al. 2022a). The breakdown of large or small plastic polymers takes place through the aerobic or anaerobic action of microorganisms. In aerobic degradation of MPs, oxygen acts an electron acceptor with water and carbon dioxide as major by-products. In anaerobic mode of MP-biodegradation, bacteria utilize iron, sulphate, manganese, and carbon dioxide as electron acceptors. MP polymers are not transported into the cells because of their large size and water insoluble nature. The degradation of MP polymers is driven by three major steps, i.e., microbial attachment on MP surface, utilization of MP as carbon source, and finally the degradation (Wani et al. 2023). MP polymers are depolymerized by enzymatic action both intracellularly and extracellularly. Exoenzymes degrade plastic polymers into small molecules with water, methane, and carbon dioxide as end products, the process is known as mineralization. The other processes of microbial degradation of MP include biodeterioration, biofragmentation, and assimilation (Bacha et al. 2021). Microorganisms that are capable of plastic polymer degradation have been isolated from different sites. *Staphylococcus, Streptococcus, Klebsiella, Pseudomonas*, and *Micrococcus* have been characterized with plastic-degrading potential. The biodegradability is enhanced by the addition of auto-oxidizing agents, which ensures easy degradation of the plastic polymers. *Bacillus vallismortis* has been reported to have LDPE degradation potential with 75% efficiency after 120 days of incubation (Nourollahi et al. 2019). *Bacillus amyloliquefaciens* degrades polyethylene when it is preliminary treated with heat. Besides *B. safensis* and *B. mycoides* degrade HDPE and LDPE when pretreated with sunlight and mercuric acid. Microorganisms adapt fluctuating environmental conditions through different cellular and genetic pathways. They are known to colonize MP surfaces which changes some of the properties of the polymers like strength, molecular weight, and roughness (Awasthi et al. 2017). The homoatomic and heteroatomic plastic backbone makes biodegradation of MPs a complex process. There is significantly weight loss in MP with the action of microorganisms, but process is slower as compared to the chemical degradation. The direct application of enzymes secreted by microorganisms plays an essential role in speeding up the process of MP degradation. Cutinases, isolated from *Thermobifida fusca* are well studied for their MP-degradation potential. MPs do not dissolve in water, but some of the water-soluble polymers are degraded easily and changed into ketones, acids, and alcohols. MP-biodegradation is monitored by: (1)

Table 6.2 Microbial-assisted degradation of different plastic polymers

Microorganism	Plastic type	Incubation period (weeks)	Weight loss (%)	References
Pseudozyma antarctica	Biodegradable plastic	5	8.2	Sameshima-Yamashita et al. (2019)
Rhodococcus sp.	PP	5.2	6.4	Auta et al. (2018)
B. cereus	LDPE	17.1	1.53	Zerhouni et al. (2018)
Citrobacter sp.	HDPE	12.8	7.3	Ojha et al. (2017)
Enterococcus sp.	PS	7	3.2	Hou and Majumder (2021)
Fusarium sp. and *Humicola* sp.	PET	5	2.6 and 3.1	O'Neill et al. (2007); Ronkvist et al. (2009)
P. aeruginosa	PE	4.8	6.25	Mouafo Tamnou et al. (2021)
Aspergillus niger	LDPE	18	17.7	Esmaeili et al. (2013)
Fusarium sp.	PE	4	14	Tachibana et al. (2010)
Ideonella sakaiensis	PET	5	7.2	Webb et al. (2013)
Exiguobacterium sp. *RIT 594*	PS	7	9.3	Parthasarathy et al. (2022)
Candida rugosa	PU	6.1	7.3	Gautam et al. (2007)
Cryptococcus sp.	PET	5	3.2	Masaki et al. (2005)
Paenibacillus sp.	PE	11.2	14	Bardají et al. (2019)

changes in surface properties of the polymer, (2) consumption rate of oxygen, (3) evolution of oxygen, and (4) changes in MP's physical and chemical properties (Zhang et al. 2021). Table 6.2 gives an insight into the ability of microorganisms in degrading the MP-contaminants.

Different factors play their role in plastic degradation by microorganisms. The difference in the functional groups, hydrophobicity, structure (linear/branched), bond type (ester, ether, urethane, amide), base composition, polymer form (pellet, film, powder), polymer molecular weight, density, toughness, and morphology drives the degradation rate of different polymers. The rate of MP degradation reduces with the decrease in polymer solubility (Weinstein et al. 2016). Plastic polymers are less susceptible to microbial degradation with the decrease in the degree of solubility. Amorphous polymers are vulnerable to enzyme-mediated degradation than the crystalline polymers. Thus, MPs can restrict microbial action in hydrophobic environment by inhibiting water absorption process (Santo et al. 2013).

6.8 Conclusion and Future Prospects

The limitations in standard sampling protocols, extraction methods, detection procedures, and analytical tools bring a lot of hurdles in management, surveillance, and degradation processes. The heterogeneity of MPs is a serious matter of concern in understanding their physiochemical parameters. The lack of established, uniform, and standard policies for the assessment of MPs and NPs is challenging. The collection of inadequate amounts of MP from the contaminated sites makes it difficult to detect the plastic polymers. The available analytical tools are not good enough to detect low concentrations (Veerasingam et al. 2020; Hossain et al. 2022). There is a paramount need for highly efficient and multi-residual tools to solve the challenges confronted in MP analysis and characterization. The low rate of biodegradation driven by bacteria and fungi is also very challenging. Since most of the microorganisms are unculturable, and it is difficult to study the process of bioremediation in laboratory conditions. Metagenomics, an unculturable study method, is one of the recent approaches to detect the novel microorganisms and genes capable of degrading plastic polymers (Handelsman 2004; Wani et al. 2022b, d). Metagenomics in association with other meta-omics approaches can be useful in predicting the efficient plastic-degrading bacteria. The recent genome editing and genetic engineering technologies can be useful in amplifying the specific genes and subsequently cloning and expressing them in a suitable vector and the host (Wani et al. 2022c, e; Mir et al. 2022). In this chapter, we have given a detailed account of the plastisphere along with the different analytical tools used in its characterization and analysis. We have highlighted the applicability of microorganisms in degrading different types of plastic polymers.

References

Alimi OS, Fadare OO, Okoffo ED (2021) Microplastics in African ecosystems: current knowledge, abundance, associated contaminants, techniques, and research needs. Sci Total Environ 755:142422

Amann RI, Ludwig W, Schleifer K-H (1995) Phylogenetic identification and in situ detection of individual microbial cells without cultivation. Microbiol Rev 59:143–169

Andler R, Tiso T, Blank L, Andreeßen C, Zampolli J, D'Afonseca V, Guajardo C, Díaz-Barrera A (2022) Current progress on the biodegradation of synthetic plastics: from fundamentals to biotechnological applications. Rev Environ Sci Biotechnol 2022 214 21(4):829–850. https://doi.org/10.1007/S11157-022-09631-2

Andrady AL (2011) Microplastics in the marine environment. Mar Pollut Bull 62:1596–1605

Arias-Andres M et al 2019 undefined (2018) Collateral effects of microplastic pollution on aquatic microorganisms: an ecological perspective. Elsevier

Arpia AA, Chen W-H, Ubando AT et al (2021) Microplastic degradation as a sustainable concurrent approach for producing biofuel and obliterating hazardous environmental effects: a state-of-the-art review. J Hazard Mater 418:126381

Auta HS, Emenike CU, Jayanthi B, Fauziah SH (2018) Growth kinetics and biodeterioration of polypropylene microplastics by Bacillus sp. and Rhodococcus sp. isolated from mangrove sediment. Mar Pollut Bull 127:15–21

Awasthi S, Srivastava P, Singh P et al (2017) Biodegradation of thermally treated high-density polyethylene (HDPE) by Klebsiella pneumoniae CH001. 3 Biotech 7:332

Azzarello MY, Van Vleet ES (1987) Marine birds and plastic pollution. Mar Ecol Prog Ser 37:295–303

Bacha A-U-R, Nabi I, Zhang L (2021) Mechanisms and the engineering approaches for the degradation of microplastics. ACS EST Eng 1:1481–1501

Bardají DKR, Furlan JPR, Stehling EG (2019) Isolation of a polyethylene degrading Paenibacillus sp. from a landfill in Brazil. Arch Microbiol 201:699–704
Barnes DKA (2002) Invasions by marine life on plastic debris. Nature 416(6883):808–809
Browne MA, Dissanayake A, Galloway TS et al (2008) Ingested microscopic plastic translocates to the circulatory system of the mussel, Mytilus edulis (L.). Environ Sci Technol 42:5026–5031
Browne MA, Galloway TS, Thompson RC (2010) Spatial patterns of plastic debris along estuarine shorelines. Environ Sci Technol 44:3404–3409
Bryant JA, Clemente TM, Viviani DA et al (2016) Diversity and activity of communities inhabiting plastic debris in the North Pacific Gyre. mSystems 1:1234
Carlén A, Nikdel K, Wennerberg A et al (2001) Surface characteristics and in vitro biofilm formation on glass ionomer and composite resin. Biomaterials 22:481–487
Carpenter EJ, Smith K Jr (1972) Plastics on the Sargasso Sea surface. Science 175:1240–1241
Carson HS, Nerheim MS, Carroll KA, Eriksen M (2013) The plastic-associated microorganisms of the North Pacific Gyre. Mar Pollut Bull 75:126–132
Caruso G (2020) Microbial colonization in marine environments: overview of current knowledge and emerging research topics. J Mar Sci Eng 8:78
Cole M, Lindeque P, Halsband C, Galloway TS (2011) Microplastics as contaminants in the marine environment: a review. Mar Pollut Bull 62:2588–2597
Cózar A, Echevarría F, González-Gordillo JI et al (2014) Plastic debris in the open ocean. Proc Natl Acad Sci 111:10239–10244
Da Costa JP, Nunes AR, Santos PSM et al (2018) Degradation of polyethylene microplastics in seawater: insights into the environmental degradation of polymers. J Environ Sci Health Part A 53:866–875
Dang H, Li T, Chen M, Huang G (2008) Cross-ocean distribution of Rhodobacterales bacteria as primary surface colonizers in temperate coastal marine waters. Appl Environ Microbiol 74:52–60
Delacuvellerie A, Cyriaque V, Gobert S et al (2019) The plastisphere in marine ecosystem hosts potential specific microbial degraders including Alcanivorax borkumensis as a key player for the low-density polyethylene degradation. J Hazard Mater 380:120899
DeLong EF, Wickham GS, Pace NR (1989) Phylogenetic stains: ribosomal RNA-based probes for the identification of single cells. Science 243:1360–1363
Deng H, Fu Q, Li D, Zhang Y, He J, Feng D, Zhao Y, Du G, Yu H, Ge C (2021) Microplastic-associated biofilm in an intensive mariculture pond: Temporal dynamics of microbial communities, extracellular polymeric substances and impacts on microplastics properties. J Clean Prod 319:128774. https://doi.org/10.1016/j.jclepro.2021.128774
Derraik JG (2002) The pollution of the marine environment by plastic debris: a review. Mar Pollut Bull 44:842–852
Dudek KL, Cruz BN, Polidoro B, Neuer S (2020) Microbial colonization of microplastics in the Caribbean Sea. Limnol Oceanogr Lett 5:5–17
Eich A, Mildenberger T, Laforsch C, Weber M (2015) Biofilm and diatom succession on polyethylene (PE) and biodegradable plastic bags in two marine habitats: early signs of degradation in the pelagic and benthic zone? PLoS One 10:14312
Eriksen M, Mason S, Wilson S et al (2013a) Microplastic pollution in the surface waters of the Laurentian Great Lakes. Mar Pollut Bull 77:177–182
Eriksen M, Maximenko N, Thiel M et al (2013b) Plastic pollution in the South Pacific subtropical gyre. Mar Pollut Bull 68:71–76
Esmaeili A, Pourbabaee AA, Alikhani HA et al (2013) Biodegradation of low-density polyethylene (LDPE) by mixed culture of Lysinibacillus xylanilyticus and Aspergillus niger in soil. PLoS One 8:e71720
Feng L, He L, Jiang S et al (2020) Investigating the composition and distribution of microplastics surface biofilms in coral areas. Chemosphere 252:126565
Fok L, Cheung PK (2015) Hong Kong at the Pearl River Estuary: a hotspot of microplastic pollution. Mar Pollut Bull 99:112–118

Foulon V, Le Roux F, Lambert C et al (2016) Colonization of polystyrene microparticles by vibrio crassostreae: light and electron microscopic investigation. Environ Sci Technol 50:10988–10996
Frère L, Maignien L, Chalopin M et al (2018) Microplastic bacterial communities in the Bay of Brest: influence of polymer type and size. Environ Pollut 242:614–625
Gallowaya TS, Lewisa CN (2016) Marine microplastics spell big problems for future generations. Proc Natl Acad Sci USA 113:2331–2333
Gautam R, Bassi AS, Yanful EK (2007) Candida rugosa lipase-catalyzed polyurethane degradation in aqueous medium. Biotechnol Lett 29:1081–1086
Goldstein MC, Rosenberg M, Cheng L (2012) Increased oceanic microplastic debris enhances oviposition in an endemic pelagic insect. Biol Lett 8:817–820
Gregory MR (1996) Plastic 'scrubbers' in hand cleansers: a further (and minor) source for marine pollution identified. Mar Pollut Bull 32:867–871
Gregory MR (2009) Environmental implications of plastic debris in marine settings—entanglement, ingestion, smothering, hangers-on, hitch-hiking and alien invasions. Philos Trans R Soc B Biol Sci 364:2013–2025
Guo X-P, Niu Z-S, Lu D-P et al (2017) Bacterial community structure in the intertidal biofilm along the Yangtze Estuary, China. Mar Pollut Bull 124:314–320
Hamdan LJ, Coffin RB, Sikaroodi M, Greinert J, Treude T, Gillevet PM (2012) Ocean currents shape the microbiome of Arctic marine sediments. ISME J 2013 74 7(4):685–696. https://doi.org/10.1038/ismej.2012.143
Handelsman J (2004) Metagenomics: application of genomics to uncultured microorganisms. Microbiol Mol Biol Rev 68:669–685
Harrison JP, Schratzberger M, Sapp M, Osborn AM (2014) Rapid bacterial colonization of low-density polyethylene microplastics in coastal sediment microcosms. BMC Microbiol 14:1–15
Harrison J, Hoellein T, Sapp M et al (2018) Microplastic-associated biofilms: a comparison of freshwater and marine environments. library.oapen.org
Heistad A, Scott T, Skaarer AM et al (2009) Virus removal by unsaturated wastewater filtration: effects of biofilm accumulation and hydrophobicity. Water Sci Technol 60:399–407
Hirai N, Mun MK, Masuda T et al (2015) Atomic force microscopy analysis of biofilms formed on different plastics. Mater Technol 30:B57–B60. https://doi.org/10.1179/1753555714Y.0000000238
Hook AL, Chang CY, Yang J et al (2012) Combinatorial discovery of polymers resistant to bacterial attachment. Nat Biotechnol 309(30):868–875
Hossain R, Islam MT, Ghose A, Sahajwalla V (2022) Full circle: challenges and prospects for plastic waste management in Australia to achieve circular economy. J Clean Prod 368:133127
Hou L, Majumder EL-W (2021) Potential for and distribution of enzymatic biodegradation of polystyrene by environmental microorganisms. Materials 14:503
Hurley R, Woodward J, Rothwell JJ (2018) Microplastic contamination of river beds significantly reduced by catchment-wide flooding. Nat Geosci 114(11):251–257
Jiang P, Zhao S, Zhu L, Li D (2018) Microplastic-associated bacterial assemblages in the intertidal zone of the Yangtze Estuary. Sci Total Environ 624:48–54
Kasmuri N, Tarmizi NAA, Mojiri A (2022) Occurrence, impact, toxicity, and degradation methods of microplastics in environment—a review. Environ Sci Pollut Res 29:30820–30836
Kerfahi D, Harvey BP, Kim H et al (2022) Whole community and functional gene changes of biofilms on marine plastic debris in response to ocean acidification. Microb Ecol 1:1–13
Kirstein IV, Kirmizi S, Wichels A et al (2016) Dangerous hitchhikers? Evidence for potentially pathogenic Vibrio spp. on microplastic particles. Mar Environ Res 120:1–8
Konopka A (2009) What is microbial community ecology? ISME J 311(3):1223–1230
Koutny M, Sancelme M, Dabin C et al (2006) Acquired biodegradability of polyethylenes containing pro-oxidant additives. Polym Degrad Stab 91:1495–1503
Krueger MC, Harms H, Schlosser D (2015) Prospects for microbiological solutions to environmental pollution with plastics. Appl Microbiol Biotechnol 99(21):8857–8874. https://doi.org/10.1007/S00253-015-6879-4

Lacerda ALDF, Rodrigues LDS, van Sebille E et al (2019) Plastics in sea surface waters around the Antarctic Peninsula. Sci Rep 9:3977

Lagarde F, Olivier O, Zanella M, Daniel P, Hiard S, Caruso A (2016) Microplastic interactions with freshwater microalgae: Hetero-aggregation and changes in plastic density appear strongly dependent on polymer type. Environ Pollut 215:331–339. https://doi.org/10.1016/j.envpol.2016.05.006

Law KL, Morét-Ferguson S, Maximenko NA et al (2010) Plastic accumulation in the North Atlantic subtropical gyre. Science 329:1185–1188

Lebreton L-M, Greer S, Borrero JC (2012) Numerical modelling of floating debris in the world's oceans. Mar Pollut Bull 64:653–661

Li R, Hsieh CL, Young A et al (2015) Illumina synthetic long read sequencing allows recovery of missing sequences even in the "finished" C. elegans genome. Sci Rep 5:1–15

Li WC, Tse H, Fok L (2016) Plastic waste in the marine environment: a review of sources, occurrence and effects. Sci Total Environ 566:333–349

Li W, Zhang Y, Wu N, Zhao Z, Xu W, Ma Y, Niu Z (2019) Colonization Characteristics of Bacterial Communities on Plastic Debris Influenced by Environmental Factors and Polymer Types in the Haihe Estuary of Bohai Bay. China Environ Sci Technol 53(18):10763–10773. https://doi.org/10.1021/ACS

Lobelle D, Cunliffe M (2011) Early microbial biofilm formation on marine plastic debris. Mar Pollut Bull 62:197–200

Lopez Lozano R, Mouat J (2009) Marine litter in the Northeast Atlantic Region: assessment and priorities for response. MAP Library 27: 210–215.

MacHado MC, Cheng D, Tarquinio KM, Webster TJ (2010) Nanotechnology: pediatric applications. Pediatr Res 675(67):500–504

Masaki K, Kamini NR, Ikeda H, Iefuji H (2005) Cutinase-Like Enzyme from the Yeast Cryptococcus sp. Strain S-2 Hydrolyzes Polylactic Acid and Other Biodegradable Plastics. Appl Environ Microbiol 71(11):7548–7550. https://doi.org/10.1128/AEM.71.11.7548-7550.2005

Masó M, Garcés E, Pagès F, Camp J (2003) Los plásticos flotantes son potenciales vectores de dispersión de especies formadoras de proliferaciones algales nocivas. Sci Mar 67:107–111

McCauley SJ, Bjorndal KA (1999) Conservation implications of dietary dilution from debris ingestion: sublethal effects in post-hatchling loggerhead sea turtles. Conserv Biol 13:925–929

McCormick A, Hoellein TJ, Mason SA, Schluep J, Kelly JJ (2014) Microplastic is an Abundant and Distinct Microbial Habitat in an Urban River. Environ Sci Technol 48(20):11863–11871. https://doi.org/10.1021/es503610r

Miao L, Wang P, Hou J et al (2019) Distinct community structure and microbial functions of biofilms colonizing microplastics. Sci Total Environ 650:2395–2402

Miao L, Wang C, Adyel TM et al (2020) Microbial carbon metabolic functions of biofilms on plastic debris influenced by the substrate types and environmental factors. Environ Int 143:106007

Miao L, Yu Y, Adyel TM et al (2021) Distinct microbial metabolic activities of biofilms colonizing microplastics in three freshwater ecosystems. J Hazard Mater 403:123577

Mir TUG, Wani AK, Akhtar N, Shukla S (2022) CRISPR/Cas9: regulations and challenges for law enforcement to combat its dual-use. Forensic Sci Int 334:111274

Moore CJ (2008) Synthetic polymers in the marine environment: a rapidly increasing, long-term threat. Environ Res 108:131–139

Moore CJ, Moore SL, Weisberg SB et al (2002) A comparison of neustonic plastic and zooplankton abundance in southern California's coastal waters. Mar Pollut Bull 44:1035–1038

Mouafo Tamnou EB, Tamsa Arfao A, Nougang ME et al (2021) Biodegradation of polyethylene by the bacterium Pseudomonas aeruginosa in acidic aquatic microcosm and effect of the environmental temperature. Environ Chall 3:100056

Muyzer G, Smalla K (1998) Application of denaturing gradient gel electrophoresis (DGGE) and temperature gradient gel electrophoresis (TGGE) in microbial ecology. Antonie Van Leeuwenhoek 731(73):127–141

Naqash N, Prakash S, Kapoor D, Singh R (2020) Interaction of freshwater microplastics with biota and heavy metals: a review. Environ Chem Lett 18(6):1813–1824. https://doi.org/10.1007/S10311-020-01044-3

Ncube LK, Ude AU, Ogunmuyiwa EN et al (2021) An overview of plastic waste generation and management in food packaging industries. Recycling 6:12

Nourollahi A, Sedighi-Khavidak S, Mokhtari M et al (2019) Isolation and identification of low-density polyethylene (LDPE) biodegrading bacteria from waste landfill in Yazd. Int J Environ Stud 76:236–250

O'Neill A, Araújo R, Casal M et al (2007) Effect of the agitation on the adsorption and hydrolytic efficiency of cutinases on polyethylene terephthalate fibres. Enzym Microb Technol 40:1801–1805

Obbard RW, Sadri S, Wong YQ et al (2014) Global warming releases microplastic legacy frozen in Arctic Sea ice. Earths Future 2:315–320

Oberbeckmann S, Loeder MGJ, Gerdts G, Osborn MA (2014) Spatial and seasonal variation in diversity and structure of microbial biofilms on marine plastics in Northern European waters. FEMS Microbiol Ecol 90:478–492

Oberbeckmann S, Osborn AM, Duhaime MB (2016) Microbes on a bottle: substrate, season and geography influence community composition of microbes colonizing marine plastic debris. PLoS One 11:e0159289

Oberbeckmann S, Kreikemeyer B, Labrenz M (2018) Environmental factors support the formation of specific bacterial assemblages on microplastics. Front Microbiol 8:2709

Oberbeckmann S, Labrenz M (2020) Marine Microbial Assemblages on Microplastics: Diversity, Adaptation, and Role in Degradation. Httpsdoiorg101146annurev-Mar-010419-010633 12:209–232. https://doi.org/10.1146/ANNUREV-MARINE-010419-010633

Ogonowski M, Motiei A, Ininbergs K, Hell E, Gerdes Z, Udekwu KI, Bacsik Z, Gorokhova E (2018) Evidence for selective bacterial community structuring on microplastics. Environ Microbiol 20(8):2796–2808. https://doi.org/10.1111/1462-2920.14120

Ojha N, Pradhan N, Singh S et al (2017) Evaluation of HDPE and LDPE degradation by fungus, implemented by statistical optimization. Sci Rep 7:39515

Oni BA, Ayeni AO, Agboola O et al (2020) Comparing microplastics contaminants in (dry and raining) seasons for Ox-Bow Lake in Yenagoa, Nigeria. Ecotoxicol Environ Saf 198:110656

Onyena AP, Aniche DC, Ogbolu BO et al (2022) Governance strategies for mitigating microplastic pollution in the marine environment: a review. Microplastics 1:15–46

Parthasarathy A, Miranda RR, Eddingsaas NC et al (2022) Polystyrene degradation by Exiguobacterium sp. RIT 594: preliminary evidence for a pathway containing an atypical oxygenase. Microorganisms 10:1619

Phelan AA, Ross H, Setianto NA et al (2020) Ocean plastic crisis—mental models of plastic pollution from remote Indonesian coastal communities. PLoS One 15:e0236149

Pinnell LJ, Turner JW (2020) Temporal changes in water temperature and salinity drive the formation of a reversible plastic-specific microbial community. FEMS Microbiol Ecol 96:230

Pinto M, Langer TM, Hüffer T et al (2019) The composition of bacterial communities associated with plastic biofilms differs between different polymers and stages of biofilm succession. PLoS One 14:e0217165

Pompilio A, Piccolomini R, Picciani C et al (2008) Factors associated with adherence to and biofilm formation on polystyrene by Stenotrophomonas maltophilia: the role of cell surface hydrophobicity and motility. FEMS Microbiol Lett 287:41–47

Qiang L, Cheng J, Mirzoyan S, Kerkhof LJ, Häggblom MM (2021) Characterization of Microplastic-Associated Biofilm Development along a Freshwater-Estuarine Gradient. Environ Sci Technol 55(24):16402–16412. https://doi.org/10.1021/acs.est.1c04108

Reisser J, Slat B, Noble K et al (2015) The vertical distribution of buoyant plastics at sea: an observational study in the North Atlantic Gyre. Biogeosciences 12:1249–1256

Rieck A, Herlemann DPR, Jürgens K, Grossart HP (2015) Particle-associated differ from free-living bacteria in surface waters of the baltic sea. Front Microbiol 6(DEC):1297. https://doi.org/10.3389/FMICB.2015.01297

Rojas-Parrales A, Orantes-Sibaja T, Redondo-Gómez C, Vega-Baudrit J (2018) Biological degradation of plastics: polyethylene biodegradation by aspergillus and streptomyces species-a review. ACS Symp Ser 1302:69–79

Ronkvist ÅM, Xie W, Lu W, Gross RA (2009) Cutinase-catalyzed hydrolysis of poly(ethylene terephthalate). Macromolecules 42:5128–5138

Rummel CD, Jahnke A, Gorokhova E et al (2017) Impacts of biofilm formation on the fate and potential effects of microplastic in the aquatic environment. Environ Sci Technol Lett 4:258–267

Ryan PG, Moore CJ, van Franeker JA, Moloney CL (2009) Monitoring the abundance of plastic debris in the marine environment. Philos Trans R Soc B Biol Sci 364:1999–2012

Sameshima-Yamashita Y, Ueda H, Koitabashi M, Kitamoto H (2019) Pretreatment with an esterase from the yeast Pseudozyma Antarctica accelerates biodegradation of plastic mulch film in soil under laboratory conditions. J Biosci Bioeng 127:93–98

Santo M, Weitsman R, Sivan A (2013) The role of the copper-binding enzyme—laccase—in the biodegradation of polyethylene by the actinomycete Rhodococcus ruber. Int Biodeterior Biodegrad 84:204–210

Schlundt C, Mark Welch JL, Knochel AM et al (2020) Spatial structure in the "Plastisphere": molecular resources for imaging microscopic communities on plastic marine debris. Mol Ecol Resour 20:620–634

Sharma S, Sharma V, Chatterjee S (2021) Microplastics in the Mediterranean Sea: sources, pollution intensity, sea health, and regulatory policies. Front Mar Sci 8:15?23

Shen M, Zhu Y, Zhang Y et al (2019) Micro(nano)plastics: unignorable vectors for organisms. Mar Pollut Bull 139:328–331

Sogin ML, Morrison HG, Huber JA et al (2006) Microbial diversity in the deep sea and the underexplored "rare biosphere". Proc Natl Acad Sci USA 103:12115–12120

Stoodley P, Cargo R, Rupp CJ et al (2002) Biofilm material properties as related to shear-induced deformation and detachment phenomena. J Ind Microbiol Biotechnol 296(29):361–367

Sun X, Chen B, Xia B, Li Q, Zhu L, Zhao X, Gao Y, Qu K (2020) Impact of mariculture-derived microplastics on bacterial biofilm formation and their potential threat to mariculture: A case in situ study on the Sungo Bay. China Environ Pollut 262:114336. https://doi.org/10.1016/j.envpol.2020.114336

Tachibana K, Hashimoto K, Yoshikawa M, Okawa H (2010) Isolation and characterization of microorganisms degrading nylon 4 in the composted soil. Polym Degrad Stab 95:912–917

Thompson RC (2006) Plastic debris in the marine environment: consequences and solutions. Mar Nat Conserv Eur 193:107–115

Tu C, Chen T, Zhou Q et al (2020) Biofilm formation and its influences on the properties of microplastics as affected by exposure time and depth in the seawater. Sci Total Environ 734:139237

Valm AM, Mark Welch JL, Borisy GG (2012) CLASI-FISH: principles of combinatorial labeling and spectral imaging. Syst Appl Microbiol 35:496–502. https://doi.org/10.1016/J.SYAPM.2012.03.004

Veerasingam S, Ranjani M, Venkatachalapathy R et al (2020) Microplastics in different environmental compartments in India: analytical methods, distribution, associated contaminants and research needs. TrAC Trends Anal Chem 133:116071

Venkataraman A, Bassis CM, Beck JM, Young VB, Curtis JL, Huffnagle GB, Schmidt TM (2015) Application of a neutral community model to assess structuring of the human lung microbiome. Am Soc Microbiol 6(1). https://doi.org/10.1128/mBio.02284-14

Wani AK, Akhtar N, Naqash N et al (2022a) Bioprospecting culturable and unculturable microbial consortia through metagenomics for bioremediation. Clean Chem Eng 7:11234

Wani AK, Akhtar N, Singh R et al (2022b) Prospects of advanced metagenomics and meta-omics in the investigation of phytomicrobiome to forecast beneficial and pathogenic response. Mol Biol Rep 12:3241

Wani AK, Akhtar N, Singh R et al (2022c) Genome centric engineering using ZFNs, TALENs and CRISPR-Cas9 systems for trait improvement and disease control in Animals. Vet Res Commun 47:1–16
Wani AK, Hashem NM, Akhtar N et al (2022d) Understanding microbial networks of farm animals through genomics, metagenomics and other meta-omic approaches for livestock wellness and sustainability. Ann Anim Sci 22:234–255
Wani AK, Rahayu F, Kadarwati FT et al (2022e) Metagenomic screening strategies for bioprospecting enzymes from environmental samples. IOP Conf Ser Earth Environ Sci 974:012003
Wani AK, Akhtar N, Naqash N et al (2023) Discovering untapped microbial communities through metagenomics for microplastic remediation: recent advances, challenges, and way forward. Environ Sci Pollut Res 30:1–24
Webb HK, Arnott J, Crawford RJ, Ivanova EP (2013) Plastic degradation and its environmental implications with special reference to poly(ethylene terephthalate). Polymers 5:1–18
Weinstein JE, Crocker BK, Gray AD (2016) From macroplastic to microplastic: degradation of high-density polyethylene, polypropylene, and polystyrene in a salt marsh habitat. Environ Toxicol Chem 35:1632–1640
Wen B, Liu JH, Zhang Y, Zhang HR, Gao JZ, Chen ZZ (2020) Community structure and functional diversity of the plastisphere in aquaculture waters: Does plastic color matter? Sci Total Environ 740:140082. https://doi.org/10.1016/J.SCITOTENV.2020.140082
Wright SL, Rowe D, Thompson RC, Galloway TS (2013) Microplastic ingestion decreases energy reserves in marine worms. Curr Biol 23:R1031–R1033
Xu X, Wang S, Gao F et al (2019) Marine microplastic-associated bacterial community succession in response to geography, exposure time, and plastic type in China's coastal seawaters. Mar Pollut Bull 145:278–286
Yang Y, Liu W, Zhang Z et al (2020) Microplastics provide new microbial niches in aquatic environments. Appl Microbiol Biotechnol 104:6501–6511
Zalasiewicz J, Waters CN, Ivar do Sul JA et al (2016) The geological cycle of plastics and their use as a stratigraphic indicator of the Anthropocene. Anthropocene 13:4–17
Zerhouni K, Abbouni B, Kanoun K et al (2018) Isolation and identification of low density polythene-degrading bacteria from soil of North West of Algeria. South Asian J Exp Biol 8:76–82
Zettler ER, Mincer TJ, Amaral-Zettler LA (2013) Life in the "plastisphere": microbial communities on plastic marine debris. Environ Sci Technol 47:7137–7146
Zhang K, Hamidian AH, Tubić A et al (2021) Understanding plastic degradation and microplastic formation in the environment: a review. Environ Pollut 274:116554

Chapter 7
Detoxification of Biomedical Waste

Abhishek Singh, Neha Chakrawarti, Vishnu D. Rajput, Karen Ghazaryan, Tatiana Minkina, Abdel Rahman Mohammad Said Al-Tawaha, Abdel Razzaq Al-Tawaha, Marwa Adel Qotb, and Arun Karnwal

Abstract Biomedical-related industrial waste which has toxic effect on plant and animal life is a major concern for environmentalists. Due to their long persistence in the environment, they can get into the food chain via food and water, causing more harm. Since improper management of biomedical waste poses a risk to medical service providers, patients, and the environment, it has received a lot of attention from the treatment technology community. Despite the implementation of legit waste administration training programs for all healthcare service staff, there is still room for improvement in existing protocols, plans, and policies. Due to their long existence, biomedical wastes are tough to degrade in nature, and even if degraded, they may not be completely removed. This implies that the prolonged permanence of biomedical waste may have adverse impacts on both animal and human popula-

A. Singh (✉) · K. Ghazaryan
Faculty of Biology, Yerevan State University, Yerevan, Armenia

N. Chakrawarti · T. Minkina
Govind Ballabh Pant University of Agriculture and Technology, Pantnagar, Uttarakhand, India

V. D. Rajput
Academy of Biology and Biotechnology, Southern Federal University, Rostov-on-Don, Russia

A. R. Mohammad Said Al-Tawaha
Department of Biological Sciences, Al Hussein Bin Talal University, Ma'an, Jordan

A. R. Al-Tawaha
Department of Crop Science, Faculty of Agriculture, Universiti Putra Malaysia, Serdang, Selangor, Malaysia

National Agricultural Research Center, Baqa, Jordan

M. Adel Qotb
Soil Fertility and Plant Nutrition Department, Soils, Water and Environmental Research Institute, ARC, Giza, Egypt

A. Karnwal
Department of Microbiology, Lovely Professional University, Phagwara, Punjab, India

A. Karnwal, A. R. Mohammad Said Al-Tawaha (eds.), *Microbial Applications for Environmental Sustainability*, https://doi.org/10.1007/978-981-97-0676-1_7

tions. Photocatalysis has evolved as an environmentally sustainable approach to address this issue, and it is receiving growing recognition for its ability to eliminate pollutants and improve the safety and transparency of the surrounding environment. Nanostructured photocatalysts exhibit greater performance compared to traditional photocatalysts in the process of photodegradation. This superiority may be attributed to their beneficial characteristics, including nontoxicity, cost-effectiveness, and high absorption efficiency over a wide range of solar wavelengths. The purpose of this chapter is to critically assess various aspects of biomedical waste along with new method, nanophotocatalysts for biomedical waste management and to highlight their significance in healthcare service settings due to their unique properties.

Keywords Biomedical waste · Nanostructure · Nanophotocatalysts · Environment

7.1 Introduction

One of the most pressing problems that developing countries must now address is the disposal of biomedical waste (Ali et al. 2017). With the increase in global population, the amount of biomedical waste generated has also increased, and the resources at our disposal are inadequate to deal with it (Azuma et al. 2016). The potential consequences of post-treatment and disposal of waste generated inside the hospital system may inadvertently expose patients to health risks, as it might result in the release of pathogens and hazardous substances into the surrounding environment (Baguma 2017). According to Balasubramanian et al. (2006), inadequate construction of landfills may result in the potential contamination of surface, drinking, and groundwater resources when untreated medical wastes are disposed of at such facilities. Furthermore, it has been observed that the improper handling, storage, and disposal of chemical disinfectants used in the treatment of medical waste may lead to the release of chemical compounds into the surrounding environment (Bohara and Pawar 2015). Burning garbage is a prevalent method; however, inadequate combustion or use of inappropriate materials may result in air pollution and the accumulation of residual ash. The primary challenges associated with healthcare waste include limited awareness regarding the health hazards associated with such waste, insufficient training in appropriate waste management practices, inadequate waste management infrastructure and disposal systems, inadequate financial and human resources, and a lack of prioritization of the issue (Bohara et al. 2017). Many nations lack the necessary laws or fail to enforce them. To handle biomedical waste effectively and efficiently, a variety of treatment techniques and processes, including burning, autoclaving, hydroclaving, and microwave treatments, must be used (Ugwuishiwu et al. 2016). It is essential for every innovation to be implemented to ensure the conservation of the environment and the overall well-being of society. The significance of using cost-effective remedies has escalated in light of the substantial increase in the pace and variety of waste generation, encompassing both its quantity and quality (Huy et al. 2017). The importance of advancements in waste management and waste engineering cannot be overstated. The selection of waste treatment methods has gained significant importance in the field of waste management due to the inherent inability to completely eliminate waste products (Huy et al. 2017).

In addition to the aforementioned techniques, there are many more commonly used approaches for managing biomedical waste, such as granulation, pulverization, shredding, grinding, mixing, agitation, and crushing. This particular treatment method has the benefit of achieving a minimum of a 60% decrease in the bulk volume of the waste materials. Although mechanical treatment does not effectively eliminate pathogens or thoroughly clean equipment, it does effectively reduce the amount of waste, creating space for further treatment or disposal. Crushers, millers, shattering machines, and splinters represent a limited selection of the machinery used in mechanical treatment processes.

These treatment strategies have the potential to alleviate the psychological impact of garbage on human observers by modifying its visual characteristics. Mechanical treatment could increase the surface area of solid fragments before any further chemical or thermal treatment. This could reduce the amount of bulk disposal needed at the same time. Nanophotocatalysts are widely used in environmental and ecological safety because they are cost-effective, very stable, very good at photocatalysis, and safe for humans (Mehta and Chopra 2022).

Various methods have been used for the treatment of sewage, including ion exchange, microbes, and adsorption, in order to achieve effective purification. Nevertheless, these techniques have several limitations that need to be considered. These constraints include the significant financial investment required, the intricate technical nature of the processes involved, the potential risk of secondary contamination, and the limited efficacy of degradation (Jiang et al. 2012). According to Rehman et al. (2009), nanophotocatalysts provide a highly targeted approach to address the challenges associated with biomedical waste management, particularly in terms of energy efficiency, environmental consequences, and ecological considerations. There are several advantages associated with the use of a photocatalysis methodology: (a) Photocatalysis offers a feasible alternative to conventional energy-intensive treatment methods, such as ultrafiltration and reverse osmosis, by utilizing pollution-free and renewable solar energy. (b) Unlike conventional treatment methods, which simply transfer pollutants from one phase to another, photocatalysis promotes the generation of nonhazardous products. (c) Furthermore, photocatalysis demonstrates its potential in the elimination of various potentially harmful substances, as supported by studies conducted by Rehman et al. (2009) and Keshavarz et al. (2020). The effectiveness of photocatalysts has been impeded by the dependence of their activity on many key factors: (a) the separation of charges, (b) the enhancement of interfacial charge transfer, and (c) the inhibition of charge carrier recombination (Rehman et al. 2009; Keshavarz et al. 2020).

Nanostructured catalytic membranes, nanosorbents, and nanophotocatalyst-based approaches for pollutant removal may be efficient and friendly to the environment, but waste purification using these technologies still necessitates significant resources and time (Bohara et al. 2015). The treatment of biomedical waste presents many difficulties; special care must be taken to prevent hazardous waste from posing environmental and health risks (Bystrzejewska-Piotrowska et al. 2009). Flexible, low-cost, and efficient new waste treatment equipment is needed for commercialization (Datta et al. 2018). Nanophotocatalysts, nanomotors, nanomembranes, nanosorbents, and imprinted polymers are just a few examples of cutting-edge nanomaterials that have revolutionized the decontamination of biomedical waste in recent years (Chen and Mao 2007; Çalıs and Arkan 2014; Carraro et al. 2016).

7.2 Biomedical Waste Problem in COVID-19 Era

Biomedical wastes are defined by the World Health Organization (WHO) as any and all waste products from human or animal healthcare settings, including hospitals, clinics, laboratories, operating rooms, and immunization clinics (Fig. 7.1) (Datta et al. 2018). Likewise categorized are the byproducts derived from biomedical research, specifically referring to the residual materials generated during laboratory experimentation. The category of personal protection equipment (PPE) encompasses various wearables, such as gloves and masks, that are used by healthcare professionals and patients alike in the context of regular or public community health services.

The World Health Organization (WHO) has issued guidelines for the classification of various forms of municipal solid waste (MSW), including anatomical and animal waste, microbiological waste, waste containing sharp objects and pharmaceuticals, as well as solid waste, liquid waste, waste resulting from incineration, and waste containing cytotoxic chemicals (Singh et al. 2001). The recent worldwide spread of the COVID-19 pandemic has greatly enhanced biomedical wastes (BMWs). Even though the World Health Organization (WHO) has waste

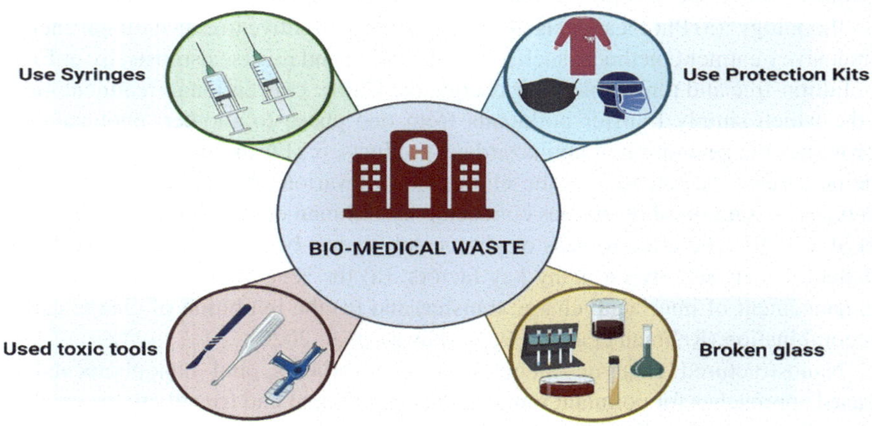

Fig. 7.1 Different types of biomedical waste

management guidelines for both normal health situations and pandemics, these guidelines become extremely difficult to implement in the current state of emergency. However, the developed world's already-in-place waste management systems are a major source of difficulty for the developing world (Matete and Trois 2008). BMWs in India can be divided roughly between two groups: infected vehicles and clean vehicles. Cytotoxic chemicals and pharmaceuticals made up the bulk of the clean categories. The infected garbage is divided into four categories, each with its own color code: red, yellow, blue, and white.

The red bin is used to dispose of objects that have been contaminated after coming in touch with the patient, a wound, or an infected component (such as dressings and POP casts). Placentas and other pathological wastes are considered anatomical wastes and are therefore yellow. Colors blue and white designate medical equipment like syringes and gloves as well as common household items like plastic and knives (Haque et al. 2021).

7.3 System for Managing Biomedical Waste

7.3.1 *Chemical-Based Waste Management*

As most of the chemical disinfection has been applied to inaccessible surfaces and pieces of equipment, the situation is hopeless (Slaughter et al. 2019). Some MWs are too cumbersome to transport to an incinerator or landfill, so chemical disinfection is the method of choice. Most types of liquid medical waste must undergo chemical disinfection before being released into the environment (Emmanuel et al. 2004). Chlorine compounds are the primary ingredient for use in chemical disinfectants. One common chemical disinfectant is sodium hypochlorite.

7.3.2 *Incineration-Based Waste Management*

BMWs are destroyed via incineration-based waste management system technique by being subjected to high temperatures. Incinerators designed to burn solid waste into a mixture of byproducts are essential to this waste management strategy (Datta et al. 2018). Toxic substances can be rendered harmless through rapid incineration in the presence of heat and a steady supply of oxygen. More than 90% of MW in most developed countries have been disposed of using this method, making it widely regarded as the safest scientific procedure for disposing of MW worldwide. Following the initial implementation of BMW rules in the early 1990s, India began using incineration (Datta et al. 2018). Incomplete combustion results in the production of several toxic substances, known as products of incomplete combustion, and while this is the preferred and most efficient technique for managing BMWs, it has been observed to produce them (PIC). Additionally, dioxins are a class of chemicals that have been shown to be produced during incineration (Datta et al. 2018). There is a class of toxic substances known as dioxins and dioxin-like compounds that pose serious risks to human health. Humans' aryl hydrocarbon receptors (AhR) are known to be activated by aryl hydrocarbons, with this activation resulting in impacts such as immune system illnesses, hormonal imbalances, reproductive issues, and neurological abnormalities (Schuhmacher and Domingo 2006).

7.3.3 Steam- and Thermal-Based Waste Management

Autoclaving is a tried-and-true method of sanitizing a wide variety of MWs (Jang et al. 2006; Nemathaga et al. 2008). Decontaminating various types of reusable medical equipment is another common application of this method (Hossain et al. 2012). Some developing countries have turned to thermal treatment as an alternative to incineration (Oweis et al. 2005).

7.3.4 Land Disposal Based Waste Management

The large quantity of BMW has been disposed of successfully through land disposal or burial. Landfills are used to store the ash generated by incinerators, along with some biodegradable hospital wastes (Oweis et al. 2005; Blenkharn 2005). Pre-processed BMW landfilling is an environmentally sound solution for toxic waste management (Lakshmikanthan et al. 2017). These days, BMWs are disposed of with the help of bioreactor simulators. In comparison to dry simulators, leachate recirculation has a higher potential for reducing carbon emissions (Reddy et al. 2009). When it comes to managing trash at hospitals, this disposal method is crucial.

7.4 Photocatalytic Reactions: An Overview of Their Basic Principles and Working Mechanism for Waste Management

When electromagnetic radiation serves as the catalyst for a series of reactions, the process is termed as "photocatalysis" (Ugwuishiwu et al. 2016). The two primary steps of photocatalysis are reduction and oxidation. When photons with energies over the bandgap of a material excite electrons in the conduction band (CB), they jump to the valence band (VB), where they produce positive holes. Reactive oxygen species (ROS) like O_2 and OH are created from the resulting electrons and holes (oxidation) (Rajeshwar et al. 2008). The nature of the reactive oxygen species (ROS) produced in response to a given set of conditions (material type and photon flux) is material and context-specific (Fujishima et al. 2000). In a typical photocatalytic process, the significant result of photocatalysis is the formation of ROS. This may have a variety of consequences, such as antibacterial activity and dye degradation (Khan et al. 2015). Changes in the environment, such as those caused by degradation, reduction, adoption, or antibacterial activity, can provide a measure of photocatalysis' efficacy (Lee and Abd Hamid 2015). Typically, by comparing the amounts of the target chemicals before and after the reaction, one may determine the effectiveness of a photocatalytic process. The rate at which electrons and holes can be extracted from a photon is a measure of the photocatalyst's efficiency. One of the primary factors that reduce photocatalytic activity is electron–hole recombination (Singh et al. 2020).

7.5 Features and Classification of Nanophotocatalysts

Compared to larger particles or bulk materials, nanosized photocatalyst particles exhibit greatly amplified reactivity (Keshavarz et al. 2018). Under the right conditions, novel nanophotocatalysts have been developed that can use solar energy to synthesize organic compounds (Sudhaik et al. 2018). Nanophotocatalysts that can speed up different chemical reactions in living things through photocatalysis can be roughly put into three groups. Some of these groups are photocatalysts that work with surface plasmon resonance, those that work with metal-organic charge transfer, and those that work with semiconductors (see Fig. 7.2). Graphene semiconductors, core–shell composites, non-metal-doped semiconductors, and metal-modified semiconductors are some of the many types of semiconductors that can be broken down into smaller groups (see Fig. 7.2). A lot of different things have been done with metal-free catalysis, such as photocatalytic water purification, breaking down oxytetracycline, and using carbon quantum dots to support graphitic carbon nitride (Huy et al. 2017; Sudhaik et al. 2018). Metal oxide semiconductors, such as TiO_2, ZnO, Fe_2O_3, and WO_3, are considered the most favorable choice among the many types of photocatalysts due to their broad bandgap energy and their ability to withstand photo-corrosion.

TiO_2 is now the most used photocatalyst due to its capacity to accelerate the oxidation of organic molecules as well as the fact that it is thermally stable, nontoxic, economical, and chemically and biologically inert. The photocatalytic activity of a catalyst can be affected by its structural and surface properties (Ahmed et al. 2011). These include the catalyst's surface area, porosity, crystal composition, particle size distribution, and bandgap energy. When compared to bulk materials, nanophotocatalysts typically exhibit the following characteristics (Radhika et al. 2019). Because of their enormous surface area to volume ratio, they have a high particle fraction, which in turn causes a high division of active sites on the catalyst surface. Furthermore, the energy gap between the valence and conduction bands of these nanoparticles is highly size-dependent (Townsend et al. 2012). And the absorbance wavelength can be modified by varying the nanocatalyst's size. Also, by controlling their aggregate size, the nanocatalyst's optical and electronic properties can be

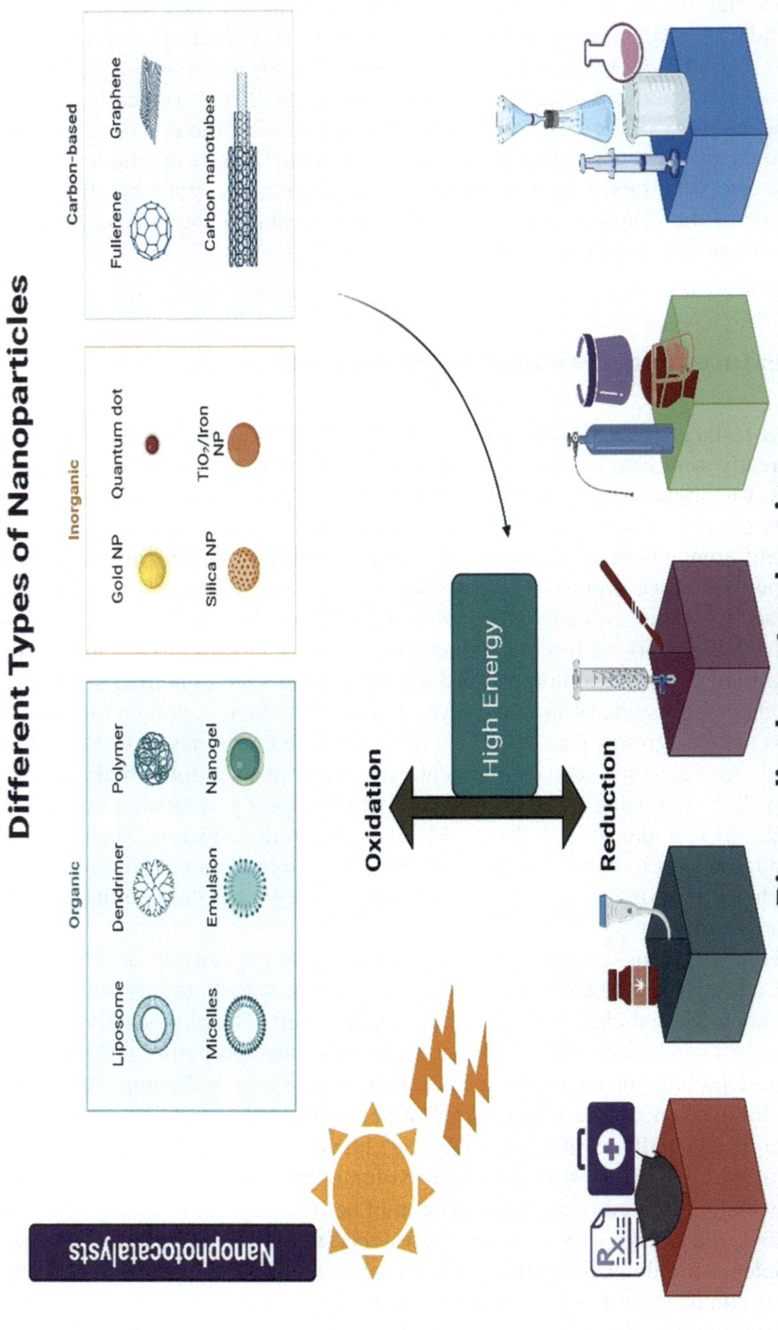

Fig. 7.2 Schematic representation of the primary categories of nanophotocatalysts and the general procedure for disinfecting biomedical and biological waste

altered (Townsend et al. 2012). These advantages have led to the use of nanophotocatalysts in a wide variety of reactions, such as organic synthesis, water splitting for hydrogen fuel generation, cancer cell inactivation, and dye degradation (Keshavarz et al. 2017). Nanophotocatalysts' degradation of pollutants and dyes has been the primary focus of research into their development. Because of these characteristics, nanophotocatalysts are gaining popularity as catalysts for a variety of organic reactions, opening up green chemistry by providing alternatives to the conventional methods typically employed in academic and commercial settings, which rely on the application of heat to accomplish the same ends (Kaur et al. 2013).

7.6 Using Nanophotocatalysts in the Biomedical Field

As photocatalysts are more effective than traditional methods (such as chlorination) at killing off harmful microorganisms, they could be used instead of those methods (Kumar et al. 2018). A variety of air and water purification systems can use photocatalysis, which is a flexible and efficient technique (Zhang and Gamage 2010). Furthermore, the use of photocatalytic surfaces has been employed as a result of their inherent purity. Recent developments in photocatalytic applications have been seen in several disciplines, including environmental and indoor air quality, plant protection, effluents, wastewater, drinking water disinfection, dye removal, the pharmaceutical and food sectors, laboratories, and hospitals, as well as biology and medicine. The photodegradation of hazardous chemicals and pollutants is considered to have a favorable cost–benefit ratio due to its minimal energy requirements, convenient use of solar energy, and decreased treatment duration (Ramos-Delgado et al. 2016). Photocatalytic processes provide disinfecting qualities that make them valuable in biological applications. The photocatalytic activity of *Staphylococcus aureus* (S. aureus), a prevalent pathogenic bacterium associated with implant-related infections, has been investigated in relation to the TiO2 layer on titanium substrates and stainless steel.

7.7 Conclusion

Biomedical waste disposal has now become a global issue. Many different methods which are generally used for safe disposal of biomedical waste are chemical based and are resulting in various types of pollution. Thus, these methods are not environmentally safe. These harmful effects of chemical solution can be reversed by photocatalysts. Here, the electromagnetic radiation catalyzes the various steps in safe BMW disposal. These photocatalysts as they are naturally derived from renewable sources such as solar radiations provide a cost-effective solution. Within photocatalysts, nanophotocatalysts being a nanosized substances have great catalyzing activity. Among these nanophotocatalysts, TiO_2 is the widely used substance due to its thermostability and other various benefits. Other substances used as photocatalysts may be Fe_2O_3 and WO_3. The use of these photocatalysts in process of BMW degradation is very new. Seeing the benefit of photocatalysts, there needs to be dissemination of this new technology among the biomedical community and also there is need to include nanophotocatalysts in routine BMW disposal. This in turn will result in maintaining a clean environment through safe, ecofriendly, and cost-effective BMW disposal methods.

Acknowledgements KG is supported by internal grant of YSU. AS is supported by the 23PostDoc-4D007 grant provided by the Science Committee of the Republic of Armenia. VDR and TM acknowledge the financial support of the Ministry of Science and Higher Education of the Russian Federation, agreement no. 075-15-2022-1122, and the Strategic Academic Leadership Program of the Southern Federal University ("Priority 2030").

Conflict of Interest

Ethics Approval and Consent to Participate Not applicable.

Consent for Publication The article contains no such material that may be unlawful, defamatory, or which would, if published, in any way whatsoever, violate the terms and conditions as laid down in the agreement.

Availability of Data and Materials Not applicable.

Competing Interests The authors declare that they have no conflict of interest in the publication.

References

Ahmed S, Rasul MG, Brown R, Hashib MA (2011) Influence of parameters on the heterogeneous photocatalytic degradation of pesticides and phenolic contaminants in wastewater: a short review. J Environ Manag 92(3):311–330. https://doi.org/10.1016/J.JENVMAN.2010.08.028

Ali M, Wang W, Chaudhry N, Geng Y (2017) Hospital waste management in developing countries: a mini review. Waste Manag Res 35(6):581–592. https://doi.org/10.1177/0734242X17691344

Azuma T, Arima N, Tsukada A, Hirami S, Matsuoka R, Moriwake R, Ishiuchi H, Inoyama T, Teranishi Y, Yamaoka M, Mino Y, Hayashi T, Fujita Y, Masada M (2016) Detection of pharmaceuticals and phytochemicals together with their metabolites in hospital effluents in Japan, and their contribution to sewage treatment plant influents. Sci Total Environ 548–549:189–197. https://doi.org/10.1016/J.SCITOTENV.2015.12.157

Baguma D (2017) Public health safety and environment in inadequate hospital and healthcare settings: a review. Public Health 144:23–31. https://doi.org/10.1016/J.PUHE.2016.11.014

Balasubramanian J, Sabumon PC, Lazar JU, Ilangovan R (2006) Reuse of textile effluent treatment plant sludge in building materials. Waste Manag 26(1):22–28. https://doi.org/10.1016/J.WASMAN.2005.01.011

Blenkharn JI (2005) Safe disposal and effective destruction of clinical wastes. J Hosp Infect 60(4):295–297. https://doi.org/10.1016/j.jhin.2005.01.029

Bohara RA, Pawar SH (2015) Innovative developments in bacterial detection with magnetic nanoparticles. Appl Biochem Biotechnol 176(4):1044–1058. https://doi.org/10.1007/S12010-015-1628-9

Bohara RA, Yadav HM, Thorat ND, Mali SS, Hong CK, Nanaware SG, Pawar SH (2015) Synthesis of functionalized Co0.5Zn0.5Fe2O4 nanoparticles for biomedical applications. J Magn Magn Mater 378:397–401. https://doi.org/10.1016/J.JMMM.2014.11.063

Bohara RA, Throat ND, Mulla NA, Pawar SH (2017) Surface-modified cobalt ferrite nanoparticles for rapid capture, detection, and removal of pathogens: a potential material for water purification. Appl Biochem Biotechnol 182(2):598–608. https://doi.org/10.1007/S12010-016-2347-6

Bystrzejewska-Piotrowska G, Golimowski J, Urban PL (2009) Nanoparticles: their potential toxicity, waste and environmental management. Waste Manag 29(9):2587–2595. https://doi.org/10.1016/J.WASMAN.2009.04.001

Çalıs S, Arkan B (2014) The views of the nursing students about the medical wastes and their effects on the environmental and human health. Proc Soc Behav Sci 116:1472–1476. https://doi.org/10.1016/J.SBSPRO.2014.01.419

Carraro E, Bonetta S, Bertino C, Lorenzi E, Bonetta S, Gilli G (2016) Hospital effluents management: chemical, physical, microbiological risks and legislation in different countries. J Environ Manag 168:185–199. https://doi.org/10.1016/J.JENVMAN.2015.11.021

Chen X, Mao SS (2007) Titanium dioxide nanomaterials: synthesis, properties, modifications and applications. Chem Rev 107(7):2891–2959. https://doi.org/10.1021/CR0500535/ASSET/CR0500535.FP.PNG_V03

Datta P, Mohi G, Chander J (2018) Biomedical waste management in India: critical appraisal. J Lab Physicians 10(1):006–014. https://doi.org/10.4103/JLP.JLP_89_17

Emmanuel E, Keck G, Blanchard JM, Vermande P, Perrodin Y (2004) Toxicological effects of disinfections using sodium hypochlorite on aquatic organisms and its contribution to AOX formation in hospital wastewater. Environ Int 30(7):891–900. https://doi.org/10.1016/J.ENVINT.2004.02.004

Fujishima A, Rao TN, Tryk DA (2000) Titanium dioxide photocatalysis. J Photochem Photobiol C: Photochem Rev 1(1):1–21. https://doi.org/10.1016/S1389-5567(00)00002-2

Haque MS, Uddin S, Sayem SM, Mohib KM (2021) Coronavirus disease 2019 (COVID-19) induced waste scenario: a short overview. J Environ Chem Eng 9(1):104660. https://doi.org/10.1016/J.JECE.2020.104660

Hossain MS, Balakrishnan V, Rahman NNNA, Sarker MZI, Kadir MOA (2012) Treatment of clinical solid waste using a steam autoclave as a possible alternative technology to incineration. Int J Environ Res Public Heal 9(3):855–867. https://doi.org/10.3390/IJERPH9030855

Huy ND, Tien NTT, Huyen LT, Quang HT, Tung TQ, Luong NN, Park SM (2017) Screening and production of manganese peroxidase from Fusarium sp. on residue materials. Mycobiology 45(1):52–56. https://doi.org/10.5941/myco.2017.45.1.52

Jang YC, Lee C, Yoon OS, Kim H (2006) Medical waste management in Korea. J Environ Manag 80(2):107–115. https://doi.org/10.1016/J.JENVMAN.2005.08.018

Jiang L, Wang Y, Feng C (2012) Application of photocatalytic technology in environmental safety. Proc Eng 45:993–997. https://doi.org/10.1016/J.PROENG.2012.08.271

Kaur P, Bansal P, Sud D (2013) Heterostructured nanophotocatalysts for degradation of organophosphate pesticides from aqueous streams. J Korean Chem Soc 57(3):382–388. https://doi.org/10.5012/JKCS.2013.57.3.382

Keshavarz M, Tan B, Venkatakrishnan K (2017) Cell selective apoptosis induced by polymorphic alteration of self-assembled silica nanowebs. ACS Appl Mater Interfaces 9(7):6292–6305. https://doi.org/10.1021/ACSAMI.6B14836/SUPPL_FILE/AM6B14836_SI_001.PDF

Keshavarz M, Tan B, Venkatakrishnan K (2018) Multiplex photoluminescent silicon nanoprobe for diagnostic bioimaging and intracellular analysis. Adv Sci 5(3):1700548. https://doi.org/10.1002/ADVS.201700548

Keshavarz M, Wales DJ, Seichepine F, Abdelaziz MEMK, Kassanos P, Li Q, Temelkuran B, Shen H, Yang GZ (2020) Induced neural stem cell differentiation on a drawn fiber scaf-

fold—toward peripheral nerve regeneration. Biomed Mater 15(5):055011. https://doi.org/10.1088/1748-605X/AB8D12
Khan MM, Adil SF, Al-Mayouf A (2015) Metal oxides as photocatalysts. J Saudi Chem Soc 19(5):462–464. https://doi.org/10.1016/J.JSCS.2015.04.003
Kumar A, Thakur PR, Sharma G, Naushad M, Rana A, Mola GT, Stadler FJ (2018) Carbon nitride, metal nitrides, phosphides, chalcogenides, perovskites and carbides nanophotocatalysts for environmental applications. Environ Chem Lett 172 17(2):655–682. https://doi.org/10.1007/S10311-018-0814-8
Lakshmikanthan P, Santhosh LG, Sivakumar BGL (2017) Evaluation of bioreactor landfill as sustainable land disposal method. pp 243–254. https://doi.org/10.1007/978-981-10-1930-2_14
Lee KM, Abd Hamid SB (2015) Simple response surface methodology: investigation on advance photocatalytic oxidation of 4-chlorophenoxyacetic acid using UV-active ZnO photocatalyst. Materials 8(1):339–354. https://doi.org/10.3390/MA8010339
Matete N, Trois C (2008) Towards Zero Waste in emerging countries—a South African experience. Waste Manag 28(8):1480–1492. https://doi.org/10.1016/J.WASMAN.2007.06.006
Mehta M, Chopra L, Manikanika (2022) Applications of nano photocatalysts in the degradation of biomedical waste: a short review. Mater Today Proc 68:695–700. https://doi.org/10.1016/J.MATPR.2022.05.563
Nemathaga F, Maringa S, Chimuka L (2008) Hospital solid waste management practices in Limpopo Province, South Africa: a case study of two hospitals. Waste Manag 28(7):1236–1245. https://doi.org/10.1016/J.WASMAN.2007.03.033
Oweis R, Al-Widyan M, Al-Limoon O (2005) Medical waste management in Jordan: a study at the King Hussein Medical Center. Waste Manag 25(6):622–625. https://doi.org/10.1016/J.WASMAN.2005.03.011
Radhika NP, Selvin R, Kakkar R, Umar A (2019) Recent advances in nano-photocatalysts for organic synthesis. Arab J Chem 12(8):4550–4578. https://doi.org/10.1016/J.ARABJC.2016.07.007
Rajeshwar K, Osugi ME, Chanmanee W, Chenthamarakshan CR, Zanoni MVB, Kajitvichyanukul P, Krishnan-Ayer R (2008) Heterogeneous photocatalytic treatment of organic dyes in air and aqueous media. J Photochem Photobiol C: Photochem Rev 9(4):171–192. https://doi.org/10.1016/J.JPHOTOCHEMREV.2008.09.001
Ramos-Delgado NA, Gracia-Pinilla M, Mangalaraja RV, O'Shea K, Dionysiou DD (2016) Industrial synthesis and characterization of nanophotocatalysts materials: titania. Nanotechnol Rev 5(5):467–479. https://doi.org/10.1515/NTREV-2016-0007/ASSET/GRAPHIC/J_NTREV-2016-0007_CV_005.JPG
Reddy KR, Hettiarachchi H, Parakalla NS, Gangathulasi J, Bogner JE (2009) Geotechnical properties of fresh municipal solid waste at Orchard Hills Landfill, USA. Waste Manag 29(2):952–959. https://doi.org/10.1016/J.WASMAN.2008.05.011
Rehman S, Ullah R, Butt AM, Gohar ND (2009) Strategies of making TiO_2 and ZnO visible light active. J Hazard Mater 170(2–3):560–569. https://doi.org/10.1016/J.JHAZMAT.2009.05.064
Schuhmacher M, Domingo JL (2006) Long-term study of environmental levels of dioxins and furans in the vicinity of a municipal solid waste incinerator. Environ Int 32(3):397–404. https://doi.org/10.1016/J.ENVINT.2005.09.002
Singh Z, Bhalwar R, Jayaram J, Tilak VW (2001) An introduction to essentials of bio-medical waste management. Med J Armed Forces India 57(2):144–147. https://doi.org/10.1016/S0377-1237(01)80136-2
Singh P, Shandilya P, Raizada P, Sudhaik A, Rahmani-Sani A, Hosseini-Bandegharaei A (2020) Review on various strategies for enhancing photocatalytic activity of graphene based nanocomposites for water purification. Arab J Chem 13(1):3498–3520. https://doi.org/10.1016/J.ARABJC.2018.12.001
Slaughter RJ, Watts M, Vale JA, Grieve JR, Schep LJ (2019) The clinical toxicology of sodium hypochlorite. Clin Toxicol (Phila) 57(5):303–311. https://doi.org/10.1080/15563650.2018.1543889

Sudhaik A, Raizada P, Shandilya P, Jeong DY, Lim JH, Singh P (2018) Review on fabrication of graphitic carbon nitride based efficient nanocomposites for photodegradation of aqueous phase organic pollutants. J Ind Eng Chem 67:28–51. https://doi.org/10.1016/J.JIEC.2018.07.007
Townsend TK, Browning ND, Osterloh FE (2012) Nanoscale strontium titanate photocatalysts for overall water splitting. ACS Nano 6(8):7420–7426. https://doi.org/10.1021/NN302647U/ASSET/IMAGES/MEDIUM/NN-2012-02647U_0009.GIF
Ugwuishiwu BO, Owoh IP, Udom IJ (2016) Solar energy application in waste treatment—a review. Niger J Technol 35(2):432–440. https://doi.org/10.4314/njt.v35i2.27
Zhang Z, Gamage J (2010) Applications of photocatalytic disinfection. Int J Photoenergy. https://doi.org/10.1155/2010/764870

Chapter 8
Microbial Processes for Reducing E-Waste Pollution

Abdel Rahman Mohammad Said Al-Tawaha, Riham F. Zahalan, Arun Karnwal, Nataliia Nesterova, Iftikhar Ali, Ibrahim Al Rawashdeh, Abdel Razzaq Al-Tawaha, Roghayeh Karimirad, and Arun Karnwal

Abstract Millions of tonnes of E-waste are dumped annually worldwide. The rapid growth of the industrial and technological sectors is increasing the produced materials of electrical and electronic equipment (EEE). Changes in the modern lifestyles and the population affordability of having new technologies enhanced the EEE industry. Several components and raw materials are involved in the production process. Shipping and exporting change the final destinations of produced materi-

A. R. Mohammad Said Al-Tawaha (✉) · I. Al Rawashdeh
Department of Biological Sciences, Al Hussein Bin Talal University, Ma'an, Jordan
e-mail: Abdel-al-tawaha@ahu.edu.jo

R. F. Zahalan
Soil Maintenance and Reclamation Department, The General Commission for Scientific Agricultural Research, GCSAR, Damascus, Syria

A. Karnwal
Department of Microbiology, Lovely Professional University, Phagwara, Punjab, India

N. Nesterova
National University of Life and Environmental Sciences of Ukraine, Kyiv, Ukraine

I. Ali
School of Life Science, Chinese University of Hong Kong, Shatin, Hong Kong

Centre for Plant Science and Biodiversity University of Swat, Charbagh, Pakistan

A. R. Al-Tawaha
Department of Crop Science, Faculty of Agriculture, Universiti Putra Malaysia, Serdang, Selangor, Malaysia

National Agricultural Research Center, Baqa, Jordan

R. Karimirad
Department of Horticultural Sciences, Faculty of Agriculture and Natural Resources, University of Mohaghegh Ardabili, Ardabīl, Iran

A. Karnwal
Department of Microbiology, School of Bioengineering and Biosciences, Phagwara, Punjab, India

A. Karnwal, A. R. Mohammad Said Al-Tawaha (eds.), *Microbial Applications for Environmental Sustainability*, https://doi.org/10.1007/978-981-97-0676-1_8

als. Developing and under-developing countries are bearing increased burdens of this industry and mostly receiving the low quality of the produced EEE products. The low-cost labor increased the established factories in these countries, leaving tonnes of E-waste resulting from manufacturing a wide range of E-products that were then shipped to other parts of the world. Despite the high benefits of the information revolution, the negative sides of accumulated toxic materials are increasing. New research that investigated the related risks and threats of EEE industry had shown a high potential of threatening both the environment and public health with their high loads of toxic metals and pollutants. The recently increased awareness of public health started to identify possible risks and develop an eco-friendly solution for accumulated E-waste. This chapter describes the capacity of microorganisms and the potential of bioremediation methods in decomposing and removing different types of dumped E-waste: the related risks and the limitations of the process.

Keywords Bioremediation · EEE · E-wastes · Organisms · Toxic metals

8.1 Introduction

The drastically increased pollution of our natural resources has become a global concern with an urgent need to rapidly act for the environment and public health. A variety of pollutants from different sources are threatening soil, water, and atmosphere, leading to higher rates of human diseases and environmental pollution, and decreasing the quality of natural resources. Soil and water are considered fundamental pillars of food production. The direct contact between cultivated plants and components of the growing environment, soil and water, facilitates the movement of the set of pollutants toward the produced food, whether by staking on outer surfaces or reaching plants' tissues through plants' absorbance. Eventually, pollutants will reach our tables and invisibly threaten our health causing several health disorders from simple illnesses to become carcinogenic agents with time and accumulation.

Several attempts are currently being made to reduce environmental pollution using all available methods and materials. Yet, the process is time, effort, and cash-consuming. Most importantly, sustainable results cannot be ensured under the unwise human usage of these natural resources.

8.2 E-Wastes, Sources, and Impacts

Although there is no comprehensive definition of ***E-waste,*** the term is used to refer to a set of complex wastes that are mainly defined as electronic wastes. It includes a variety of plastics, metals and other materials usually used in manufacturing electronic devices and equipment. The abbreviation WEEE is globally common to refer to this type of waste, which stands for wastes of electrical and electronic equipment.

Discarded parts of mobile phones, computers, TV sets, and batteries are the most known for this type of waste.

After the huge and rapid global technological and industrial revolution, E-waste started to pose increased threats to natural resources, the environment, and public health. According to Plepiene and Youhanan (2019), the globally generated E-waste was about of 44.7 million tons in 2016. A new report by UNITAR estimated an average of 7.3 kg/capita and about 53.6 million tons of E-waste by 2019 (Balde et al. 2022). The most terrifying fact is that generated E-wastes are expected to increase up to 74.7 million tons by 2030 if it is not recycled, and controlled and human practices do not change.

Even though information technology and advanced electronic industries have together facilitated the modern life of humans, E-wastes represent the negative side of these technologies. This fact was widely recognized after the increased and accumulated E-wastes with a variety of highly toxic components that are difficult to treat. Toxic materials include toxic metals such as cadmium and nickel, toxic gases, chlorinated substances, plastics, brominated substances, etc. (Rao 2014).

The main source of E-wastes is the electrical, information, and communication industries. Table 8.1 shows examples of E-wastes' sources and related risks. Large

Table 8.1 Types of hazardous material, sources, and related risks

Types of hazardous material	Sources	Related risks
Cadmium	Batteries, semiconductor chips, coating materials, infrared detectors, UV stabilizers	Carcinogenic agent, kidney diseases, bone brittleness, lung damages
Mercury	Batteries, fluorescent lamps	Nervous system disorders, kidney diseases
Chromium	Cables, plastic computer housing, hard disks, magnetic, floppy disks, tapes, pigments	DNA damages, asthma, cough, skin, nose and eye irritations, and allergic reactions
Lead	Batteries, cable sheathing, glass of monitors, cathode ray tubes, X-ray shielding devices	Kidney damages, anemia, brain and nervous system disorders, blood pressure, and toxic symptoms
Arsenic	LEDs[a], semiconductor, light emitting diodes	Carcinogenic agent especially for lungs, liver, skin, and bladder, skin diseases, and nerve disorders
Antimony	Computer pieces such as plastics and glass	Carcinogenic agent, stomach ulcers
Nickel	Batteries, thermostat, light bulbs and lamps, panel displays	Kidney diseases, tremors, brain damages and memory problems, blood pressure and heart disorders, senses irritations
Silver	Electronic pieces, conductors	Respiratory and breathing disorders, skin disease
Barium	Lamps and sparkplugs, vacuum tubes	Brain, liver, heart, and spleen damages

[a] *LEDs* Light-emitting diodes

amounts of this produced equipment are transported from the origin country of production to different destinations. This movement reshapes the accumulated amounts of E-waste in each country. The level of livelihood and education are key factors that affect generated E-waste amounts, especially in developing countries (Moletsane and Venter 2018). Impacts of E-waste on the environment and natural resources are mainly due to the toxic materials themselves and the current methods of E-waste removal and disposal that may maximize their risks on natural resources. Landfilled E-wastes, for instance, may pose massive risks on water resources through the deep leaching of polluted landfill wastewater toward groundwater. Excessive heavy metal concentrations from electronic wastes (E-Wastes) can infiltrate the tissues of cultivated plants in contaminated soils via plant uptake, posing significant health risks to humans as potential carcinogenic agents. Further, significant amounts of toxic substances pollute the air when using thermal disposal methods such as incineration. These pollutants may move with winds to threat local environment and even consider a cross boarder pollution.

Methods of E-waste disposal have changed with time. As the world developed, claims of using eco-friendly solutions have increased. The produced amounts of generated E-wastes also increased. Formal and informal recycling have become more used. The biological removal has started to attract higher attention since they are environmentally friendly methods. Yet the biological process requires longer time compared with thermal and dumping methods. The traditional ways of incineration, open dumping, and using of landfills are used at larger scale, especially in developing and under-developing countries. Figure 8.1 illustrates the used methods for E-waste disposal and remediation.

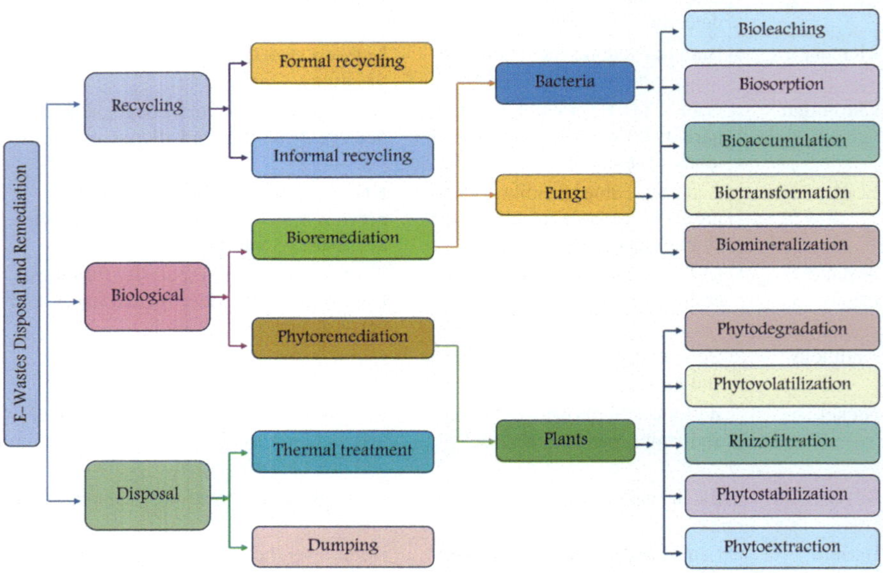

Fig. 8.1 Methods of E-waste disposal and remediation

8.3 Soil Fauna and Role of Environment and Bioremediation

Several types of microorganisms have been used for soil and water bioremediation; the most common types include:

Bacteria play a crucial role in bioremediation due to their significant capacity as biosorbents for pollution reduction. They possess the ability to disperse under regulated circumstances and exhibit resilience in the face of harsh stress conditions. Coelho et al. (2015) and Balamurugan et al. (2014) have shown the efficacy of several bacterial species, including Bacillus subtilis, Bacillus cereus, and Pseudomonas putida, in the process of chromium removal. According to previous studies conducted by Vullo et al. (2008) and Coelho et al. (2015), it has been shown that Pseudomonas veronii, Sporosarcina ginsengisoli, and Kocuria flava can effectively eliminate arsenic, copper, cadmium, and zinc from contaminated environments.

Bacteria are spread in the biosphere because of their metabolic capability and are simply grown under a broad range of environmental conditions and secrete enzymes. Different enzymes from aerobic bacteria including Alcaligenes, Pseudomonas, Rhodococcus, Sphingomonas, and Mycobacterium have mostly degraded hydrocarbons and pesticides. In comparison to aerobic bacteria, enzymes from anaerobic bacteria have been applied for the bioremediation of dechlorination of trichloroethylene, polychlorinated biphenyls, and chloroform. The most typical enzymes from microbial and genetically engineered microbial are responsible for the degradation of the various classes of pollutants such as cytochrome P450, hydrolases, laccases, dehydrogenases, dehalogenases, lipases, and proteases.

8.3.1 *Fungal*

The utilization of fungal microorganisms in the procedure of bioremediation has been commonly known as mycoremediation or fungal bioremediation. Fungi can break down chemical compounds and biomass leftovers via the secretion of various enzymes including protease, amylase, nuclease, and lipase. These enzymes facilitate the decomposition of cell walls, lipids, proteins, DNA, and other organic molecules. According to the study conducted by Balabanova et al. (2018), these fungi have been used in the process of biotransforming organic waste compounds and removing contaminants from the surrounding environment. These microorganisms have the potential to break down organic compounds, metalloids, metals, and radionuclides by the production of enzymes and other substances through a chemical amendment, thereby influencing the chemical bioavailability.

Furthermore, it has adapted to endure in many environments consistently. Fungi can metabolize many ambient substances and employ the resulting products to sustain themselves, hence eliminating the need for additional sustenance. The degradation of lignocelluloses and other contaminants is facilitated by the action of numerous extracellular oxidoreductases. The microorganisms in question can digest and immobilize hazardous metals, radionuclides, and metalloids. Furthermore, they possess the capability to store these substances in various cellular compartments or transport them via hyphae. Additionally, it successfully breaks down certain compounds that are not susceptible to bacterial degradation, such as 2,4,6-trinitrotoluene, dioxins, and pharmaceutical substances.

The previously mentioned attributes provide fungi a viable candidate for the removal of metal or organic pollutants in the atmosphere, soil, and water, hence enabling their use in cost-effective and environmentally sustainable bioremediation practices (Harms et al. 2011). The enzymes possessed by several fungi have a distinctive characteristic that enables the fungal strain to concurrently metabolize multiple molecules of diverse contaminants, even those that vary in structure. According to Kües (2015), Phanerochaete chrysosporium can degrade a wide range of harmful substances, including xylene, ethylbenzene, benzene, toluene, N-heterocyclic explosives, organochlorines, nitroaromatic compounds, polycyclic aromatic hydrocarbons, synthetic dyes, pesticides, polychlorinated dibenzo-p-dioxins, and synthetic polymers. This capacity remains effective even when these substances are present together in a mixture. The primary categories of enzymes involved in the degradation of pollutants and waste materials include cell-associated enzymes, extracellular oxidoreductases, and various transferase enzymes.

White-rot fungus biodegrades lignin in carbon recycling environments. Bjerkandera adusta, Phanerochaete chrysosporium, and Pleurotus sp., which generate laccases and peroxidases, are effective bioremediators (Yang et al. 2013; Voberkova et al. 2016).

Marine fungus provides useful secondary metabolites, enzymes, polysaccharides, biosurfactants, and fatty acids for heavy metal and hydrocarbon bioremediation. These fungi are more effective in harsh settings than others since they can tolerate high pH and salinity. Marine fungus can remediate heavy metal ions in nanotechnology, demonstrating extremophilic bacteria' bioremediation potential. Marine fungi's variety, ecological benefits, and biotechnological applications for enzymes, medicines, biopesticides, biodiesel, and bioremediation show its promise (Thatoi et al. 2013). The biotechnological use of this fungus' isolated enzymes has been discovered recently (Bonugli-Santos et al. 2015). These high-tolerance bacteria may interact with marine environments' heavy metal ions to synthesize metal nanoparticles.

The extremophilic fungi with such attributes are better microorganisms for cost-effectiveness and environmental-friendly together with biotransformation of raw materials in many industries such as food industries, animal fodder preparation, and textiles. Sinha et al. (2011) reported the implications of metallophilic microbes in bioremediation to generate problems with heavy metals from the environment that is produced by the synthesis of nanoparticles for effective bioremediation.

Cryptococcus sp. is a psychrophilic fungus occurring in deep-sea sediments, which can tolerate high concentrations of heavy metals $CdCl_2$, $CuSO_4$, $ZnSO_4$, and Pb $(CH_3COO)_2$ (100 mg/L). Different hydrolytic enzymes also show activity in extremophilic conditions and are involved in remediation techniques under stress conditions including extra-heavy crude oil contamination and high salinity The laccases activity under stress conditions is applied in the bioremediation mechanism in *Pestalotiopsis palmarum,* and lignin peroxidases are secreted from extra-heavy crude oil as an original of energy and carbon.

Fungi can uptake heavy metals into their spores and mycelium, so serving as potent biocatalysts in the process of metal bioremediation. The approach used for the remediation of heavy metals involves many mechanisms, including agglomeration, both extracellular and intracellular, precipitation, and valence conversion. According to Gupta et al. (2017), it was shown that the mycelium of fungus exhibited the secretion of potent extracellular enzymes with the ability to degrade cellulose and lignin.

8.3.2 Green Algae

Green algae such as Cladophora sp. have a high bioaccumulation ability for toxic metals and can be applied as a strong and alternative waste management factor. Although algae are also capable to degrade various kinds of hydrocarbon, their participation in hydrocarbon biodegradation is still not well introduced (Das and Chandran 2011). Some algae are capable of up-taking and decomposition of polycyclic aromatic hydrocarbons including *Scenedesmus quadricauda, S. capricornutum, S. platydiscus, and Chlorella vulgaris*. Species like Monoraphidium braunii and Selenastrum capricornutum have the potential to eliminate benzene, bisphenol, naphthalene, and toluene from a polluted environment.

8.4 E-Waste Remediation and Mechanism of Bioremediation

Chemical, physical, and biological methods are used for E-waste disposal. Incineration is among the old physical methods. Hydrometallurgical, pyrometallurgical, and biometallurgical processes in addition to extractive metallurgy are the main practiced chemical methods worldwide. With the developed standards of living and global awareness about public health and increased demands of saving the environment, biological methods were preferred and practiced widely. This type of eco-friendly method includes micro-remediation, phytoremediation, and vermiremediation (Patel and Kasture 2014).

Bioremediation refers to the usage of microbes to detoxify and decompose environmental pollutants. Microbes have different mechanisms for removing heavy metals ions from the environments, such as complexation of exopolysaccharides, adsorption to cell surfaces, biosynthesis of metallothioneins, intracellular accumulation, and other proteins that trap ions and convert them to volatile compounds. Recently, research is being concentrated on the expansion of consortia or genetically modified microbes for removing pollutants from the environment.

Environmental preservation and sustainability are significant concerns in contemporary society. The issue of environmental pollution refers to the entry of several hazardous substances into the natural surroundings. One of the most pressing challenges of the twenty-first century is the global concern and prioritization of decision-making. Over time, the presence of hazardous waste has resulted in substantial and permanent harm to the Earth, thereby elevating it to a topic of considerable worldwide concern (Orloff and Falk 2003). The agricultural sector produces a variety of hazardous waste, including pesticides, fertilizers, and veterinary product waste. The overutilization of pesticides, fertilizers, and other substances used within this sector, along with the resultant generation of trash, has resulted in contamination of both water and land resources. Numerous chemical substances often used in agricultural practices, such as endosulfan, DDT, atrazine, aldrin, toxaphene, heptachlor, chlordane, dieldrin, and lindane, are disposed of as hazardous waste. The use of this substance has mostly been employed within the field of agriculture to enhance productivity. The category of domestically produced hazardous waste encompasses several substances, including rat poisons, paints, insect repellents, and disinfectants, which possess poisonous properties. Certain categories of toxic hazardous wastes are specifically acknowledged as persistent pollutants due to their unique physical and chemical characteristics and elevated levels of toxicity. The majority of the existing research indicates that synthetic compounds possess an intrinsic resistance to degradation mechanisms, rendering them environmentally persistent. These materials possess characteristics that make them suitable for extended transportation, exhibit resistance to degradation, have adverse toxicological effects, and have a prolonged half-life. Bioremediation refers to the use of microorganisms to reduce or eliminate the presence of toxic hazardous wastes in the natural environment.

Bioremediation is applied constantly as a good clean-up, economic, environmental-friendly, and energy-efficient, alternative approach over the physical and chemical methods. The two great mechanisms to perform bioremediation are either by improving the growth of the native microbial community that exists at the site such as biostimulation or via presenting microbes posing better bioremediation capabilities (Agnello et al. 2016). During the bioremediation process, microbes make use of hazardous waste as an origin of energy and carbon to change them into nonpoisonous or less poisonous by-products. It is an economic method, which commonly has public acceptance and can often be performed on-site (Boopathy 2000). One of the main benefits of bioremediation is the production of nontoxic elements including nitrogen (N), carbon (C), and hydrogen (H) which simply get re-entry into the environment. It is, nevertheless, not always appropriate as the range of contaminants on which it is efficient is restricted, the time duration involved is almost long, and the remaining contaminant levels accessible may not always be suitable. Although the techniques applied are not technically intricate, remarkable knowledge and skill may be needed to design and perform a favored bioremediation scheme, because of the requirement to thoroughly evaluate a site for suitability and to improve conditions to attain an acceptable outcome. It also has been reported that by utilizing the indigenous microbial community that exists at the contaminated place, the bioremediation speed is entirely slow. In these times, several progresses have been performed in the bioremediation process to improve its speed. This advancement includes transgenic-based bioremediation, utilization of exogenous biosurfactants generating microbes which improve the accessibility of contaminants to the microorganism, the application of mixed culture, and the utilization of rhizoremediation despite bioremediation (Kumari and Singh 2016). The utilization of bacterial and algal consortia causes increased removal of multiple persistent organic contaminants that exist in landfill leachate over single algal and bacterial treatment. It also has been reported that the bacterial and algal consortia are more capable of a decrease in cytotoxicity over individual algal and bacterial cultures (Kumari and Singh 2016). Nevertheless, the bioremediation method has got some restrictions. It is a slow process, and only a few species of microbe capable of generating special enzymes have shown their potent capability to degrade pollutants. So, many researchers prefer genetically engineered microorganisms because of the large generation of favorable enzymes at optimized conditions. Many experiments utilize the typical features of microbial or genetically engineered microbial capable of generating special enzymes to metabolize or catalyze the xenobiotic pollutant to acquire biomass and energy, such as agrochemical, polyhalogenated compounds, microplastic, and hydrocarbons. This method converts toxic kinds to nontoxic kinds and sometimes new products. The bioremediation and degradation of wastes are challenging tasks due to there being no reliable approach to date that is sustainable about the complete taking away of these pollutants. Eventually, various groups of microorganisms that are already existing in nature may supply solutions for the bioremediation and degradation of toxic hazardous wastes. Okpokwasili (2007) demonstrated that microorganisms are being applied for the transformation and bioremediation of pollutants for a long time.

8.5 Types of E-Wastes' Bioremediation Methods

8.5.1 Bioleaching

In this method, solvable pollutants are adsorbed on the surfaces of cellular structure. The larger the number of absorption groups, the higher the predicted effectiveness of biosorption method. These groups include phosphate and carboxyl. The main mechanisms of E-waste bioleaching are acidolysis, complexolysis, and redoxolysis processes, and the main involved organisms are heterotrophic and autotrophic bacteria in addition to heterotrophic fungi. Bioremediation by bioleaching could be achieved directly and indirectly. Direct bioleaching happens through oxidizing insoluble toxic metals such as heavy metals by microorganisms that produce organic acids. Oxidation helps in transforming these toxins into soluble form that could be then removed and controlled. The indirect process mainly happens through oxidizing the metal surrounding the microbe. As a result, free metal ions are produced from oxidized metal compounds (Patel and Kasture 2014).

Bioleaching process is affected mainly by temperature and pH of the medium. Higher effectiveness is usually obtained at about 40° C and low pH values (>6) for the majority of used organisms. According to Ahmed et al. (2019), several heavy metals could be disposed of using bioleaching method such as copper, zinc, lead, aluminum, nickel, sliver, cadmium, and gold.

The research of Stephen and Macnaughtont (1999) identified a set of bacteria and fungi effective for bioleaching: *Aspergillus fumigatus* and *Aspergillus niger fungi and Acidithiobacillus thiooxidans, Micrococcus roseus, and Thiobacillus ferrooxidans* bacteria for arsenic and cadmium and lead.

8.5.2 Biosorption

Sulfate and amnio groups: Findings of previous research approved the effectiveness of using bacterial species such as *Arthrobacter sp., Pseudomonas sp., Enterobacter sp.* J, and *Enterobacter sp. J* for heavy metals removal, copper, chromium, cadmium, nickel, and zinc.

Aspergillus niger, saccharomyces cerevisiae, and penicillium chrysogenum are common fungal specious for biosorption of lead, copper, mercury, cadmium, and zinc. For algal species, Sargassum sp. and Spirogyra sp. are among the most used strains (Ahmed et al. 2019). *Bacillus sphaericus, myxococcus xanthus, pseudomonas aeruginosa, and uranium streptoverticillium cinnamoneum* are commonly used bacteria for biosorption method and the bio-removal of chromium, uranium, cadmium, according to the findings of Hu et al. (1996); Atkinson et al. (1998); Ahalya et al. (2003).

8.5.3 Bioaccumulation

As the absorbed pollutants enter the organism's cellular structure, they become concentrated and accumulate within the biomass of the organism. In this mechanism, toxic metals are usually sequestrated within the organisms through active metabolism (Prakash et al. 2012). For organic pollutants, the completion of the process requires catalyzing contaminants inside the cells' cytoplasm to transforming them into less toxic contaminants. Different bacteria and fungi specious were identified for bioaccumulation method. *Bacillus circulans, bacillus megaterium, deinococcus radiodurans,* and *micrococcus luteus* bacteria are used for chromium and uranium bioaccumulation, while *Aspergillus niger* and *Monodictys pelagica* fungi are used for chromium and lead (Demirbas 2001; Srinath et al. 2002; Malik 2004).

8.5.4 *Biotransformation*

From the name, this mechanism refers to the change of pollutants' chemical forms through removal or addition of electrons (Prakash et al. 2012). Biotransformation happens through enzymatic reduction, thereby giving additional electrons from the exterior of the involved enzymes to the targeted toxic metal. Another available biotransformation happens indirectly by metal-reducing and sulfate-reducing bacteria that transform toxic metals into immobilized and unavailable forms. The findings of recent research indicated effective breakdown of plastics into less harmful or harmless forms through biotransformation using specific microbes of Rhodococcus ruber. Some bacterial strains have shown effective degrading of ε-caprolactone such as Psychrobacter, Moritella, Shewanella, and Pseudomonas (Kathiresan 2003). Anaeromyxobacter sp., Clostridium sphenoides, and Halomonas sp. Bacteria are used in the biotransformation of uranium pollutions (Lovley and Coates 1997; Francis 1998; Malik 2004).

8.5.5 *Biomineralization*

Mineralization as a natural process usually refers to the bioactivity of microorganisms in which minerals are produced as a result of series of interactions with mineralized materials. In biomineralization mechanism, ions from toxic pollutants and ions related to used organisms are combined, interact, and led to precipitation of toxic compounds with unavailable forms. The findings of Patel and Kasture (2014) approved the capacity of some fungal and bacterial strains in lead, cadmium, and arsenic removal through biomineralization mechanism. Bacillus fusiformis, Cupriavidus metallidurans, Desulfotomaculum, Auripigmentum, and Sporosarcina ginsengisoli Bacteria in addition to Aspergillus flavus fungi are good examples of organisms used for biomineralization of heavy metals (Tabak et al. 2005; Benzerara et al. 2011; Achal et al. 2012; Govarthanan et al. 2012).

8.6 Limitations and Conclusion

Although several physical, chemical, and biological methods have showed good results in E-waste removal with their loads of highly toxic compounds and metals, yet all these methods are considered not adequate to control the annual increased accumulated tones of these toxic wastes worldwide. Adopting the eco-friendly solutions requires time, cash, and experience. Most importantly, it requires strong belief in the importance of the sustainable use of available natural resources and the need to act for the environment.

The integrated solution of E-waste management and removal starts by reducing the yearly produced and dumped amounts of E-wastes. This could be achieved through the wise use of resources, the controlled use of technology, recycling, and increasing the awareness about related risks among consumers. Scientific research regarding E-waste disposal, methods, effectiveness, and applicability should be governmentally supported. The combined effects of previously mentioned points with the available removal mechanisms can with time minimize the negative impacts of E-waste worldwide. It will allow the world to benefit from technology without increasing pollution or causing any harms.

References

Achal V, Pan X, Fu Q, Zhang D (2012) Biomineralization based remediation of As (III) contaminated soil by Sporosarcina ginsengisoli. J Hazard Mater 201:178–184

Agnello AC, Bagard M, van Hullebusch ED, Esposito G, Huguenot D (2016) Comparative bioremediation of heavy metals and petroleum hydrocarbons co-contaminated soil by natural attenuation, phytoremediation, bioaugmentation and bioaugmentation-assisted phytoremediation. Sci Total Environ 563–564:693–703

Ahalya N, Ramachandra TV, Kanamadi RD (2003) Biosorption of heavy metals. Res J Chem Environ 7(4):71–79

Ahmed T, Liaqat I, Murtaza R, Rasheed A (2019) Bioremediation: approaches for E-waste management: a step toward sustainable environment. Springer Nature, Cham, pp 267–289

Atkinson BW, Bux F, Kasan HC (1998) Considerations for application of biosorption technology to remediate metal-contaminated industrial effluents. Water SA 24(2):129–135

Balabanova L, Slepchenko L, Son O, Tekutyeva L (2018) Biotechnology potential of marine fungi degrading plant and algae polymeric substrates. Front Microbiol 9(1527):1–15

Balamurugan D, Udayasooriyan C, Kamaladevi B (2014) Chromium (VI) reduction by Pseudomonas putida and Bacillus subtilis isolated from contaminated soils. Int J Environ Sci Technol 5(3):522

Balde CP, D'Angelo E, Luda V, Kuehr R (2022) Transboundary E-waste flows. United Nations Institute for Training and Research (Unitar), Bonn, pp 1–66

Benzerara K, Miot J, Morin G, Ona-Nguema G, Skouri-Panet F, Férard C (2011) Significance, mechanisms, and environmental implications of microbial biomineralization. Compt Rendus Geosci 343(2):160–167

Bonugli-Santos RC, dos Santos Vasconcelos MR, Passarini MR (2015) Marine-derived fungi: diversity of enzymes and biotechnological applications. Front Microbial 6:1–15

Boopathy R (2000) Factors limiting bioremediation technologies. Bioresour Technol 74(1):63–67

Coelho LM, Rezende HC, Coelho LM, de Sousa PAR, Melo DFO, Coelho NMM (2015) Bioremediation of polluted waters using microorganisms. In: Shiomi N (ed) Advances in bioremediation of wastewater and polluted soil, vol 74(3). InTech, Shanghai, pp 293–301

Das N, Chandran P (2011) Microbial degradation of petroleum hydrocarbon contaminants: an overview. Biotechnol Res Int 2011:941810

Demirbas A (2001) Heavy metal bioaccumulation by mushrooms from artificially fortified soils. Food Chem 74(3):293–301

Francis AJ (1998) Biotransformation of uranium and other actinides in radioactive wastes. J Alloys Compd 271:78–84

Govarthanan M, Lee KJ, Cho M, Kim JS, Kamala-Kannan S, Oh BT (2012) Significance of autochthonous Bacillus sp. KK1 on biomineralization of lead in mine tailings. Chemosphere 90(8):2267–2272

Gupta A, Joia J, Sood A, Sood R, Sidhu C, Kaur G (2017) Microbes as potential tool for remediation of heavy metals: A review. J. Microb. Biochem. Technol 8:364–372. https://doi.org/10.4172/1948-5948.1000310

Harms H, Schlosser D, Wick LY (2011) Untapped potential: exploiting fungi in bioremediation of hazardous chemicals. Nat Rev Microbiol 9(3):177–192

Hu MZC, Norman JM, Faison BD, Reeves ME (1996) Biosorption of uranium by Pseudomonas aeruginosa strain CSU: characterization and comparison studies. Biotechnol Bioeng 51(2):237–247

Kathiresan K (2003) Polythene and plastic-degrading microbes in an Indian mangrove soil. Rev Biol Trop 51:629–633

Kües U (2015) Fungal enzymes for environmental management. Curr Opin Biotechnol 33:268–278

Kumari B, Singh DP (2016) A review on multifaceted application of nanoparticles in the field of bioremediation of petroleum hydrocarbons. Ecol Eng 97:98–105

Lovley DR, Coates JD (1997) Bioremediation of metal contamination. Curr Opin Biotechnol 8(3):285–289

Malik A (2004) Metal bioremediation through growing cells. Environ Int 30(2):261–278

Moletsane RI, Venter C (2018) Electronic waste and its negative impact on human health and the environment. IEEE. 978-1-5386-3060

Okpokwasili GC (2007) Biotechnology and clean environment. In: Proceedings of the 20th Annual conference of the Biotechnology Society of Nigeria (BSN), Abakaliki, Ebonyi State University

Orloff K, Falk H (2003) An international perspective on hazardous waste practices. Int J Hyg Environ Health 206(4–5):291–302

Patel A, Kasture A (2014) E (Electronic) waste management using biological systems-Overview. Int J Curr Microbiol App Sci 3(7):495–504

Plepiene JM, Youhanan L (2019) E-waste and raw materials: from environmental issues to business models. IVL Swedish Environmental Research Institute, Stockholm, pp 2–44

Prakash A, Satyanarayana T, Johri BN (2012) Microorganisms in environmental management: microbes and environment. Springer, New York, Dordrecht

Rao LN (2014) Environmental impact of uncontrolled disposal of E-wastes. Int J Chem Tech Res 6(2):1343–1353

Sinha A, Sinha R, Khare SK (2011) Heavy metal bioremediation and nanoparticle synthesis by metallophiles. In: Parmar N, Singh A (eds) Geomicrobiology and biogeochemistry, soil biology. Springer, Berlin, pp 101–118

Srinath T, Verma T, Ramteke PW, Garg SK (2002) Chromium (VI) biosorption and bioaccumulation by chromate resistant bacteria. Chemosphere 48(4):427–435

Stephen JR, Macnaughtont SJ (1999) Developments in terrestrial bacterial remediation of metals. Curr Opin Biotechnol 10(3):230–233

Tabak HH, Lens P, van Hullebusch ED, Dejonghe W (2005) Developments in bioremediation of soils and sediments polluted with metals and radionuclides 1. Microbial processes and mechanisms affecting bioremediation of metal contamination and influencing metal toxicity and transport. Rev Environ Sci Biotechnol 4(3):115–156

Thatoi H, Behera BC, Mishra RR et al (2013) Biodiversity and biotechnological potential of microorganisms from mangrove ecosystems: a review. Ann Microbiol 63:1–19. https://doi.org/10.1007/s13213-012-0442-7

Voberkova S, Vaverkova MD, Buresova A, Adamcova D, Vrsanska M, Kynicky J, Brtnicky M, Adam V (2016) Effect of inoculation with white-rot fungi and fungal consortium on the compositing efficiency of municipal solid waste. Waste Manag 61:157–164

Vullo DL, Cretti HM, Daniel MA, Rairez SAM, Zalts A (2008) Cadmium, zinc, and copper biosorption mediated by Pseudomonas veronii 2E. Bioresour Technol 99(13):5574–5581. Elsevier

Yang S, Hai FI, Nghiem LD (2013) Understanding the factors controlling the removal of trace organic contaminants by white-rot fungi and their lignin-modifying enzymes: a critical review. Bioresour Technol 141:97–108

Chapter 9
Microbial Enzymes for Eco-Friendly Recycling of Wastepaper by Deinking

Sapna Rawat, Abhishek Singh, Vishnu D. Rajput, Karen Ghazaryan, Tatiana Minkina, Abdel Rahman Mohammad Said Al-Tawaha, Abdel Razzaq Al-Tawaha, Marwa Adel Qotb, and Arun Karnwal

Abstract The papermaking industry relies heavily on forests for the raw material wood. In order to preserve forests, the practice of deinking and bleaching (recycling) used paper has become increasingly popular. As conventional chemical deinking processes are harmful to the environment due to the use of toxic chemicals, there is a need for more effective and eco-friendly deinking methods. Enzymatic deinking, with its high efficiency and low environmental impact, is a promising solution that utilizes cellulolytic, hemicellulolytic, and ligninolytic enzymes. During bleaching, the pulp is made more accessible to the chemicals and requires less of them to achieve the same levels of brightness by means of enzymatic action (both singly and in combination) and physical treatment. These processes enhance the pulp's strength properties and brightness. This analysis details the many options available for recycling paper in a way that has the smallest possible ecological footprint.

S. Rawat
University of Delhi, Delhi, India

A. Singh (✉) · K. Ghazaryan
Faculty of Biology, Yerevan State University, Yerevan, Armenia

V. D. Rajput · T. Minkina
Academy of Biology and Biotechnology, Southern Federal University, Rostov-on-Don, Russia

A. R. Mohammad Said Al-Tawaha
Department of Biological Sciences, Al Hussein bin Talal University, Ma'an, Jordan

A. R. Al-Tawaha
Department of Crop Science, Faculty of Agriculture, Universiti Putra Malaysia, Serdang, Selangor, Malaysia

National Agricultural Research Center, Baqa, Jordan

M. Adel Qotb
Soil Fertility and Plant Nutrition Department, Soils, Water and Environmental Research Institute, ARC, Giza, Egypt

A. Karnwal
Department of Microbiology, Lovely Professional University, Phagwara, Punjab, India

A. Karnwal, A. R. Mohammad Said Al-Tawaha (eds.), *Microbial Applications for Environmental Sustainability*, https://doi.org/10.1007/978-981-97-0676-1_9

Keywords Enzymes · Deinking · Bleaching · Cellulolytic · Hemicellulolytic · Ligninolytic

9.1 Introduction

The harvesting of plants by paper manufacturing industries is a major contributor to pollution in the environment, which is becoming an increasingly urgent and serious problem. Due to the rapid expansion of the paper industry, forest reserves are being rapidly depleted. Therefore, the development of novel approaches to these challenges is a top priority. Plant cutting should be outlawed, and paper recycling should be promoted. The practice of recycling used paper is a very beneficial endeavor that may effectively mitigate the environmental burden of toxins. The use of recycled paper offers a viable means of mitigating the need to engage in deforestation practices, hence yielding significant environmental benefits. In order to facilitate recycling efforts, the deinking sector now utilizes chemical deinking methods that need the use of potentially dangerous chemicals, including sodium hydroxide (NaOH), sodium carbonate (Na_2CO_3), sodium silicate (Na_2SiO_3), magnesium sulfate ($MgSO_4$), ethylenediaminetetraacetic acid (EDTA), and hydrogen peroxide (H_2O_2). The use of chemical techniques within the recycling process not only influences the caliber of the recovered pulp fiber but also engenders detrimental consequences on the environment and aquatic organisms. In order to address these consequences, it is essential to use biological approaches that include the utilization of microbial enzymes such as cellulase, pectinase, xylanase, laccase, and lipase for the purpose of separating and eliminating ink from wastepaper pulp (Chauhan et al. 2012). These biological agents possess the exclusive ability to mitigate the ecotoxic burden on the environment. The recycling of mixed office waste (MOW) is hindered by the conventional deinking procedure because of the inclusion of fused noncontact ink, often known as toner. At elevated temperatures necessary for printing, the toner undergoes polymerization and then adheres to the paper fibers by fusion. In order to liberate the ink particles from the fibers, the enzymes produced by microorganisms catalyze the hydrolysis of the link between the fiber and ink in a specific region. The effectiveness of the deinking process is influenced by several parameters, including the origin and specificity of microbial enzymes, the composition of the wastepaper being deinked, and the kind of ink present in the wastepaper (Lee et al. 2007, 2013). Biological deinking methods are gaining popularity as a more environmentally sustainable option compared to the prevalent use of chemical deinking techniques in the wastepaper deinking sector (Gübitz et al. 1998; Kaur et al. 2010). The wastepaper recycling sector necessitates the attainment of whiter and brighter outcomes, a goal that may be accomplished via the use of chemical deinking methods. Nevertheless, this approach is not without its limitations, since it compromises the structural integrity of the recycled paper and amplifies the discharge of biological oxygen demand (BOD) and chemical oxygen demand (COD) constituents. The high levels of these contents generated after various deinking treatments result in substantial effluent treatment costs (Wang et al. 2018; Shankar et al. 2018). To reduce chemical use and alleviate environmental stress, chemical and biological deinking can be used in combination. In a recent study, the efficiency of xylano-pectinolytic enzymes and alkaline chemical deinking was compared for mixed office waste (MOW) paper pulp. The study found that the synergistic action of xylanase and pectinase enzymes hydrolyzed the fiber-ink bonding region of the pulp, exposing the lignin and making it easier to remove the ink particle (Ibarra et al. 2012). The effectiveness of integrating biological deinking agents, derived from a solitary bacterial strain, into a deinking process for MOW paper was effectively shown (Imamoglu 2006). Enzymatic deinking is a method that offers significant benefits and demonstrates environmental sustainability by minimizing the use of harsh alkaline deinking chemicals. By utilizing microbial enzymes for the proper recycling of wastepaper, it is possible to reduce environmental pollution caused by activities like cutting down trees for paper and using toxic alkaline chemicals for deinking. Ultimately, enzymatic deinking is a sustainable method that supports environmental conservation efforts.

9.2 Worldwide Utilization of Recycled Papers

Paper recycling has become increasingly important in the pulp and paper industry as a means of conserving natural resources in light of growing demand and falling supplies of timber (Vyas and Lachke 2003). Paper recycling is now a sustainable business that can help protect forests for future wood and paper products, thanks to innovations like deinking. There has been a dramatic rise in the use of recycled paper in the papermaking industry (Chee and Drelich 2000). In India, for instance, the market for recycled paper is anticipated to grow between four and five million tonnes annually. In 2011, the USA saw a 66.8% increase in paper recovery or 1.2 million tonnes. Global consumption of recycled paper increased from around 200 million tonnes in 2006 to around 250 million tonnes in 2012 (Bajpai et al. 1999). The consumption of recycled paper has increased the most in China (Bajpai et al. 1999). The ink, printing method, and fiber type must all be considered when attempting to remove contaminants from wastepaper. Several factors, including ink particle hydrophobicity, fiber type, ink particle size, pH, ink type, and temperature, can affect the efficacy of a deinking process. Traditional deinking methods have proven to be insufficient for nonimpact printed papers, making them notoriously difficult to recycle. Consequently, their share of the recovered paper market continues to grow. The deinking process for water-based flexographic newsprint is particularly challenging, and removing cross-linked colored inks used in offset lithography poses similar difficulties (Lee et al. 2007).

9.3 Basic Principle of Deinking Wastepaper

Paper recycling entails the processes of defibering, deinking, and reforming. Defibering is the process of separating the wood fibers from the recycled material, and reforming is the process of making the final product. Although defibering and reforming are relatively straightforward processes, deinking plays a crucial role in the production of certain recycled paper products, such as printing paper of paperboard (Keong Lee et al. 2011). The deinking process plays a critical role in recycling used paper by separating ink from fibers and producing ink particles of the appropriate size and shape for further ink removal stages. When the ink particles are small, they have a higher tendency to reattach to the fibers because they can get inside the fiber lumen. The rate of ink detachment and redeposition during pulping are both first-order processes. At high consistency, the shear stress between the fibers helps to detach ink particles from the fibers, enabling the ink to be rinsed away (Sondhi et al. 2015). Flotation and wash deinking are the two primary deinking processes used in the paper industry. Wash deinking is a filtration separation process that keeps the useful wood fibers and allows the unwanted ink to pass through the screens. Separation is independent of chemical properties, so it is useful for separating inks with wide variations in both. However, yield is drastically reduced (yield can be reduced to as low as 70%) as a result of paper fines and ash being lost in the process. Flotation deinking, however, uses surfactants and air bubbles to selectively remove hydrophobic particles from solution, with yields of greater than 90%. Chemical, physical, and enzymatic deinking can be used together to extract aged and strongly bound hydrophilic inks, while the removal of loosely bound ink is straightforward even without these additional steps.

9.4 Need of Microbial Enzymes for Wastepaper by Deinking

Paper is essential in so many aspects of daily life, including the workplace, educational institutions, and private homes. Production is impacted by the daily occurrence of paper waste. This increases demand for the plant's raw materials, which in turn reduces the forest's ability to provide those materials (Fig. 9.1) (Siqueira et al. 2020). During the papermaking process, cellulase is used primarily as a raw material derived from wood (tree). While paper waste (including newspapers, photocopies, and laser printer output) is biodegradable and good for the environment, it is also the leading cause of tree cutting. Paper products like old newspapers, photocopies, and laser prints are easily recyclable, thanks to the deinking process (Qian and Goodell 2005). Removing the ink from recycled paper is called "deinking," and it is a crucial part of the paper recycling process. Currently, harmful chemicals are used in the deinking process used by the paper industries (Qian and Goodell 2005). The wastewater from most existing deinking processes must be treated, which can be expensive because it may contain large amounts of chemicals that are harmful to the environment (Ibarra et al. 2012). Flotation is the most common traditional method for removing ink (Bhat 2000). Many pulp and paper companies are investing in enzymes that break down lignocellulosic materials (Bhat 2000). Traditional deinking uses surfactants to facilitate toner flotation and high temperatures to expedite the formation of toner aggregates. In addition, bleaching chemicals are used in conventional deinking, whereas only a small number of enzymes are required for enzymatic deinking (Lee et al. 2013). Enzymatic deinking is the most effective

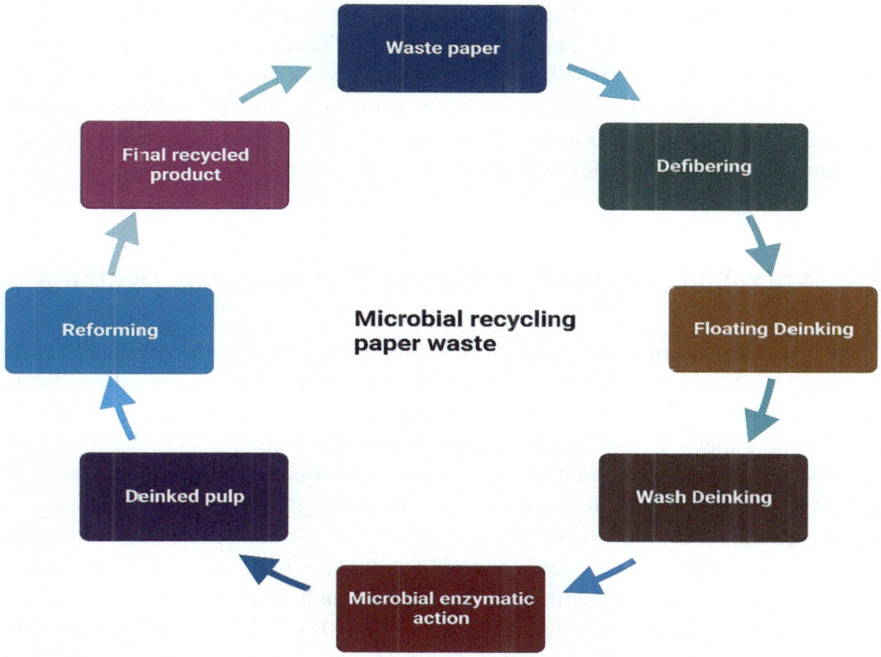

Fig. 9.1 Diagrammatic representation of microbial enzyme-based paper recycling

method for removing inks from paper waste, as it not only removes the inks but also has a low environmental impact and a low cost per tonne of paper waste (Lee et al. 2013). The potential of a biological approach to deinking has only just begun to be explored. Traditional enzymes and chemicals can be swapped out for cellulase (Vinod Kumar et al. 2018). When it comes to deinking paper pulp, the two most powerful microbial enzymes are cellulase and xylanase. Due to its low cost and low impact on the environment, enzymatic deinking has recently gained widespread attention (Ibarra et al. 2012; Siqueira et al. 2020). It also results in effluents that are less toxic, making their cleanup a breeze. Cellulase, hemicellulase, laccase, pectinase, esterase, -amylase, and ligninolytic enzymes from a wide variety of bacteria and fungi have all been used, singly or in combination, for enzymatic deinking of a wide variety of wastepapers. The commercially available enzyme would reduce costs compared to traditional deinking chemicals, but it is too expensive to compete the use of a crude cellulase enzyme to achieve deinking on a biological level. The bacteria are grown in the Mettupalayam forest, and the crude enzyme is then prepared from them. The enzyme was isolated and used in place of the synthetic deinking agent. The primary goal of implementing a biological method for deinking without the use of harmful chemicals is to eliminate ink particles from photocopied documents through an enzymatic ink removal process. This process involves utilizing crude cellulase enzyme extracted from isolated bacterial strains. The utilization of *Bacillus wiedmannii* for efficient ink removal from wastewater is a new development, requiring only a small quantity of chemicals. This approach is an area that has not been fully explored yet (Kumar and Satyanarayana 2014). Bacterial cellulase accomplishes the primary goal of recycling paper by eliminating the ink (Kumar and Satyanarayana 2014). It also highlights eco-friendly methods of improving the environment to create a better tomorrow.

9.5 Microbial Enzymatic Action for Wastepaper by Deinking

As people have become more aware of sustainability issues, the paper and pulp industry has steadily used more microbial enzymes to reduce their negative effects on the ecosystem. Enzymes cut down on processing times, energy requirements, and chemical loads. Enzymes can be advantageous in deinking and bleaching procedures for the paper and pulp sector, as well as in waste management, since they can increase the levels of biological oxygen demand (BOD) and chemical oxygen demand (COD) (Ma et al. 2011). The removal of lignin and hemicelluloses by xylanases and ligninases in the paper and pulp industries increases the pulp's value. Amylases have many applications in these fields, including starch coating, deinking, cleaner paper, and better drainage (Pala et al. 2004). Cellulases are employed to enhance deinking, soften fibers, and improve drainage, whereas lipases are utilized to aid in deinking and improve pitch control (Pala et al. 2006). Used paper can be recycled through a bioprocess that has been developed using cellulase (Pan and Nguyen 2007). Replacing chlorine with laccase in the chemical pulping process diminishes the quantity of waste that contributes to the depletion of the ozone layer and the acidification of the oceans. Furthermore, the utilization of mannases to degrade glucomannan enhances brightness in the paper industry (Pan and Nguyen 2007).

9.5.1 Cellulase

Cellulase is a widely studied and important enzyme with numerous practical applications in industry (Kaur et al. 2007). Cellulase requires a combination of different types of cellulase to degrade cellulose, which is the major substrate of the cellulase enzyme (Linder and Teeri 1997). Cellulase enzymes can be degraded down into two distinct groups: those that cleave cellulose from within (endoglucanase, E.C. 3.2.1.4) and those that cleave cellulose from outside (exo-glucanase, E.C. 3.2.1.91), the latter of which produces cellobiose (Prasad et al. 1993). The active sites of enzymes interact with substrates at their carbohydrate-binding domains. Cellulase without carbohydrate-binding sites is used for enzymatic bleaching in the paper pulp industries (Zuo and Saville 2005).

9.5.2 *Xylanases*

Xylanase free from cellulase has demonstrated its effectiveness in enhancing delignification without interfering with the pulp's cellulose backbone prior to bleaching (Viikari et al. 1994; Blanco et al. 1995). The utilization of several xylanases, each with distinct catalytic efficiencies, is beneficial for achieving complete hydrolysis of hemicellulosic materials (Chauhan et al. 2014). Purified xylanases are utilized in biobleaching as they can both deink and enhance the quality of deinked laser-printed paper simultaneously (Maity et al. 2012).

9.5.3 *Amylase*

The degradation of the starch layer on paper's surface, resulting in the alteration of various pulp properties, can be achieved by amylase.

9.5.4 *Laccases*

Combining laccase with specific low molar mass chemical mediators can deink ONP, a paper substrate consisting of lignin-rich mechanical pulp, leading to pulps that are lighter and simpler to bleach (Bajpai 1999). The enzymatically deinked pulp had greater bleachability than the chemically deinked pulp. The addition of laccase to pulp and paper treated with ONP results in changes in their properties (Xu et al. 2009). Studies have shown that laccase obtained from *Myceliophthora thermophila* and *Trametes versicolor* effectively removes water-soluble textile and ink jet ink dyes by up to 94%, and aqueous dye-based ink jet inks by 40–80%, ultimately increasing the ISO brightness by 5% (Xu et al. 2009).

9.5.5 *Lipase*

Lipase oil-based inks can be cleaned up with alkaline lipases, and inks made from vegetable oils can be broken down with esterase. Treatment of a mixture of ONP and magazine with lipase preparations resulted in alterations in brightness and ink concentration. Deinking of soybean oil-based ink printed paper using lipase preparations led to a decrease in ink residue and dirt counts (Mørkbak et al. 1999). Treatment of various papers with lipases and esterases led to an increase in deinking. Amphiphilic lipases have a surfactant effect, allowing deinked paper to be recovered (Leduc et al. 2011).

9.5.6 *Pectinases*

Pectinases have the capacity to alter the fiber surface or bonds situated close to ink particles. The breakdown of secondary fibers by pectinases during washing or flotation can release ink particles, thereby reducing the need for conventional deinking chemicals.

9.6 Enzymes' Impact on Paper Quality and Production Volumes

Enzymatic deinking pulp outperforms chemically deinked pulp in terms of brightness, impurity removal, and strength, as evidenced by an increase in Canadian Standard Freeness (CSF) from 510 to 570 mL and 440 to 490 mL through enzyme treatment (Singh et al. 2012). In all the enzyme-treated samples, freeness increased when compared with the control. Black-and-white hemicellulase-treated newsprint increased freeness by 14%, while colored flexo-printed newsprint treated with cellulase increased freeness by 50% (Soni et al. 2010). Enzymatic deinking of pulp results in improved drainage and wet web strength, making it more suitable for paper machine operations. In a study on deinking office papers, recycled fibers were enzymatically pretreated with cellulase or a combination of cellulase and xylanase to evaluate their impact on various properties, such as freeness, sheet strength, optical characteristics, and surface properties (Tatsumi et al. 2000). The mixture of cellulase and xylanase treated pulps showed a substantial increase in burst strength, while tensile strength remained almost constant, indicating the effectiveness of the enzymatic treatment. The enzymatic deinking of old newspaper (ONP) results in the release of reducing sugars; however, the losses incurred during this process remain unchanged. The enzymatic activity facilitated the liberation of microfibrils from their fibrous structures, ultimately resulting in their dispersion. According to Singh et al. (2019), the enzymatic treatment resulted in better yields compared to traditional deinking. Additionally, the observed changes in characteristics suggested a favorable reaction to the enzyme treatment.

9.7 The Impact of Different Enzymes on Effluent Quality

Enzymatic deinking results in lower chemical oxygen demand (COD) in the wastewater compared to chemical deinking, thereby reducing the burden on the wastewater treatment systems and producing clearer water (Dutt et al. 2012). The enzyme treatment reduced the COD load by 50% compared to the standard deinking method (Lee et al. 2007). Enzyme reactions produced process water that, upon entering the clarifier, had a higher total COD and lower dissolved BOD compared to the control. Effluents from laboratory scale experiments suggest that enzyme reactions are safer than the baseline run. Deinking offset newsprint without using any chemicals results in the lowest COD load in the white water from the pulp (Pala et al. 2004). The COD load of enzyme-treated pulp is proportional to the amount and type of enzymes used. After adding either 0.2% or 1.0% of the enzyme, the average CODs were 11 and 20 kg/ton, respectively. Treatment with 0.2% enzymes reduced COD load by 50% compared to standard deinking (Pala et al. 2004). The switch from chemical to enzymatic deinking did not result in any statistically significant difference in COD entering the biological treatment plant, nor did it affect the sludge volume index (Pala et al. 2004). The transfer of calcium ions from the deinking plant to the biological treatment plant resulted in a doubling of the ash content, which only required a minor adjustment to the dewatering equipment but had a significant impact on the dewatering properties of the bio-sludge.

9.8 Conclusion

Microbial enzymes used in the deinking process were chosen after careful consideration of the effects of factors like pH, temperature, pulp consistency, and mechanical stress on their performance and longevity. The ink, printing procedure, and type of wastepaper all play a role in determining which enzymes to use in the enzymatic deinking process. The use of microbial enzymes can increase the brightness of used paper and reduce the size of ink particles regardless of the paper or ink used. This allows for more efficient waste management and chemical cost savings during the deinking process of discarded paper.

Acknowledgements KG is supported by internal grant of YSU. AS is supported by the 23PostDoc-4D007 grant provided by the Science Committee of the Republic of Armenia. VDR and TM acknowledge the financial support of the Ministry of Science and Higher Education of the Russian Federation, agreement no. 075-15-2022-1122, and the Strategic Academic Leadership Program of the Southern Federal University ("Priority 2030").

Conflict of Interest

Ethics Approval and Consent to Participate Not applicable.

Consent for Publication The article contains no such material that may be unlawful, defamatory, or which would, if published, in any way whatsoever, violate the terms and conditions as laid down in the agreement.

Availability of Data and Materials Not applicable.

Competing Interests The authors declare that they have no conflict of interest in the publication.

References

Bajpai P (1999) Application of enzymes in the pulp and paper industry. Biotechnol Prog 15(2):147–157. https://doi.org/10.1021/BP990013K

Bajpai P, Bajpai PK, Kondo R (1999) Enzymatic deinking. Biotechnol Environ Prot Pulp Pap Ind 91–108. https://doi.org/10.1007/978-3-642-60136-1_6

Bhat MK (2000) Cellulases and related enzymes in biotechnology. Biotechnol Adv 18(5):355–383. https://doi.org/10.1016/S0734-9750(00)00041-0

Blanco A, Vidal T, Colom JF, Pastor FIJ (1995) Purification and properties of xylanase A from alkali-tolerant Bacillus sp. strain BP-23. Appl Environ Microbiol 61(12):4468–4470. https://doi.org/10.1128/AEM.61.12.4468-4470.1995

Chauhan PS, Puri N, Sharma P, Gupta N (2012) Mannanases: microbial sources, production, properties and potential biotechnological applications. Appl Microbiol Biotechnol 93(5):1817–1830. https://doi.org/10.1007/S00253-012-3887-5/METRICS

Chauhan PS, Sharma P, Puri N, Gupta N (2014) A process for reduction in viscosity of coffee extract by enzymatic hydrolysis of mannan. Bioprocess Biosyst Eng 37(7):1459–1467. https://doi.org/10.1007/S00449-013-1118-9/METRICS

Chee K, Drelich J (2000) Interfacial effects of a polyalkylene oxide/fatty acid surfactant blend in flotation deinking of mixed office papers rehydroxylation dating of fired-clay ceramics view project flotation fundamentals view project

Dutt D, Tyagi CH, Singh RP, Kumar A (2012) Effect of enzyme concoctions on fiber surface roughness and deinking efficiency of sorted office paper. Cellul Chem Technol 46(9):611–623

Gübitz GM, Mansfield SD, Böhm D, Saddler JN (1998) Effect of endoglucanases and hemicellulases in magnetic and flotation deinking of xerographic and laser-printed papers. J Biotechnol 65(2–3):209–215. https://doi.org/10.1016/S0168-1656(98)00130-8

Ibarra D, Monte MC, Blanco A, Martínez AT, Martínez MJ (2012) Enzymatic deinking of secondary fibers: cellulases/hemicellulases versus laccase-mediator system. J Ind Microbiol Biotechnol 39(1):1–9. https://doi.org/10.1007/S10295-011-0991-Y

Imamoglu S (2006) Deinkability efficiency of waste office paper printed using a duplicating machine. J Appl Sci 6(9):2006–2009. https://doi.org/10.3923/JAS.2006.2006.2009

Kaur J, Chadha BS, Kumar BA, Saini HS (2007) Purification and characterization of two endoglucanases from Melanocarpus sp. MTCC 3922. Bioresour Technol 98(1):74–81. https://doi.org/10.1016/J.BIORTECH.2005.11.019

Kaur A, Mahajan R, Singh A, Garg G, Sharma J (2010) Application of cellulase-free xylano-pectinolytic enzymes from the same bacterial isolate in biobleaching of kraft pulp. Bioresour Technol 101(23):9150–9155. https://doi.org/10.1016/J.BIORTECH.2010.07.020

Keong Lee C, Che Omar I, Rosli Wan Daud W (2011) Enzymatic and chemical deinking of mixed office wastepaper and old newspaper: paper quality and effluent characteristics Comparative properties of soda pulp stalk, bast and core Malaysian grown kenaf View project Anti-biofilm activity of endophytic fungus against foodborne bacteria View project. https://doi.org/10.15376/biores.6.4.3859-3875

Kumar V, Satyanarayana T (2014) Production of endoxylanase with enhanced thermostability by a novel polyextremophilic Bacillus halodurans TSEV1 and its applicability in waste paper deinking. Process Biochem 49(3):386–394. https://doi.org/10.1016/J.PROCBIO.2013.12.005

Leduc C, Lanteigne-roch L, Daneault C (2011) Use of enzymes in deinked pulp bleaching. Cellul Chem Technol 45(9):657–663

Lee CK, Darah I, Ibrahim CO (2007) Enzymatic deinking of laser printed office waste papers: some governing parameters on deinking efficiency. Bioresour Technol 98(8):1684–1689. https://doi.org/10.1016/J.BIORTECH.2006.05.052

Lee CK, Ibrahim D, Che Omar I (2013) Enzymatic deinking of various types of waste paper: efficiency and characteristics. Process Biochem 48(2):299–305. https://doi.org/10.1016/J.PROCBIO.2012.12.015

Linder M, Teeri TT (1997) The roles and function of cellulose-binding domains. J Biotechnol 57(1–3):15–28. https://doi.org/10.1016/S0168-1656(97)00087-4

Ma Y, Wang L, Sun X (2011) Deinking technology and deinking agent of waste paper. Adv Mater Res 335–336:1340–1343. https://doi.org/10.4028/WWW.SCIENTIFIC.NET/AMR.335-336.1340

Maity C, Ghosh K, Halder SK, Jana A, Adak A, Mohapatra PKD, Pati BR, Mondal KC (2012) Xylanase isozymes from the newly isolated bacillus sp. CKBx1D and optimization of its deinking potentiality. Appl Biochem Biotechnol 167(5):1208–1219. https://doi.org/10.1007/S12010-012-9556-4/METRICS

Mørkbak AL, Degn P, Zimmermann W (1999) Deinking of soy bean oil based ink printed paper with lipases and a neutral surfactant. J Biotechnol 67(2–3):229–236. https://doi.org/10.1016/S0168-1656(98)00179-5

Pala H, Mota M, Gama FM (2004) Enzymatic versus chemical deinking of non-impact ink printed paper. J Biotechnol 108(1):79–89. https://doi.org/10.1016/J.JBIOTEC.2003.10.016

Pala H, Mota M, Gama FM (2006) Factors influencing MOW deinking: laboratory scale studies. Enzym Microb Technol 38(1–2):81–87. https://doi.org/10.1016/J.ENZMICTEC.2005.04.022

Pan J, Nguyen KL (2007) Energy required for detachment of inks from printed newsprint in fibre slurries—effects of vehicles and aging. Colloids Surfaces A Physicochem Eng Asp 302(1–3):354–359. https://doi.org/10.1016/J.COLSURFA.2007.02.065

Prasad DY, Heitmann JA, Joyce TW (1993) Enzymatic deinking of colored offset newsprint. Nord Pulp Pap Res J 8(2):284–286. https://doi.org/10.3183/NPPRJ-1993-08-02-P284-286

Qian Y, Goodell B (2005) Deinking of laser printed copy paper with a mediated free radical system. Bioresour Technol 96(8):913–920. https://doi.org/10.1016/J.BIORTECH.2004.08.015

Shankar S, Shikha, Bhan C, Chandra R, Tyagi S (2018) Laccase based de-inking of mixed office waste and evaluation of its impact on physico-optical properties of recycled fiber. Environ Sustain 131(3):233–244. https://doi.org/10.1007/S42398-018-0021-3

Singh A, Yadav RD, Kaur A, Mahajan R (2012) An ecofriendly cost effective enzymatic methodology for deinking of school waste paper. Bioresour Technol 120:322–327. https://doi.org/10.1016/J.BIORTECH.2012.06.050

Singh A, Kaur A, Yadav RD, Mahajan R (2019) An efficient eco-friendly approach for recycling of newspaper waste. 3 Biotech 9(2):1–7. https://doi.org/10.1007/S13205-019-1590-2/METRICS

Siqueira JGW, Rodrigues C, Vandenberghe LP d S, Woiciechowski AL, Soccol CR (2020) Current advances in on-site cellulase production and application on lignocellulosic biomass conversion to biofuels: a review. Biomass Bioenergy 132:105419. https://doi.org/10.1016/J.BIOMBIOE.2019.105419

Sondhi S, Sharma P, George N, Chauhan PS, Puri N, Gupta N (2015) An extracellular thermo-alkali-stable laccase from Bacillus tequilensis SN4, with a potential to biobleach softwood pulp. 3 Biotech 5(2):175–185. https://doi.org/10.1007/S13205-014-0207-Z/TABLES/4

Soni R, Nazir A, Chadha BS (2010) Optimization of cellulase production by a versatile Aspergillus fumigatus fresenius strain (AMA) capable of efficient deinking and enzymatic hydrolysis of Solka floc and bagasse. Ind Crop Prod 31(2):277–283. https://doi.org/10.1016/J.INDCROP.2009.11.007

Tatsumi D, Higashihara T, Kawamura S, Matsumoto T (2000) Ultrasonic treatment to improve the quality of recycled pulp fiber. J Wood Sci 46(5):405–409. https://doi.org/10.1007/BF00776405/METRICS

Viikari L, Kantelinen A, Sundquist J, Linko M (1994) Xylanases in bleaching: from an idea to the industry. FEMS Microbiol Rev 13(2–3):335–350. https://doi.org/10.1111/J.1574-6976.1994.TB00053.X

Vinod Kumar N, Rani ME, Gunaseeli R, Kannan ND (2018) Paper pulp modification and deinking efficiency of cellulase-xylanase complex from Escherichia coli SD5. Int J Biol Macromol 111:289–295. https://doi.org/10.1016/J.IJBIOMAC.2017.12.126

Vyas S, Lachke A (2003) Biodeinking of mixed office waste paper by alkaline active cellulases from alkalotolerant Fusarium sp. Enzym Microb Technol 32(2):236–245. https://doi.org/10.1016/S0141-0229(02)00273-9

Wang F, Zhang X, Zhang G, Chen J, Sang M, Long Z, Wang B (2018) Studies on the environmentally friendly deinking process employing biological enzymes and composite surfactant. Cellulose 25(5):3079–3089. https://doi.org/10.1007/S10570-018-1778-3/METRICS

Xu Q, Fu Y, Gao Y, Qin M (2009) Performance and efficiency of old newspaper deinking by combining cellulase/hemicellulase with laccase-violuric acid system. Waste Manag 29(5):1486–1490. https://doi.org/10.1016/J.WASMAN.2008.10.007

Zuo Y, Saville BA (2005) Efficacy of immobilized cellulase for deinking of mixed office waste. J Pulp Pap Sci 31(1):3–6

Chapter 10
The Potential and Reality of Agricultural Wastewater Reuse

Nataliia Nesterova, Abdel Rahman Mohammad Said Al-Tawaha, and Arun Karnwal

Abstract The soils of arid regions face a range of environmental and sometimes man-made problems, but mainly water scarcity (usually <200 mm of precipitation per year^{-1}) and land devastation. The modern development of agriculture is a fundamental strategy for responding to significant problems in an arid climate, in which the severity of global warming is even stronger. It is important to analyze the types of climate change manifestations and their impact on agricultural productivity in arid climates, to understand possible methods of sustainable intensification of agriculture that can help respond to climate change in such conditions. Yet extensive agriculture, unsustainable use, and global climate change are causing an irrecoverable damage to arid lands, with far-reaching consequences to the species, groundwater resources, ecosystem productivity, and ultimately the communities dependent on these systems. However, adequate research and effective policies to protect arid land biodiversity and sustainability are lacking because a large proportion of arid areas are in developing countries, and the unique diversity in these systems is frequently overlooked. However, certain manifestations of climate change and their consequences should be considered in the context of the positive aspects of sustainable agriculture in arid lands. Thus, agricultural productivity in arid lands can be increased by 21–33% through actions aimed at increasing productivity and sustainability, such as improved crop genetics, better management of natural resources, climate-friendly agriculture, etc.

N. Nesterova (✉)
National University of Life and Environmental Sciences of Ukraine, Kyiv, Ukraine
e-mail: natalianesterova@nubip.edu.ua

A. R. Mohammad Said Al-Tawaha
Department of Biological Sciences, Al Hussein Bin Talal University, Ma'an, Jordan
e-mail: Abdel-al-tawaha@ahu.edu.jo

A. Karnwal
Department of Microbiology, Lovely Professional University, Phagwara, Punjab, India

A. Karnwal, A. R. Mohammad Said Al-Tawaha (eds.), *Microbial Applications for Environmental Sustainability*, https://doi.org/10.1007/978-981-97-0676-1_10

Keywords Arid climate agriculture · Sustainable agriculture arid regions · Climate change impact arid agriculture · Arid land biodiversity · Water scarcity agriculture · Climate-friendly farming arid zones

10.1 Introduction

Water is an irreplaceable natural resource for life, industrial, and production spheres and anthropogenic human activities as a whole. Global warming, water scarcity, and urban population growth lead to increased demand for water resources for domestic, industrial, and agricultural needs. Water resources can be used in various aspects, such as urban water supply, industrial reuse, agricultural growth, ecological recycling, replenishment of grounded water levels, etc., which requires a detailed approach, considering that reclaimed water replaces traditional sources of water supply. The modernization of existing methods and processes for cleaning used water, which were significantly improved in the twentieth century, contributed to the potentially wider production of wastewater reutilization technology and continue to improve the course of the twenty-first century. The environmental aspect of water reuse may include the improvement and restoration of wetlands, although water reuse in the agricultural aspect will be increased mostly by the implementation of recycling in irrigation. Reusing wastewater after its treatment can definitely be a good alternative for regions that are in need of clean water or have limited access to water resources. At the same time, effective water treatment requires the possible connection of advanced cleaning technologies and available financial resources.

At the same time, many new effective technologies for industrial wastewater treatment have recently been developed. The complexity of the cleaning task is determined by the nature of pollution—usually a number of substances that require a different approach act as undesirable components. It is characteristic of water treatment facilities that they line up in a certain sequence. Such a complex is called a line of sewage treatment plants. Some systems and types of equipment include different wastewater treatment methods. Due to the diverse content of wastewater entering treatment and the high requirements for water treatment, it is most appropriate to use mixed methods of treatment. For more complex pollution, it is necessary to use devices with a specific effect and special water purification technologies. Each method of cleaning wastewater and technical liquids is carried out using different devices. Such actions will make it possible to increase the efficiency of wastewater treatment and achieve the highest level of water quality that meets the norms of the main technological processes for reuse methods and structures that allow improving the quality of wastewater discharged into open water bodies, to reduce costs for purified water.

Developing a strategy for sustainable development of semiarid and arid regions is especially relevant in the 21st century. Increasingly, irreversible environmental problems such as devastation, loss of land from agricultural use, and, as a consequence, loss of economic support from society, are emerging. In such vulnerable ecosystems, the successful implementation of sustainable management strategies depends on a well-founded understanding of the processes at different scales that underlie the dynamics of the complex system. There is ample evidence that, instead of traditional approaches, only interdisciplinary research works effectively to solve the problem and manage the conservation of natural resources.

The purpose of this material is to present the possibilities of wastewater reuse in the field of agricultural and industrial production, characterize ways and experiences of water resource reuse that can be used in arid and semiarid regions in the context of sustainable agriculture, and model proposals for deepening the scope of wastewater application.

10.2 Wastewater Reuse

Industrial waste and sediments containing salts of heavy metals and so on represent environmental hazard, as they have high toxic, carcinogenic, and mutagenic effects on living organisms. Getting into bodies of water, ions of heavy metals have an adverse effect on their biocenoses and limit the possibility of their economic and recreational use (Baskar et al. 2022). Getting into human body via water and food products, heavy metal ions do not only influence human health but also destroy the gene pool (Kesari et al. 2021; Ostad-Ali-Askari and Eslamian 2021).

At the same time, waste and sediments containing such salts can serve as sources of valuable ferrous and nonferrous metals, if a method for their processing is known. Moreover, prices for copper, nickel, tin, and other metals have been rising steadily over the past few years. Waste and sediments, in particular, galvanic waste and galvanic sludge contain up to 30% of heavy metals and are disposed at best in landfills and often on the terrain or in sewers. Thus, waste can be divided into two main types—liquid and solid. Liquid is wastewater contaminated with all components process solutions, and solid waste is the actual sediments formed in various technological processes (Nikolaos 2018).

Waste neutralization is a technological operation or a combination of them, as a result of which a primary toxic substance or group substances are converted into neutral, nontoxic, and nondegradable compounds (Al-Harahsheh et al. 2017).

In simple words, recycling or waste neutralization is using waste for good. This process is a set of technological operations, as a result of which one or more types of products are produced from waste or they are used to generate heat, energy, etc.

According to the degree of impact on the environment and people, waste is divided into five hazard classes (Fig. 10.1), so most galvanic sludge has a hazard class of III—moderately dangerous (Burger et al. 1997; Stamford and Azapagic 2019).

To reduce the environmental hazard of waste from galvanic production, in particular galvanic sludge, methods of chemical fixation are most often used, produced by ferritization of the solid phase of waste, silicatization, solidification of waste

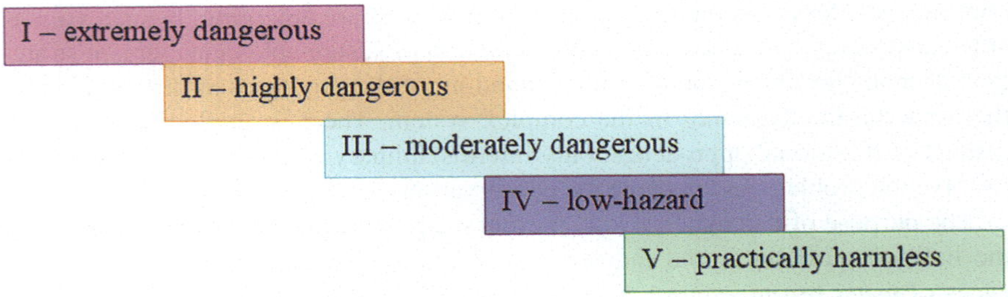

Fig. 10.1 Hazard classes of waste according to the degree of impact on the environment and people

using inorganic and organic binders, and sintering (Kochetov et al. 2020; Kesari et al. 2021; Mymrin et al. 2023).

The chemical fixation method is generally used for the disposal of chromium-containing galvanic sludge, which, after drying to hygroscopic humidity (if there is a significant content of organic components in them—after calcination until ashing) can be used as a dye for production, for example, decorative glass, depending on the chemical composition, galvanic sludge treated in this way produces bright blue glass, blue-green, and dark brown to black. It has also been established that it is possible to add galvanic deposits to the asphalt mixture in amounts of up to 20% without harming the environment. Sludge enriched with iron when wastewater is treated with a reagent obtained from waste electrolytic method is used to obtain highly valuable complex ferrites, which are used in the electrical industry, radioengineering, and chemical industry as catalysts (Lieberman et al. 2018).

Today, completely eliminating environmental pollution is the method of firing galvanic sludge in the presence of silicates. At temperatures above 1100 °C, more than 95% of hexavalent chromium is converted into trivalent chromium. Fusion of galvanic sludge with silicates in a 1:1 ratio at 800–1000 °C makes it possible to obtain sparingly soluble compounds of heavy metals They can be used without further processing for various economic needs, for example, as a facing material in construction. When introducing heavy metals into a ceramic mixture not only their reliable neutralization and burial occur but also many properties of ceramics are improved, such as mechanical strength, brightness, etc. (Brozova et al. 2013).

The most appropriate methods for recycling galvanic sludge are those that allow the extraction of valuable components—method of extraction separation of valuable components from heat-treated galvanic sludge. Leaching of nonferrous metals was carried out using sulfuric acid, and the sulfuric acid solutions obtained after adjusting the acidity and filtration contain mainly copper, cobalt, chromium, cadmium, and alkali metals. Selective separation of such solutions is very difficult, and for this purpose, a method of selective extraction separation of cobalt with salts of quaternary ammonium bases has been developed. In addition, one of the promising recycling products is calcium ferrite, synthesized on the basis of sediment-sludge after electrocoagulation and reagent cleaning methods and has high anticorrosion efficiency compared to analogues. Sediment-sludge in combination with spent etching solutions can be used as raw materials for the production of magnetic fluids. Magnetic fluids can be used to extract oil and oils from wastewater, as well as to clean up oil spills in the seas (Vilarinho et al. 2017).

Until recently, galvanic sludge was clearly classified as a nonrecyclable type of industrial waste, and the only way to neutralization was burial at specialized landfills. Moreover, due to the high toxicity of galvanic sludge, especially stringent requirements are imposed on the construction of landfills: the construction of reliable chemical protection, waterproofing, and mandatory concreting. Since the construction of such landfills requires vast areas without a visible prospect of their subsequent release in favor of agriculture and significant investment, it is obvious that on the Eurasian and Asian continents such landfills in the required quantities were not available, and in the near foreseeable future, it is unlikely to happen.

Therefore, a rather interesting way of recycling galvanic sludge is to add it as one of the components to the charge in the manufacture of pigments, dyes, glazed ceramic tiles, since during the processing heavy metals are converted into slightly soluble compounds. However, not all types of sludge are suitable for processing using this technology and obviously, the method cannot solve the problem in the context of constantly generating new volumes of galvanic sludge in developing countries.

The problem of recycling chromate production sludge, despite its relevance, has still not found a satisfactory solution. The proposed methods of processing into building materials or raw materials for agricultural practices cannot be implemented in industrial scale, mainly due to the presence of toxic chromium (VI) compounds. In addition, they are ineffective economically (Al-Harahsheh et al. 2017).

Along with chromium, solid waste contains large amounts of magnesium. Only complex extraction of chromium and magnesium can significantly increase economic efficiency of disposal and will allow the return of treated wastewater to the practical chain of the agricultural sector (Baskar et al. 2022).

At the same time, it is worth noting that exclusively the burial of solid and liquid waste from galvanic production in building materials, according to a number of scientists is unacceptable, since building materials (brick, tiles, etc.) over time collapse and become a secondary source of pollution environment. Their use as mineral dyes and microfertilizers in the agricultural sector, first proposed about 15 years ago, is extremely difficult due to the variable composition of precipitation, although the content of nitrogen (1.5–4.6%) and phosphorus (0.53–1.2%) is high in them. But the application in metallurgy is hampered by the high content of zinc, copper, and cadmium (Kesari et al. 2021).

10.3 Water Conservation and Management in Semiarid and Arid Lands for Sustainable Agriculture

The development strategy for sustainable agriculture has become more relevant than ever over the past 50 years. The modern agricultural sector provides fiber for the textile industry, food and feed for livestock, and much more. Due to the growing population of the planet, the demand for raw materials is also increasing, and the constantly emerging problems of global climate change even lead to the threat of a shortage of resources, primarily in the arid regions of the planet (Brown 1997; Singh et al. 2023).

Sustainable farming technologies promote the stable and continuous production of crops, especially in water-scarce environments, to ensure sufficient food and raw materials in the future. According to the Food and Agriculture Organization of the United Nations (FAO) (Kingra and Misra 2021), sustainable agriculture practices are based on the following **five principles**:

1. food chain optimization,
2. protection and saving of natural resources,
3. improving the welfare and economic situation of the population,
4. promoting the sustainability of ecosystems and natural communities,
5. support of government initiatives and regulations.

Successful agriculture largely depends on efficient irrigation, which requires huge amounts of water and energy resources. Sustainable development aims to optimize water and energy consumption to meet the moisture needs of plants.

In sustainable agriculture, optimizing the use of water and energy resources is achieved primarily by growing types of crops that use less water, as well as using smart irrigation methods. In particular, the drip irrigation method, compared to furrow irrigation, requires 20–40% less water and at the same time allows increasing crop yields by 20–50%.

Climate change and opportunities for sustainable agriculture will clearly affect the water resources of semiarid and arid regions. The danger comes from the assumption that the temperature change is only a few degrees, and corresponding changes in the structure of precipitation distribution can lead to an increase in the average long-term river flow and water availability by 10–40% in some regions and their reduction by 10–30% in others. This "leverage" effect will cause even small changes in temperature to significantly affect river flow conditions, which will certainly affect the water supply of growing urban populations, infrastructure, industry and much of agriculture, which will be especially vulnerable if their water supply is unstable. In addition, the global climate is projected to be more prone to floods and droughts (Kingra and Misra 2021).

However, attempts to solve the problem of conserving water resources not only failed but also led to catastrophic consequences. Thus, the desire to increase the production of biofuels has already led to water stress and hunger in many regions. Also engineering approaches to flood protection can protect the population from "normal" floods, but at the same time can make it very vulnerable to catastrophic destruction of infrastructure, as happened in Nepal and India in 2008. But here are approaches to mitigate water shortages, such as energy-intensive technologies desalination could, in turn, exacerbate the impacts of climate change if they are used on a large scale.

It is therefore logical that water conservation and management will become a key aspect of the risks and responses of climate change adaptation and the adoption of sustainable agriculture technologies in dry climate countries.

Addressing the problem of increasing water scarcity can begin with making a number of fundamental changes in the way societies manage and develop their water resources. Unlike energy, water is difficult to transport over long distances, and its use patterns are largely tied to specific terrain and conditions that vary significantly from country to country, as well as within countries. It is obvious that different water sources are often connected to each other through the water cycle. Planting forests on slopes can deplete groundwater supplies in valleys; excessive pumping of groundwater in a particular area can lead to drying up of watercourses in its surroundings; using rivers to generate hydroelectric power can negatively impact fish populations and the incomes of fishermen downstream.

Therefore, water resources need to be properly managed and reused, taking into account the variability of the water cycle, existing uncertainties, scarcity, or abundance, and the interrelationships between different users locally, regionally, and globally.

Achieving water security requires technical and institutional innovation and can create opportunities to improve service delivery as well as initiate new economic activities. For decades now, Singapore has been actively implementing an innovative policy of integrated water resources management, collection and nonstandard use of water resources, which has ensured almost complete water security in the region. Even though the country is located in a tropical rainfall area, due to small catchment area, agriculture has limited water resources and traditionally depends on supplies from neighboring countries. And so, Singapore has provided a powerful and sustainable water supply system based on four different sources known as "four national taps" (local river basin water, imported water, purified water, known as "NEWater" and desalinated water). By introducing high-tech changes to the system and thereby achieving maximum efficiency in the use of each of the four sources, this will ensure a sustainable water supply that can support further economic growth. Singapore has achieved sustained reductions in vulnerability levels in its agricultural sector to climate change because the country is able to take advantage of even unusual rainfall events by storing rainwater in artificial coastal lagoons, which reduces the volume of water that needs to be desalinated, thereby minimizing CO_2 emissions (Guven et al. 2023).

It is obvious that the technical solutions adopted in Singapore can be implemented in other countries, either in whole or in part, but this technology alone cannot solve global water problems, especially in arid zones. Lack of water or incorrect distribution of points for returning water resources to the practice of sustainable agriculture creates a number of specific problems and difficulties in arid climates.

An objective example of a successful combination of infrastructural and institutional measures for water resources management can be identified in the Central Asian region, which is located in a semiarid and arid zone, where natural water deficit has existed for many thousands of years. Sustainable farming technologies cover about 85% of the region's water resources used for irrigation. This technology forms an assessment of water consumption in the irrigation sector specifically in conditions of drought or new air temperature conditions, and also develops adaptation measures based on changing the types of crops grown and the introduction of water and energy-saving technologies at the on-farm level (Guven et al. 2023).

In the countries of the Central Asian region, investment projects have already been introduced within the framework of the Kyoto Protocol on mechanisms of environmentally friendly development, which involve a combination of several aspects aimed at optimizing the agricultural sector and at the same time the competent use of water resources (Shan et al. 2020).

From an economic perspective, it is important to note the reduction in energy costs and raw materials in the production of a unit of output, as well as increasing the efficiency of production or use of natural resources through the introduction of advanced technologies in the private sector. The broadest environmental block provides active promotion of environmental protection and prevention of its degradation by minimizing the consumption of natural resources and production waste. The widespread introduction of technologies aimed at reusing resources and using renewable natural resources to reduce negative impacts on the environment has become absolutely obvious. Of course, the reuse of water in arid regions is inextricably linked with technologies for removing pollutants, heavy metals, as well as water desalination, which requires new effective technological approaches and investments. An incredibly important step in the adoption of these criteria for environmentally friendly development was the increase in the level of public understanding of the problems of management and conservation of natural resources in the context of sustainable agricultural practices.

However, the practices of water management technologies under the Kyoto Protocol are just beginning to be introduced into agricultural practice in some countries and also have a long-term perspective; today, it is important to use such a mechanism as "hard" options for water abstraction and water resource management. These options range from large dams to household rainwater harvesting structures. At the same time, important hydraulic structures are canals, tunnels, pipelines, which serve not only to directly satisfy human needs but also, less obviously, create interconnected systems that, due to using of different water sources, are less affected by climate variability, provide greater flexibility and a higher level of security of water supply.

Climate change and water scarcity will impact not only the volumes of water supplies that meet the planet's current requirements but also the appropriate types of water storage infrastructure (natural, artificial, small or large reservoirs). It is quite promising to organize and preserve natural systems of water reserves, such as aquifers (naturally or artificially recharged), wetlands, and lakes, as well as artificial water storage of all types and scales, including rainwater collection in households, traditional ponds in villages, small dams and dams, and large-scale reservoirs (Lacombe et al. 2019).

Another aspect is the possibility of transferring runoff between different watersheds. In a situation of increasing variability and increasing pressure on water resources, the ability to transfer water resources from one river basin to another may prove promising, especially where water demand urban conglomerates and key economic sectors are growing rapidly. For example, China has already transferred large volumes of water from the Yangtze River basin to the Yellow River basin; India, Australia, and Spain are actively discussing the effectiveness of such a scheme, but so far they cannot offer anything else. South Africa, Mexico, and the USA are already actively using such transfers, which have shown an increase in the economic efficiency of water use by up to 20% (Gulamussen et al. 2019; Goyal and Kumar 2021).

There is already a danger that efforts to manage water resources in semi-arid and arid regions will be sidelined as planning for adaptation rather than environmental management is more important. Therefore, it is very important that the steps for introducing technologies for the efficient use of natural resources are that, firstly, concentrated efforts may be required to develop and initiate adaptation strategies; and second, they should be integrated, and not carried out in parallel, with ongoing water management work.

Both of these areas must be effectively integrated into national development strategies, and the parallelism of these processes with the dominance of one strategy must be avoided in the context of sustainable agriculture.

10.4 Conclusion

Recycled water can and should be an important alternative to freshwater, especially in regions with significant water scarcity. This chapter aims to show ways to effectively expand this reuse of water resources as a promising solution to meet the consumer and agricultural water supply needs of the regions of the future. It is shown that expanding the spectrum of wastewater reuse can increase the percentage of supply due to the creation of alternative sources of water for industrial and domestic use and also reduces pressure on natural resources and supports a sustainable approach to providing water to the population.

However, as noted, it is the lack of funding for research into modern waste processing technologies that significantly affects the state of the water use problem. At the same time, the formation in a number of countries of public-private partnerships for the construction and operation of treatment facilities with minimal resources minimizes the stage of repeated water purification. Not considering low funding, technologies for reincorporating water into the agricultural sector are limited both by natural cataclysms, prolonged droughts, water and salt stress, and by environmental fluctuations, the increase in the population of individual states, natural diseases and pathogens, increased energy consumption, etc. Introducing the field of innovative wastewater management technologies will give states the opportunity to improve their wastewater disposal skills at minimal costs.

Therefore, the reuse of wastewater in households and agriculture can regulate and close the water cycle, thereby ensuring the effective reuse of existing water resources. Of course, this can solve the problem of water security in regions with arid and semi-arid climates, and in the future—throughout the planet.

References

Al-Harahsheh M, Hussain YA, Al-Zoubi H, Batiha M, Hammouri E (2017) Hybrid precipitation-nanofiltration treatment of effluent pond water from phosphoric acid industry. Desalination 406:88–97. https://doi.org/10.1016/j.desal.2016.06.014

Baskar AV, Bolan N, Hoang SA, Sooriyakumar P, Kumar M, Singh L, Vinu A, Sarkar B, Kirkham MB, Rinklebe J, Wang S, Wang H, Balasubramanian R, Siddique KH (2022) Recovery, regeneration and sustainable management of spent adsorbents from wastewater treatment streams: a review. Sci Total Environ 822:153555. https://doi.org/10.1016/j.scitotenv.2022.153555

Brown LR (1997) The agricultural link: how environmental deterioration could disrupt economic progress. World watch. World Watch Institute, Washington, DC, p 136

Brozova S, Ingaldi M, Sperlin I (2013) Economical aspects of high-temperature heating utilization for industrial waste treatment, 22nd edn. METAL, Ostrava

Burger J, Martin M, Cooper K, Gochfeld M (1997) Attitudes toward environmental hazards: where do toxic wastes fit? J Toxicol Environ Health 51(2):109–121. https://doi.org/10.1080/00984109708984015

Goyal K, Kumar A (2021) Development of water reuse: a global review with the focus on India. Water Sci Technol 84(10-11):3172–3190. https://doi.org/10.2166/wst.2021.359

Gulamussen NJ, Arsénio AM, Matsinhe NP, Rietveld LC (2019) Water reclamation for industrial use in sub-Saharan Africa—a critical review. Drink Water Eng Sci 12:45–58. https://doi.org/10.5194/dwes-12-45-2019

Guven H, Ersahin ME, Ozgun H, Ozturk I, Koyuncu I (2023) Energy and material refineries of future: wastewater treatment plants. J Environ Manag 329:117130. https://doi.org/10.1016/j.jenvman.2022.117130

Kesari KK, Soni R, Jamal QMS, Tripathi P, Lal JA, Jha NK, Siddique KH, Kumar P, Tripathi V, Ruokolainen J (2021) Wastewater treatment and reuse: a review of its applications and health implications. Water Air Soil Pollut 232:1–28. https://doi.org/10.1007/s11270-021-05154-8

Kingra PK, Misra AK (2021) Agricultural input use efficiency and climate change: ways to improve the environment and food security. In: Bhatt R, Meena RS, Hossain A (eds) Input use efficiency for food and environmental security. Springer, Singapore. https://doi.org/10.1007/978-981-16-5199-1_2

Kochetov G, Kovalchuk O, Samchenko D (2020) Development of technology of utilization of products of ferritization processing of galvanic waste in the composition of alkaline cements. Eastern-European J Enterp Technol 5(10-107):6–13. https://doi.org/10.15587/1729-4061.2020.215129

Lacombe G, Chinnasamy P, Nicol A (2019) Background paper 1. Climate risks and solutions: adaptation frameworks for water resources planning, development and Management in South Asia Background Paper. In: Review of climate change science, knowledge and impacts on water resources in South Asia. International Water Management Institute (IWMI), Colombo. https://doi.org/10.5337/2019.202

Lieberman RN, Knop Y, Palmerola NM, Goldman A, Querol X, Muñoz-Quirós C, Cohen H (2018) Fixation of treated phosphate waste and its use in concrete. J Clean Prod 178:89–97. https://doi.org/10.1016/j.jclepro.2017.12.272

Mymrin V, Ribas HE, Pedroso DE, Pedroso CL, Klitzke W, Avanci MA, Goncalves AJ, Rolim PH (2023) Physical–chemical processes of sustainable materials' production from hazardous toner waste, galvanic glass waste and spent foundry sand. J Mater Cycles Waste Manag 25(1):396–406. https://doi.org/10.1007/s10163-022-01557-9

Nikolaos V (2018) Water reuse from a circular economy perspective and potential risks from an unregulated approach. Curr Opin Environ Sci Health 2:32–45. https://doi.org/10.1016/j.coesh.2018.01.005

Ostad-Ali-Askari K, Eslamian S (2021) Water reuse in industry: necessities and possibilities. Am J Eng Appl Sci 14(1):94–102. https://doi.org/10.3844/ajeassp.2021.94.102

Shan V, Singh SK, Haritash AK (2020) Water crisis in the Asian countries: status and future trends. In: Kumar M, Munoz-Arriola F, Furumai H, Chaminda T (eds) Resilience, response, and risk in water systems . Springer transactions in civil and environmental engineering. Springer, Singapore. https://doi.org/10.1007/978-981-15-4668-6_10

Singh A, Rajput VD, Ghazaryan K, Gupta SK, Minkina T (2023) Nanopriming approach to sustainable agriculture. IGI Global, Mumbai. https://doi.org/10.4018/978-1-6684-7232-3

Stamford L, Azapagic A (2019) Environmental impacts of copper-indium-gallium-selenide (CIGS) photovoltaics and the elimination of cadmium through atomic layer deposition. Sci Total Environ 688:1092–1101. https://doi.org/10.1016/j.scitotenv.2019.06.343

Vilarinho C, Teixeira J, Araújo J, Carvalho J (2017) Effect of time and acid concentration on metal extraction from galvanic sludges. In: ASME international mechanical engineering congress and exposition, vol 58493. American Society of Mechanical Engineers, p 014T11A018. https://doi.org/10.1115/IMECE2017-71370

Chapter 11
Microbial Manganese Peroxidase: Ligninolytic Enzymes for Bioremediation

Abhishek Singh, Ragini Sharma, Vishnu D. Rajput, Karen Ghazaryan, Tatiana Minkina, Abdel Rahman Mohammad Said Al-Tawaha, Shreni Agrawal, Ashi Varshney, Abdel Razzaq Al-Tawaha, and Arun Karnwal

Abstract The industrial applications of ligninolytic enzymes of microbes such as lignin peroxidase, laccase, and manganese peroxidase have increased in popularity. In the microbial ligninolytic system, manganese peroxidases play a significant role. Oxalic acid and other fungal chelators help convert the Mn (II) ions from soil and wood into Mn^{3+} form which is more reactive. The phenolic lignin is broken down and free radicals are produced, but Mn^{3+}, a less molecular weight molecule, functions as a diffusible redox intermediary. The alcohol, biofuel, food, cosmetic, agricultural, and textile industries are just a few of the many that could benefit from this

A. Singh (✉)
Faculty of Biology, Yerevan State University, Yerevan, Armenia

R. Sharma · K. Ghazaryan
Panjab Agriculture University, Ludhiana, India

V. D. Rajput · T. Minkina
Academy of Biology and Biotechnology, Southern Federal University, Rostov-on-Don, Russia

A. R. Mohammad Said Al-Tawaha
Department of Biological Sciences, Al Hussein Bin Talal University, Ma'an, Jordan

S. Agrawal
Department of Biotechnology, Parul Institute of Applied Science, Parul University, Vadodara, Gujarat, India

A. Varshney
Department of Genetics and Plant Breeding, Banaras Hindu University, Varanasi, India

A. R. Al-Tawaha
Department of Crop Science, Faculty of Agriculture, Universiti Putra Malaysia, Serdang, Selangor, Malaysia

National Agricultural Research Center, Baqa, Jordan

A. Karnwal
Department of Microbiology, Lovely Professional University, Phagwara, Punjab, India

A. Karnwal, A. R. Mohammad Said Al-Tawaha (eds.), *Microbial Applications for Environmental Sustainability*, https://doi.org/10.1007/978-981-97-0675-1_11

technology. The mechanisms for catalytic reactions, biotechnological technological applications, and sources are the primary focuses of this review. Manganese peroxidases, however, can degrade a wide variety of xenobiotic compounds and generate polymeric products that can be used as effective bioremediation instruments. Microorganisms called MnPs can break down lignin into sugars that can then be used to produce biofuels. This chapter focuses mainly on the contemporary state and recent advancements pertaining to the manganese peroxidase enzyme.

Keywords Ligninolytic enzymes · Microbial enzymes · Lignin peroxidase · Laccase · Manganese peroxidase · Fungal chelators · Mn (II) ions · Mn^{3+} form · Phenolic lignin

11.1 Introduction

You can find manganese peroxidase, an enzyme that breaks down lignin (E.C.1.11.1.13, Mn^{2+}:H_2O_2 oxidoreductase), almost everywhere in nature (Asgher et al. 2014). This enzyme's increasing popularity over the past few years can be attributed to its usefulness in a wide range of biotechnological contexts (Asgher et al. 2014; Qin et al. 2017). Because of its adaptability and the wide variety of uses it finds in industry, MnP is the enzyme with the most potential (Kersten and Cullen 2007). The MnP enzyme speeds up the change from Mn (II) to Mn (III). This process also breaks down phenolic and non-phenolic compounds (Cheng et al. 2006; Muhammad Nasir Iqbal and Asgher 2013).

Fungi form and MnPs are produced, a widely dispersed extracellular potential peroxidase (white-rot fungi). The same holds for the enzymes, laccases and LiPs, which are believed to be crucial in the depolymerization of lignin (Morgenstern et al. 2008; Hammel and Cullen 2008; Hofrichter et al. 2010; Floudas et al. 2012). The fungus P. chrysosporium was first thought to produce these heme-containing glycoproteins around 20 years ago (Qin et al. 2017). In contrast to LiP, laccase, and other peroxidases of *P. chrysosporium*, it initially received less attention (Tuomela et al. 2000).

The white-rot fungus *P. chrysosporium*, one of the best-studied organisms of its kind, produces large amounts of reactive, nonspecific free radicals via the secretion of a number of peroxidases and oxidases (Kersten and Cullen 2007; Zhang et al. 2007). There are no known microorganisms that produce enough ligninolytic enzymes for use in biotechnological processes. The BLAST (basic local alignment search tool) was used to find MnP isoenzymes. Three genes with structural makeup were found: mnp1, mnp2, and mnp3. Two more genes from P. chrysosporium were also recently isolated and added to the list. White-rot fungi typically secrete a variety of different MnP types.

MnP enzymes are various glycoproteins having molecular weights between 38 and 62.5 kDa, 350 residues of amino acids, and 43% sequence identity to LiP (Hofrichter 2002). Their formation can be attributed to the presence of two domains containing a central heminic group, one minor helix, ten major helixes, and five disulfide bridges. One of them contributes to the manganese binding site. MnP is distinguished from other reported peroxidases at this site (Sundaramoorthy et al. 1994).

Increased MnP production is crucial for lowering the price of energy and chemicals that are utilized to produce biofuel while also helping to create environmentally friendly alternatives for biomass degradation. Bio-bleaching, pollutant bioremediation/biodegradation, and bio-pulping are just a few of the industrial processes that can benefit from MnP (Asgher et al. 2013; Asgher and Iqbal 2013; Nousiainen et al. 2014).

11.2 An Overview of Enzymes

Enzymes are biological catalysts that reduce the activation energy of a reaction, making it easier for a substrate to be converted into a product. At least one polypeptide moiety is present in enzymes, which can be either proteins or glycoproteins. The areas of an enzyme that catalyze the reaction are known as the active sites. If the nonprotein part of an enzyme is needed for catalytic activity and is linked to the active sites by covalent or non-covalent bonds, it is called the prosthetic group. The protein, or glycoprotein, portion of an enzyme is referred to as the apoenzyme. The apoenzyme is linked to the prosthetic group to produce the holoenzyme.

11.3 Nomenclature of Enzymes

The term “enzyme” refers to a single catalytic entity as opposed to a series of individually catalyzed reactions. Enzyme names are according to the chemical reactions they catalyze.

11.4 Origin of MnP

MnPs have been observed by numerous researchers to be ubiquitous among bacteria, fungi, and algae (Bharagava et al. 2017; Zhang et al. 2018). To degrade lignin, fungi must produce and secrete the extracellular enzymes LiP, MnP, and laccase. *P. sordida, P. chrysosporium, P. abeled, P. rivulosus, C. elegans, calcitriol subvermispora,* as well as *Dichomeris squalens* are all generated by extracellular MnP (Hakala et al. 2006). Several medium, environmental, and nutritional growth conditions were found to result in MnP production by P. chrysosporium BKM-1767 (Urek and Pazarlioglu 2007). The search for MnP with increased tolerance to organic solvent and inorganic ions is, however, urgently needed to improve the degradation potential of xenobiotics. Substrate/mediators including veratryl alcohol, oxalic acid, and 2, 6-dimethoxyphenol catalyze MnP formation in Irpex lacteus CD2. Furthermore, the initial suspect strain was labeled as *A. baumannii* using a phylogenetic tree constructed from sequences of 16 rRNA genes (Qin et al. 2014). These results have led to the identification of several crucial microbial strains that can delignify and biodegrade resistant compounds using ligninolytic enzymes for MnP synthesis from Fusarium sp., rice straw or wood chips being a cheaper substrate of agricultural residue can be employed as the main carbon source (Huy et al. 2017). In addition, MnP was isolated from the white-rot fungus C and found to have a molecular weight of 45 kDa. MnP-BBP6 is a BBP6 strain optimized for growth at T and pH values of 7.0 and 7.0, respectively (Zhang et al. 2018). The deterioration of rice and lignin has attracted a lot of attention recently, particularly in Egypt and around the world because of the buildup of wastes from agriculture like rice straw that contribute to numerous environmental issues on a global scale. It causes mineralization of man-made lignin and gets produced extracellularly by praecox. A. niger's MnP1 isozyme has been purified. Praecox was also isolated, cloned, and studied in detail. This recently discovered MnP1 could be a distinct subset of the ubiquitous short MnP enzyme (Huy et al. 2017). Bjerkandera adusta, F. is the most common species responsible for MnP production. Flavus, C. cerrena, T. tramaeus, and F. fomentarius. It was previously discovered that nutrient limitation, specifically that of C, N, and Mn^{2+}, increases MnP production in P. chrysosporium. In contrast, mnp3 has no MREs, and it has been found that the addition of Mn^{2+} does not affect transcript levels (Pease and Tien 1992; Gettemy et al. 1998). During lignin biodegradation, the MnP ligninolytic enzymes and the laccase both are produced by Cupriavidus basilensis (Huy et al. 2017). Bio-pretreatment of lignocellulose and xenobiotic biodegradation are two promising areas where Irpex lacteus could be used. I. lacteus was shown to primarily produce MnP through laboratory testing (Salvachúa et al. 2013).

11.5 The MnPs Crystal Structure

The catalytic mechanism of an enzyme can be better understood, and more in-depth designs and transformations may be possible, all of which are important steps toward realizing enzymes' green biotechnological use (Engel et al. 2003; Chang et al. 2019; Qiu et al. 2020). MnP's heme conformation is evolutionarily conserved and is the same as that of lignin peroxidase (LiP) (Ayuso-Fernández et al. 2017). MnP in its resting state contains a Fe^{3+} penta-coordinated structure with the heme cofactor's porphyrin ring and a proximal histidine (Chang et al. 2021). The sixth coordination position is open for H_2O_2 to bind. Up to this point, various MnP's crystal structures have been reported in a number of different ways, with the highest-resolution crystal structure (0.93) depicted for MnP complexed with Mn^{2+} (Mn-MnP) (Sundaramoorthy et al. 2010). The stability of the protein is influenced by the presence of conserved Ca^{2+} ions, seen as golden spheres, and the placement of the Mn^{2+}

substrate represented in violet (George et al. 1999). The active area consists of three highly conserved amino acids (Glu39, Glu35, and Asp179) and a heme protein. The heme propionate (HEM) is located inside the hydrophobic core of the enzyme, whereas the active site is responsible for binding the Mn^{2+} substrate. The spatial arrangement of HEM is also reinforced by the presence of eight hydrogen bonds (shown by green dashed lines), two electrostatic interactions (indicated by orange dashed lines), and many other weaker interactions. Heme peroxidases have a high degree of conservation across many species, with a major alteration seen in the catalytic region of some enzymes, such as ascorbate peroxidase and cytochrome c, where phenylalanine is replaced by tryptophan (Chang et al. 2021). Furthermore, the maintenance of the Asp-His pair at positions 242 and 173 has been reported by (Chang et al. 2021).

11.6 Catalysis of MnP

At the beginning of the catalytic cycle, H_2O_2 or organic peroxide binds to the enzyme in its dormant ferric (Fe^{3+}) state. One molecule of water is produced and two oxidation equivalents of MnP-compound I (Fe^{4+}-oxo-porphyrin radical complex) are formed. This results in the synthesis of MnP-II (Fe^{4+} oxo-porphyrin radical complex) via the oxidation of Mn^{2+} to Mn^{3+}. The catalytic cycle is then finished when MnP-compound II undergoes a similar combination with Mn^{2+} to produce Mn^{3+}, generating one H_2O molecule, and is subsequently reduced to ferric MnP (Hofrichter 2002). The MnP catalytic cycle resembles other lignin and heme peroxidases when native Fe^{3+} enzymes and two reactive intermediates are present (Huy et al. 2017). MnP, in contrast to other peroxidases, utilizes Mn^{2+} as its substrate of choice, transforming it into a highly oxidized form of Mn^{3+} via multiple redox reactions (Wariishi et al. 1991).

11.7 Application of MnP in Bioremediation

11.7.1 Breakdown of Phenolic and Non-phenolic Substances

Various industrial wastewaters contain non-phenolic and phenolic substances including wastewaters from the following processes: petroleum refining, conversion of coal, paper and pulp bleaching, wood preservation, mining, dressing, textile, metal coal, dyes, and other chemical industries. Previous studies have mostly focused on the elimination of phenolic compounds in the presence of hydrogen peroxide (Mohan et al. 2005; Kim et al. 2005). The MnP enzyme is capable of producing a Mn (III) chelator that exhibits the ability to induce the oxidation of several phenolic pollutants, including amines, dyes, phenols, lignin, and dimers. According to Champagne and Ramsay (2005), it has been shown by researchers that Mn(III), which functions as a chelator, has moderate oxidizing properties and is specifically limited to the oxidation of phenolic compounds that include lignin under physiological circumstances. The Mn (III) species is dissociated from the enzyme at an elevated redox potential range of 0.8–0.9 V and thereafter stabilized by the formation of complexes with hydroxy acids. Fungi have a significant role in the production of two very effective chelating agents, namely malonate and oxalate, as shown by Kishi et al. (1994). According to Reddy et al. (2003), the introduction of a second mediator during Mn(III) oxidation results in the production of reactive radicals inside a non-phenolic substrate. A radical cation is formed via e-reduction of an aromatic ring in a LiP-catalyzed reaction. Substituted benzyl alcohols and diaryl propane structures are oxidized by Mn(III) into their respective aldehydes in the presence of thiols (R-S-H) such as glutathione (Morgenstern et al. 2008). MnP-Tra-48424, discovered by Zhang et al. and isolated from Trametes sp., decolorizes a wide variety of dyes, including triphenylmethane, azo dyes, indigo, and anthraquinone.

11.7.2 Pulp and Paper Industry

The pulp and paper industry is vital to the global economy, but it also has serious negative impacts on the natural environment and the health of humans. The pulp and paper industry consumes vast quantities of lignocellulosic materials during the papermaking process. Pulping, bleaching, and paper production are the three main stages in the production of paper, with chemical, mechanical, and combination (chemo-mechanical) pulping constituting the three types of pulping. Strong wastewater is produced by these procedures, which has a negative effect on the natural world. It has been reported that biological approaches to wastewater treatment are cost-effective, commercially viable, and environmentally safe, while chemical-based treatment methods pose risks to living organisms and are expensive and unsafe for a green environment (Iqbal et al. 2013). Diverse groups of microorganisms are playing an essential role in the proper remediation of contaminants and wastewater released. White-rot fungi are more efficient than other organisms at breaking down lignin and similar wood substances into carbon dioxide and water (Zainith et al. 2018). Many different value-added bio-products can be made by converting lignocellulosic materials or resources (Iqbal et al. 2013). Some of these materials are powerful degraders of lignin instead of hemicellulose and cellulose. Cellulosic fibers breaking and having slow reaction speeds that last for a long time are two issues that arise when using microbes directly for degrading lignocellulosic materials (Jiménez et al. 1997; Li 2003). Hatakka et al. discovered that in the presence of Mn, the chelation of MnP is aided by Tween 80, an H_2O_2 generating system, and an organic acid (II) (Hatakka et al. 2003). Conversely, Mn(II) is a potent oxidizing agent and a crucial part of the delignification process (Katagiri et al. 1995). Enzyme pulping utilizes MnP and laccases from Polyporus spp., as well as pectinases from Rhizopus spp., in the pulp and paper industry. Using ligninolytic enzymes decreases the need for sodium hydroxide (NaOH) and other chemicals in the basic pulping process (Poonpairoj et al. 2001). It has been claimed, however, that MnP and LiP are crucial in the decolorizing of kraft pulp (Moreira et al. 2003; Raj et al. 2014).

11.7.3 Application of MnPs in the Decolorization of Dye Wastewater

The concentration of dye in the beginning, the source of the enzyme, test conditions, and constituents of the medium are all factors in the decolorization rate. *Trametes pubescens* strain i8, the basidiomycete WRF fungus, was recently isolated, and its purified MnP (named MnP TP55) (Rekik et al. 2019). With the dye concentration of 50 g/mL in the starting, the MnP TP55 degraded a wide variety of dyes in 24 h including indigo dye (Indigo Carmine), polymeric dye, triphenylmethane dye (methyl green), azo dyes (Direct Red 5B and Remazol Brilliant Violet 5R), and anthraquinone dyes like Cibacet Brilliant Blue BG and Remazol Brilliant Blue Reactif. Moreover, the decolorization rate and decolorization duration are strongly correlated, i.e., the higher the decolorization rate, the longer the decolorization time. Similar studies showed that Remazol Brilliant Violet 5R, Acid Blue 158, Cibacet Brilliant Blue BG, and Poly R-478 could all be degraded by the isolated extracellular MnP BA30 from Bjerkandera adusta strain CX-9 at rates of 70%, 91%, 77%, and 80%, respectively. This means that a high decolorization rate is not always indicative of a high decolorization efficiency (Bouacem et al. 2018). Researchers tested the decolorizing potential of fungal enzymes like MnP and Lip using a variety of dyes like Congo Red, Methylene Blue, and Remazol Brilliant Blue R (RBBR). Incubation of RBBR for 7 days resulted in a decolorization rate of 17.96–44.89%, and for 15 days, the rate increased to 55.98–99.99%. In addition to decolorizing dyes, MnP can deteriorate other contaminants like polycyclic aromatic hydrocarbons (PAHs). An MnP capable of efficiently degrading various dyes was purified by Zhang et al. from the WRF Trametes sp.48424. At 18 h from dye concentration of 100 mg/L at the beginning, Remazol Brilliant Violet 5R, RBBR, methyl green, and Indigo Carmine all showed removal efficacies of 88.4, 85.0, 93.1, and 94.6%, respectively (Huy et al. 2017). It was also demonstrated that the enzyme has the capacity to break down PAHs in challenging conditions. In the aforementioned studies, pure cultures were employed for the majority of the MnP-assisted decolorization, which is useful for examining processes but typically produces inferior results compared to those of consortia of microbes. SR is a consortium of fungi that consists of Trametes species. Similar to SQ01, Chaetomium sp., it was found that R01 was more effective than using individual fungi in the decolorization of various triphenylmethane dyes (Yang et al. 2011). The consortium increased MnP production by 5.5 times compared to the Trametes monoculture, yielding 1.3 U/mL. After a dye concentration of 100 mg/L in the beginning, consortium decolorization efficiencies for Cresol Red and Coomassie Brilliant Blue G250 were 93% and 96%, respectively, within 6 days, much greater than those attained by Trametes alone (60% and 72%, respectively). Using two separate microbial consortia, ZSY (formed largely of Piscibacillus, Bacillus, and Unclassified Bacillaceae), and ZW1 (mainly formed of Clostridiisalibacter, Marinobacter, and Halomonas), more than 93% of the color of 100 mg/L Metanil Yellow G was removed within 48 h. Both groups also performed well in high-pH and high-salinity conditions, suggesting they could be applied to the treatment of alkaline textile wastewater.

11.8 Conclusion

There is still a need for extensive study of microbial MnP due to their wide variety of potential industrial and biotechnological uses. Almost all bacteria and fungi are known to possess this enzyme. Bacterial MnP is crucial in the bioremediation of industrial wastes due to its ability to oxidize a wide variety of xenobiotics compounds and nontoxic substrates. It is also widely employed in the food industry, the design of biosensors, the paper and pulp industry, the textile and distillery sectors, the creation of diagnostic kits, and environmental protection. Using small molecules known as mediators, MnP has been shown to catalyze the oxidation of a wide variety of phenolic and non-phenolic compounds.

Acknowledgements VDR and TM acknowledge the financial support by the BRICS project of the Ministry of Science and Higher Education of the Russian Federation, no. 075-15-2022-1247.

References

Asgher M, Iqbal HMN (2013) Enhanced catalytic features of sol–gel immobilized MnP isolated from solid state culture of Pleurotus ostreatus IBL-02. Chin Chem Lett 24:344–346. https://doi.org/10.1016/J.CCLET.2013.02.019

Asgher M, Ahmad Z, Iqbal HMN (2013) Alkali and enzymatic delignification of sugarcane bagasse to expose cellulose polymers for saccharification and bio-ethanol production. Ind Crop Prod 44:488–495. https://doi.org/10.1016/J.INDCROP.2012.10.005

Asgher M, Bashir F, Iqbal HMN (2014) A comprehensive ligninolytic pre-treatment approach from lignocellulose green biotechnology to produce bio-ethanol. Chem Eng Res Des 92:1571–1578. https://doi.org/10.1016/J.CHERD.2013.09.003

Ayuso-Fernández I, Martínez AT, Ruiz-Dueñas FJ (2017) Experimental recreation of the evolution of lignin-degrading enzymes from the Jurassic to date. Biotechnol Biofuels 10:1–13. https://doi.org/10.1186/S13068-017-0744-X

Bharagava RN, Chowdhary P, Yadav A, Kaithwas G (2017) Distillery wastewater: a major source of environmental pollution and its biological treatment for environmental safety. Green Technol Environ Sustain 409–435. https://doi.org/10.1007/978-3-319-50654-8_18/COVER

Bouacem K, Rekik H, Jaouadi NZ, Zenati B, Kourdali S, Elhattab M, Badis A, Annane R, Bejar S, Hacene H, Bouanane,-Darenfed, A. (2018) Purification and characterization of two novel peroxidases from the dye-decolorizing fungus Bjerkandera adusta strain CX-9. Int J Biol Macromol 106:636–646. https://doi.org/10.1016/J.IJBIOMAC.2017.08.061

Champagne PP, Ramsay JA (2005) Contribution of manganese peroxidase and laccase to dye decoloration by Trametes versicolor. Appl Microbiol Biotechnol 69:276–285. https://doi.org/10.1007/S00253-005-1964-8/METRICS

Chang M, Zhou Y, Wang H et al (2019) Crystal structure of the multifunctional SAM-dependent enzyme LepI provides insights into its catalytic mechanism. Biochem Biophys Res Commun 515:255–260. https://doi.org/10.1016/J.BBRC.2019.05.031

Chang Y, Yang D, Li R et al (2021) Textile dye biodecolorization by manganese peroxidase: a review. Molecules 26:4403. https://doi.org/10.3390/MOLECULES26154403

Cheng J, Yu SM, Zuo P (2006) Horseradish peroxidase immobilized on aluminum-pillared interlayered clay for the catalytic oxidation of phenolic wastewater. Water Res 40(2):283–290. https://doi.org/10.1016/J.WATRES.2005.11.017

Engel M, Hoffmann T, Wagner L et al (2003) The crystal structure of dipeptidyl peptidase IV (CD26) reveals its functional regulation and enzymatic mechanism. Proc Natl Acad Sci USA 100:5063–5068. https://doi.org/10.1073/PNAS.0230620100/SUPPL_FILE/0620FIG6.JPG

Floudas D, Binder M, Riley R et al (2012) The paleozoic origin of enzymatic lignin decomposition reconstructed from 31 fungal genomes. Science (80-) 336:1715–1719. https://doi.org/10.1126/SCIENCE.1221748/SUPPL_FILE/FLOUDAS.SM.PDF

George SJ, Kvaratskhelia M, Dilworth MJ, Thorneley RN (1999) Reversible alkaline inactivation of lignin peroxidase involves the release of both the distal and proximal site calcium ions and bishistidine co-ordination of the haem. Biochem J 344(1):237–244. https://doi.org/10.1042/BJ3440237

Gettemy JM, Ma B, Alic M, Gold MH (1998) Reverse transcription-PCR analysis of the regulation of the manganese peroxidase gene family. Appl Environ Microbiol 64:569–574. https://doi.org/10.1128/AEM.64.2.569-574.1998/ASSET/B316CB9A-CF8A-4134-A9BF-BE450302C2CE/ASSETS/GRAPHIC/AM0281030005.JPEG

Hakala TK, Hildén K, Maijala P et al (2006) Differential regulation of manganese peroxidases and characterization of two variable MnP encoding genes in the white-rot fungus Physisporinus rivulosus. Appl Microbiol Biotechnol 734(73):839–849. https://doi.org/10.1007/S00253-006-0541-0

Hammel KE, Cullen D (2008) Role of fungal peroxidases in biological ligninolysis. Curr Opin Plant Biol 11:349–355. https://doi.org/10.1016/J.PBI.2008.02.003

Hatakka A, Lundell T, Hofrichter M, Maijala P (2003) Manganese peroxidase and its role in the degradation of wood lignin. pp 230–243. https://doi.org/10.1021/BK-2003-0855.CH014

Hofrichter M (2002) Review: lignin conversion by manganese peroxidase (MnP). Enzym Microb Technol 30:454–466. https://doi.org/10.1016/S0141-0229(01)00528-2

Hofrichter M, Ullrich R, Pecyna MJ et al (2010) New and classic families of secreted fungal heme peroxidases. Appl Microbiol Biotechnol 87:871–897. https://doi.org/10.1007/S00253-010-2633-0/METRICS

Huy ND, Tien NTT, Huyen LT, Quang HT, Tung TQ, Luong NN, Park SM (2017) Screening and production of manganese peroxidase from Fusarium sp. on residue materials. Mycobiology 45(1):52–56. https://doi.org/10.5941/myco.2017.45.1.52

Iqbal HMN, Kyazze G, Keshavarz T (2013) Advances in the valorization of lignocellulosic materials by biotechnology: an overview. Bioresources 8:3157–3176. https://doi.org/10.15376/BIORES.8 2.3157-3176

Jiménez L, Martínez C, Pérez I, López F (1997) Biobleaching procedures for pulp from agricultural residues using Phanerochaete chrysosporium and enzymes. Process Biochem 32:297–304. https://doi.org/10.1016/S0032-9592(96)00090-8

Katagiri N, Tsutsumi Y, Nishida T (1995) Correlation of brightening with cumulative enzyme activity related to lignin biodegradation during biobleaching of kraft pulp by white rot fungi in the solid-state fermentation system. Appl Environ Microbiol 61:617–622. https://doi.org/10.1128/AEM.61.2.617-622.1995

Kersten P, Cullen D (2007) Extracellular oxidative systems of the lignin-degrading Basidiomycete Phanerochaete chrysosporium. Fungal Genet Biol 44:77–87. https://doi.org/10.1016/J.FGB.2006.07.007

Kim GY, Lee KB, Cho SH et al (2005) Electroenzymatic degradation of azo dye using an immobilized peroxidase enzyme. J Hazard Mater 126:183–188. https://doi.org/10.1016/J.JHAZMAT.2005.06.023

Kishi K, Wariiship H, Gold MH et al (1994) Mechanism of manganese peroxidase compound II reduction. Effect of organic acid chelators and pH. Biochemistry 33:8694–8701. https://doi.org/10.1021/BI00195A010/ASSET/BI00195A010.FP.PNG_V03

Li K (2003) The role of enzymes and mediators in white-rot fungal degradation of lignocellulose. ACS Symp Ser 845:196–209. https://doi.org/10.1021/BK-2003-0845.CH011

Mohan SV, Prasad KK, Rao NC, Sarma PN (2005) Acid azo dye degradation by free and immobilized horseradish peroxidase (HRP) catalyzed process. Chemosphere 58:1097–1105. https://doi.org/10.1016/J.CHEMOSPHERE.2004.09.070

Moreira MT, Feijoo G, Canaval J, Lema JM (2003) Semipilot-scale bleaching of kraft pulp with manganese peroxide. Wood Sci Technol 37:117–123. https://doi.org/10.1007/S00226-003-0175-7/METRICS

Morgenstern I, Klopman S, Hibbett DS (2008) Molecular evolution and diversity of lignin degrading heme peroxidases in the agaricomycetes. J Mol Evol 66:243–257. https://doi.org/10.1007/S00239-008-9079-3/METRICS

Muhammad Nasir Iqbal H, Asgher M (2013) Characterization and decolorization applicability of xerogel matrix immobilized manganese peroxidase produced from Trametes versicolor IBL-04. Protein Pept Lett 20(5):591–600

Nousiainen P, Kontro J, Manner H et al (2014) Phenolic mediators enhance the manganese peroxidase catalyzed oxidation of recalcitrant lignin model compounds and synthetic lignin. Fungal Genet Biol 72:137–149. https://doi.org/10.1016/J.FGB.2014.07.008

Pease EA, Tien M (1992) Heterogeneity and regulation of manganese peroxidases from Phanerochaete chrysosporium. J Bacteriol 174:3532–3540. https://doi.org/10.1128/JB.174.11.3532-3540.1992

Poonpairoj P, Peerapatsakul C, Chitradon L (2001) Trend in using fungal enzymes, lignin-and pectin-degrading enzymes. In: Improvement of the paper mulberry pulping process

Qin X, Zhang J, Zhang X, Yang Y (2014) Induction, purification and characterization of a novel manganese peroxidase from Irpex lacteus CD2 and its application in the decolorization of different types of dye. PLoS One 9:e113282. https://doi.org/10.1371/JOURNAL.PONE.0113282

Qin X, Sun X, Huang H et al (2017) Oxidation of a non-phenolic lignin model compound by two Irpex lacteus manganese peroxidases: evidence for implication of carboxylate and radicals. Biotechnol Biofuels 10:1–13. https://doi.org/10.1186/S13068-017-0787-Z/FIGURES/6

Qiu J, Wilkens C, Barrett K, Meyer AS (2020) Microbial enzymes catalyzing keratin degradation: classification, structure, function. Biotechnol Adv 44:107607. https://doi.org/10.1016/J.BIOTECHADV.2020.107607

Raj A, Kumar S, Haq I, Singh SK (2014) Bioremediation and toxicity reduction in pulp and paper mill effluent by newly isolated ligninolytic Paenibacillus sp. Ecol Eng 71:355–362. https://doi.org/10.1016/J.ECOLENG.2014.07.002

Reddy GVB, Sridhar M, Gold MH (2003) Cleavage of nonphenolic β-1 diarylpropane lignin model dimers by manganese peroxidase from Phanerochaete chrysosporium. Eur J Biochem 270:284–292. https://doi.org/10.1046/J.1432-1033.2003.03386.X

Rekik H, Zaraî Jaouadi N, Bouacem K et al (2019) Physical and enzymatic properties of a new manganese peroxidase from the white-rot fungus Trametes pubescens strain i8 for lignin biodegradation and textile-dyes biodecolorization. Int J Biol Macromol 125:514–525. https://doi.org/10.1016/J.IJBIOMAC.2018.12.053

Salvachúa D, Martínez AT, Tien M et al (2013) Differential proteomic analysis of the secretome of Irpex lacteus and other white-rot fungi during wheat straw pretreatment. Biotechnol Biofuels 6:1–14. https://doi.org/10.1186/1754-6834-6-115/TABLES/4

Sundaramoorthy M, Kishi K, Gold MH, Poulos TL (1994) The crystal structure of manganese peroxidase from Phanerochaete chrysosporium at 2.06-A resolution. J Biol Chem 269:32759–32767. https://doi.org/10.1016/S0021-9258(20)30056-9

Sundaramoorthy M, Gold MH, Poulos TL (2010) Ultrahigh (0.93 Å) resolution structure of manganese peroxidase from Phanerochaete chrysosporium: implications for the catalytic mechanism. J Inorg Biochem 104:683–690. https://doi.org/10.1016/J.JINORGBIO.2010.02.011

Tuomela M, Vikman M, Hatakka A, Itävaara M (2000) Biodegradation of lignin in a compost environment: a review. Bioresour Technol 72:169–183. https://doi.org/10.1016/S0960-8524(99)00104-2

Urek RO, Pazarlioglu NK (2007) Enhanced production of manganese peroxidase by Phanerochaete chrysosporium. Braz Arch Biol Technol 50:913–920. https://doi.org/10.1590/S1516-89132007000700001

Wariishi H, Huangs J, Dunford HB, Golds MH (1991) Reactions of lignin peroxidase compounds I and 11 with veratryl alcohol. Transient-state kinetic characterization*. J Biol Chem 266:20694–20699. https://doi.org/10.1016/S0021-9258(18)54764-5

Yang X, Wang J, Zhao X et al (2011) Increasing manganese peroxidase production and biodecolorization of triphenylmethane dyes by novel fungal consortium. Bioresour Technol 102:10535–10541. https://doi.org/10.1016/J.BIORTECH.2011.06.034

Zainith S, Chowdhary P, Bharagava RN (2018) Recent advances in physico-chemical and biological techniques for the management of pulp and paper mill waste. In: Emerging and eco-friendly approaches for waste management. pp 271–297. https://doi.org/10.1007/978-981-10-8669-4_13/COVER

Zhang Y, Zeng GM, Tang L et al (2007) A hydroquinone biosensor using modified core–shell magnetic nanoparticles supported on carbon paste electrode. Biosens Bioelectron 22:2121–2126. https://doi.org/10.1016/J.BIOS.2006.09.030

Zhang H, Zhang J, Zhang X, Geng A (2018) Purification and characterization of a novel manganese peroxidase from white-rot fungus Cerrena unicolor BBP6 and its application in dye decolorization and denim bleaching. Process Biochem 66:222–229. https://doi.org/10.1016/J.PROCBIO.2017.12.011

Chapter 12
Biofilm-Mediated Heavy Metals Bioremediation

Sufiara Yousuf and Rahul Singh

Abstract Industrial activities have generated wastes globally and are still being created and dumped, resulting in contaminated environments (terrestrial as well as aquatic). Industrial wastes include polycyclic aromatic hydrocarbons, polythene, dyes, phenols, and heavy metals which degrade the quality of soil and cause havoc to ecosystem, animal, and human welfare. In order to overcome from the problem, various remediation strategies have been initiated such as land sparing, soil excavation, crop rotation, land sharing, and offsite treatment. While these methods require trained workers and could potentially introduce harmful substances into the environment. These are capable of removing and detoxifying the heavy metal contamination. Microbial communities produce extracellular polymeric substance (biofilm) that enhances the colonization of microbial cells and shields them in harsh environmental conditions. Biofilm formation is a natural character, exhibited by microbes growing on natural and artificial surfaces. In biofilms, multiple microorganisms reside together and degrade the pollutants in polluted soil and groundwater. Biofilm functions through cell signalling and quorum sensing to remove metal contamination from the environment. These microbes include Gram-positive, Gram-negative bacteria, cyanobacteria, fungi, and microalgae. Microbial biofilms are used as biomarkers and biofilters when pollution is on large scale. Bioremediation is a developing technology and has attracted interests in research and development. This chapter provides clear idea about toxicity of heavy metals in environment and the role of microbial biofilms in bioremediation of heavy metals from environment.

Keywords Microbial communities · Biofilm · Bioremediation · Heavy metal toxicity · contaminated environments

S. Yousuf · R. Singh (✉)
Department of Zoology, School of Bioengineering and Biosciences, Lovely Professional University, Phagwara, Punjab, India

A. Karnwal, A. R. Mohammad Said Al-Tawaha (eds.), *Microbial Applications for Environmental Sustainability*, https://doi.org/10.1007/978-981-97-0676-1_12

12.1 Introduction

The enormous growth of anthropogenic activities in the field of industrial revolution (urbanization and industrialization) is seen as a severe threat to public health. Direct discard of heavy metals including chromium, cadmium, arsenic, and lead is one of the most serious environmental concerns that has wrecked nature and has led to negative consequences on various ecosystems. Heavy metals have the ability to dissolve in aquatic bodies and are easily absorbed by aquatic species, but in humans accumulation occurs by consuming aquatic foods and drinking contaminated water (Kinuthia et al. 2020). Heavy metal exposure to aquatic life especially fish increases mortality and reduces breeding potentials in adults (Olojo et al. 2005). Metal toxicity creates oxidative stress in fish that weakens the immune system and finally damages the organs (Garai et al. 2021). Heavy metal accumulation in fish is directly linked to human life and gives birth to various diseases (Has-Schön et al. 2006). Different findings have stated the detrimental effects of heavy metal accumulation in humans. Arsenic inhalation or absorption may cause renal failure, leukopenia, and neural damage, while increased exposure causes lung and skin cancers (Saha et al. 1999). Cadmium exposure causes pneumonitis, diarrhoea, decalcification, and kidney damage and is responsible for many cancers (Bernhoft 2013). Chromium exposure causes enlarged liver, increases cholesterol level, and can damage DNA (Bist and Choudhary 2022). Furthermore, lead exposure results in cognitive loss, gastrointestinal problems, depression, and heart congestion (Breton et al. 2013; Wani et al. 2015). Among urban populations, water serves as the primary pathway for exposure to heavy metals, followed by air and soil; however, exposure to rural population ensues in the order of water, soil, and air (Zhao et al. 2019). Various strategies have been developed to assure effective treatment of contaminated waters and to mitigate the effects of indiscriminate heavy metal release in water bodies. Many treatment procedures including "chemical precipitation, oxidation/reduction, ion exchange, membrane filtration, and evaporation" were developed. Nevertheless, conventional approaches for removing harmful heavy metals are inefficient and costly. As a result of these flaws, microorganisms have been employed as a sustainable and economically viable strategy for addressing the issue of heavy metal toxicity that refers to the term bioremediation.

12.2 Heavy Metal Toxicity in Environment

Heavy metals are classified as naturally existing elements with significant atomic weight and density (5 times exceeding that of water). Due to their vast environmental dissemination and diverse domestic, agricultural, medical, and technical uses, there are apprehensions about their possible impacts on ecosystem and human well-being. The toxicity of metals is influenced by various factors, including the dosage or concentration, the route of exposure, and the specific chemicals to which individuals are exposed. Public health concerns arise from the high levels of toxicity associated with priority metals such as arsenic, cadmium, chromium, lead, and mercury. These metals are recognized as posing significant risks to human health. These metals get accumulated in target tissues and enter the bloodstream following their initial interaction with the gastrointestinal tract and its ecological system (Breton et al. 2013). These metallic substances are thought to be systemic toxins that can harm various organs even at modest exposure levels. Every metal possesses a distinct set of characteristics and physicochemical properties, which in turn bestow discrete toxicological modes of action.

12.2.1 Arsenic

Arsenic is a pervasive element that can be found in trace amounts in almost all environmental matrices. Arsenic exists in two forms: organic and inorganic forms. Trivalent arsenite and pentavalent arsenate are widely recognized as the two most prevalent inorganic forms of arsenic, respectively. The organic forms of arsenic encompass monomethylarsonic acid (MMA), dimethylarsinic acid (DMA), and trimethylarsine oxide. Natural phenomena like volcanic eruption, soil erosion, as well as human activity contribute to arsenic pollution in the environment (Tchounwou et al. 2012). Various compounds that contain arsenic are manufactured commercially and have been incorporated into the production of various agricultural products, including insecticides, herbicides, fungicides, algicides, sheep dips, wood preservatives, and dyestuffs. Arsenic used in agriculture is carried into waterways, where it can accumulate in the water and the animals that live there. There is a belief that a significant number of people worldwide, estimated to be in the millions, experience chronic exposure to arsenic due to the contamination of groundwater with high levels of this toxic element. Also due to the consumption of contaminated fish or crustaceans, humans get exposed to arsenic (Chakraborti et al. 2016). Prolonged exposure to arsenic has been associated with various health issues in humans, including cancer, diabetes, hepatitis, and cardiovascular disease (Choiniere and Wang 2016). Toxic effects of arsenic include interference with metabolic processes including acetyl-CoA and succinic dehydrogenase, which can lead to poisoning. The majority of occurrences of arsenic toxicity in humans have been linked to exposure to inorganic arsenic (Rahman et al. 2013). Arsenic can inhibit 200 enzymes by the binding of trivalent arsenite to thiol or sulfhydryl groups found on protein surfaces. This is the mechanism that is most likely responsible for the extensive impacts that arsenic has on a variety of organ systems (Hughes 2002).

12.2.2 Cadmium

In accordance with the International Agency for Research on Cancer (IARC), cadmium, or Cd, is a group I carcinogenic chemical (IARC 1993). It is found in abundant quantities in the earth's crust, with an average proportion of approximately 0.1 mg/kg. Sedimentary rocks contain the highest concentration of cadmium compounds. Cadmium finds extensive use in diverse industrial applications, primarily in the production of alloys, pigments, and batteries (Wilson 1988). The metal is nephrotoxic and has the ability of causing a wide range of health problems, including osteoporosis, damage to the kidneys, brain, and testes (Järup and Åkesson 2009; Rinaldi et al. 2017). It has been hypothesized that cadmium is principally responsible for the damage to cells via producing reactive oxygen species, despite the fact that the mechanism of cadmium toxicity is not yet well comprehended (Stohs and Bagchi 1995). These ROS are known to damage single-strand DNA and to prevent the synthesis of nucleic acids and proteins (Mitra 1984). Furthermore, cadmium exposure impairs spermatogenesis, sperm motility, and hormone synthesis/release in males. Experimental and human studies have demonstrated that cadmium has the potential to interfere with female fertility by disrupting reproductive hormone balance and menstrual cycles (Kumar and Sharma 2019).

12.2.3 Chromium

Chromium (Cr) is an element present naturally in the Earth's crust and can exist in different oxidation states, commonly referred to as valence states. The most common forms of chromium are trivalent Cr (III) and hexavalent Cr (VI). Both of these forms are hazardous to humans, plants, and animals (Monalisa and Kumar 2013). Hexavalent Cr (VI) is rapidly absorbed by living cells and is more toxic, whereas trivalent Cr is less toxic as it is less soluble. Cr (VI) has higher absorption capabilities within the body due to its greater mobility within cells, enabling it to readily traverse cell membranes and accumulate in organs (Wolińska et al. 2013; Feng et al. 2020). The inhalation of the less soluble or insoluble compounds of Cr (VI) appears to be linked to carcinogenicity (Katz and Salem 1993). In a variety of animal models, it was found that exposure to Cr (VI) was connected with the development of intestinal and oral neoplasms as well as cancers of the lungs, pharynx, and buccal cavity (Deng et al. 2019). Chromium can be detected in the air, soil, and water. There are extensive numbers of natural processes and human activities that contribute to the release of chromium into the environment, but industrial enterprises are responsible for the greatest amount of this element's discharge. The metallurgical, refractory, and chemical industries are mostly responsible for the growth of chromium concentrations in the environment. The yearly environmental emission of chromium is expected to be 33 metric tonnes (Abadin et al. 2012).

12.2.4 Lead

Lead is a naturally existing metal with a bluish-grey colour that can be found in trace quantities within the Earth's crust. Human activities such as the fossil fuel burning, mining, and manufacturing allow high levels of lead to be discharged into the environment. Lead is utilized in a variety of contexts, including the industrial, agricultural, and domestic spheres. At the moment, it is put to use in the makeup of lead–acid batteries, ammunition, metal items (such as solder and pipes), and apparatus designed to shield X-rays (Ara and Usmani 2015). In several animal models, lead (Pb) exposure results in a multitude of problems. Chronic lead exposure can result in a number of different health issues, including neuropsychiatric disorders, anaemia, hypertension, and vomiting (Flora et al. 2012; Wani et al. 2015). One of the primary ways in which lead is able to exert its hazardous effect is through the participation in several biochemical processes. These processes include the ability of lead to either block or mimic the actions of calcium, as well as its ability to interact with proteins (Flora et al. 2006). Lead has the capacity to attach to sulfhydryl and amide groups of enzymes. causes the enzymes structures to change, and reduces their activity. Lead may potentially impair enzyme activity by competing with essential metallic cations for binding sites, or it may change the transport of vital cations such as calcium (Flora et al. 2007; Wani et al. 2019). Recent research has found that exposure to lead during childhood may be a factor in the development of neurological illnesses in adulthood (Reuben 2018). Furthermore, the IARC has classified lead as a potential human carcinogen, as in experiments it has shown detrimental effects, causing kidney tumours in rats and mice (IARC 1987). Moreover, it is well-established that exposure to lead can lead to gene mutations and abnormalities in sister chromatids (Yang et al. 1999). Thus, Pb exposure has negative effects on the organism.

12.2.5 Mercury

Mercury is found in the periodic table's transition element series. Mercury can exist in three different forms in nature: elemental (Hg^0), inorganic (Hg^{1+}), and organic (Hg^2). Inorganic mercury is frequently found in soil and groundwater resources, where it is consumed by bacteria and converted to methyl mercury. Methyl mercury undergoes biomagnification and severely disrupts life beneath the surface (Bernhoft 2012). Mercury exposure tends to increase along with rising industry pollution, leading to a number of major health problems (Dopp et al. 2004). Discharges, agriculture, burning of unburned fuels, mining, and outflow of industrial solid waste are all anthropogenic sources of mercury (Chen et al. 2012). Mercury exposure to humans occurs by consuming mercury contaminated fish and other aquatic organisms (Trasande et al. 2005). Exposure to high concentration of elemental mercury withdraws a number of negative effects that include degradation of mucosal lining, lung toxicity, gastrointestinal problems, nausea, skin rashes, and neurological malfunction (Bates 2003). While Hg^{2+} and MeHg deplete cellular antioxidants once they enter the cell and create covalent interactions with cysteine residues in proteins (Valko et al. 2006). Inorganic mercury enhances the generation of reactive oxygen species (ROS) by disrupting oxidative phosphorylation and electron transport within the ubiquinone-cytochrome b5 pathway (Palmeira and Madeira 1997). The study of carcinogenic effects caused by metals has attracted a lot of attention from the public health community. The association between mercury exposure and the development of cancer remains a topic of controversy and discussion. Certain studies have indicated its potential for genotoxicity, other research has not connected mercury exposure to genotoxicity (Valko et al. 2004). Studies that indicate mercury is a genotoxic substance describe that oxidative stress is referred as molecular mechanism of toxicity. As a result, it has been demonstrated that mercury causes the production of ROS, which are known to damage DNA in cells and start carcinogenic processes (Ogura et al. 1996; Valko et al. 2006). Therefore, mercury is extremely toxic in all of its forms, posing health hazards to organs and causing GI toxicity, neurotoxicity, and renal toxicity (Fig. 12.1).

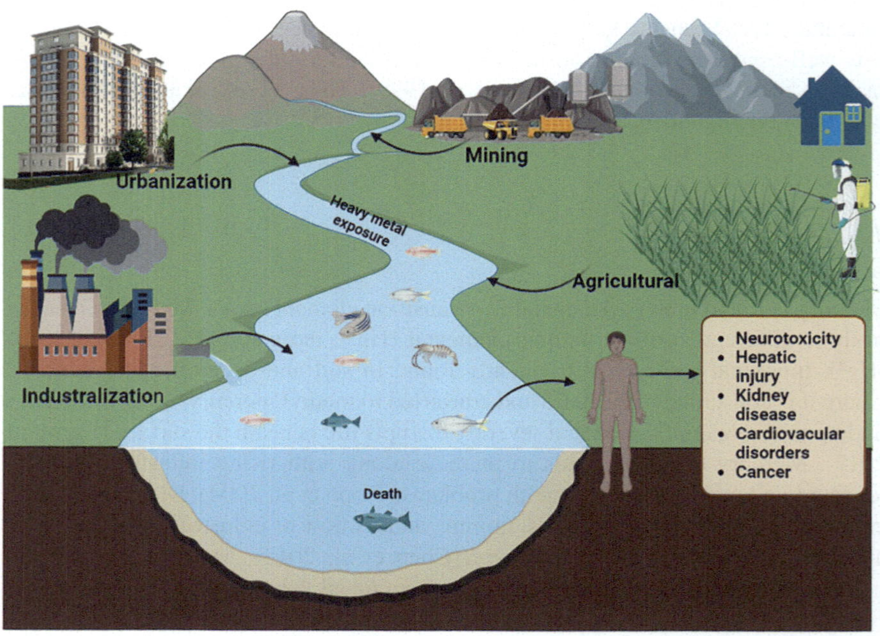

Fig. 12.1 Heavy metal toxicity in environment

12.3 Bioremediation

The word "bioremediation" describes a set of procedures that utilize biological systems to repair or clean up polluted environments. Mostly bioremediation involves bacteria, yeast, and fungi to digest contaminants and convert them into less hazardous or harmless forms. By oxidizing, immobilizing, or converting contaminants into non-toxic chemicals, the majority of native bacteria are able to successfully restore the environment. The primary objective of bioremediation is to reduce pollutant concentrations to levels that are either undetectable, non-toxic, or deemed acceptable (Vishwakarma et al. 2020). A wide spectrum of chemicals can be treated via bioremediation. Apart from municipal waste and process waters, potential microorganisms can be employed for the degradation of pesticides, industrial chemicals, components of crude oil, and even compounds that were previously thought to be non-biodegradable, including chlorinated solvents, chlorofluorocarbons, and similar substances (Kumar et al. 2011). The discovery of the bioremediation mechanism can be attributed to "George M. Robinson", who initially recognized its potential while conducting research on sewage and oil treatment (Sonawdekar 2012). The basic mechanism for bioremediation process is to stimulate the microbial population for maximum enzyme production. Microorganisms utilize pollutants as a substrate to metabolize and break down contaminants, utilizing them as a source of energy for their growth and reproduction. Energy is released when chemical bonds break and the free electrons are shifted to an electron acceptor such as O_2. The basic mechanism involves a "redox reaction" in which organic matter is oxidized by the loss of electrons, while the substance that receives the electron is reduced. The electron acceptors in the anaerobic process are nitrate, sulphate, and iron, whereas oxygen is the electron acceptor in the aerobic process (Vishwakarma et al. 2020). The majority of bioremediation systems work in aerobic conditions. Bioremediation process can be broadly categorized into two main groups: in situ and ex situ. In situ technology involves biostimulation, while ex situ technology entails manually removing contaminated material and transporting it to a different location for treatment in bioreactors, land farming, or composting (Zouboulis et al. 2019). In situ techniques are more beneficial and helpful because they cause less site disruption and transportation expense when compared to ex situ procedures.

Microbial populations have a tremendous potential for remediation of polluted sites due to their remarkable adaptability, genetic diversity, cellular simplicity, and functioning (Zhang et al. 2021). Some of the microbes employed in bioremediation are extremophiles that are able to survive in acidic, heavily metal-contaminated, or radioactive environments (Prince 2000). Bioremediation commonly involves optimizing the physical and chemical conditions, such as pH, aeration, and moisture, at the contaminated site to enhance the ability of bacteria to efficiently digest pollutants. Recently, biofilm-mediated remediation has attracted a lot of attention as cutting-edge remedial approach for the biodegradation of environmental contaminants. In comparison to using planktonic microorganisms, biofilm-mediated remediation has been recognized as a structured, effective, and reliable approach for reducing or eliminating contaminants (Abinandan et al. 2018).

12.4 Role of Biofilms in Bioremediation

The term "biofilm" refers to a group of microbial cells attached to a surface by their mutual production of extracellular polymeric molecules (EPS) (Headley et al. 1998; Flemming and Wingender 2010). In biofilms, microorganisms reside together and benefit one another through the development of several biological niches. Biofilms provide protection to microbial clusters from adverse environmental factors such as nutrition fluctuation, hazardous chemicals, pH, temperature, and salinity (Singh et al. 2006; Vaishnav et al. 2016). Biofilms have the capability to develop in any moist environment that provides sufficient nutrient availability and suitable surfaces for attachment. Biofilms consist of one or multiple species of microorganisms derived from diverse kingdoms, including bacteria, algae, fungi, and archaea (Baker et al. 2009). Each member in biofilm performs singly or collectively with different metabolic breakdown route to degrade hazardous pollutants in contaminated sites (Gieg et al. 2014; Swem et al. 2008). Clean-up of environmental pollutants through biofilm mediation is both environmentally friendly and cost-effective.

Mostly bacterial species are primarily capable of colonizing and forming biofilms to adapt and survive in challenging environmental conditions (Lima et al. 2020). In biofilm, there are 10^8–10^{11} microorganisms per g of wet weight (Flemming et al. 2016). The formation of biofilms follows a sequential series of phases:

1. Bacterial cells attach irreversibly to the surface and form matrix.
2. Quorum sensing begins and cells become layered.
3. Colonies become 3D-structured and attain maximum thickness.
4. Planktonic cell dispersion from matrix (Fig. 12.2).

Bacterial strains that adopt a biofilm lifestyle are encapsulated within an extracellular polymeric substance (EPS) matrix, which acts as a protective shield for microorganisms involved in bioremediation processes. The EPS is composed of water, proteins, lipids, nucleic acids, humic compounds, and polysaccharides. Within the biofilm community, the presence of extracellular polymeric substances (EPS) facilitates the exchange of nutrients and the removal of by-products. The extracellular enzymes found within the extracellular polymeric substances (EPS) of biofilms play a remedial role in degrading and detoxifying contaminants, including heavy metals and various organic compounds (Flemming and Wingender 2010). The abundant presence of negatively charged functional groups in the extracellular polymeric substances (EPS) allows them to effectively trap and remove heavy metals and organic contaminants. EPS can form complexes with these pollutants, facilitating their removal from contaminated sites (Pal and Paul 2008). Degradation of organic aromatic pollutants is facilitated by various genes (pah, nid, nah, bph, and ntdR) and enzymes (decarboxylase, dehydrogenase, and hydrolyase) (Mangwani

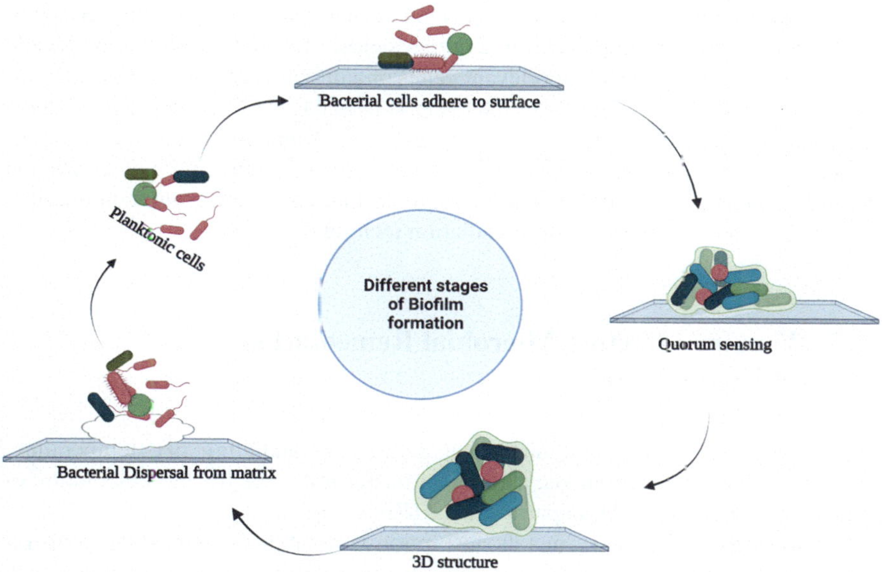

Fig. 12.2 Different stages of biofilm formation

et al. 2016). Several bacterial species, categorized within the genera, *Bacillus*, *Rhodococcus*, *Arthrobacter*, *Cycloclasticus*, *Alcanivorax,* and *Pseudomonas*, have been recognized to be capable of forming biofilms and to break down or mineralize dangerous pollutants in a variety of contaminated environments (Yoshikawa et al. 2017; Mahto and Das 2020). Paul (2015) used *Arthrobacter* sp. SUK for removal of heavy metals (Ni, Cr, and Mg), the results of study demonstrated that the species have highest removal efficiency of Cr compared to Ni and Mg. Luo et al. (2014) applied *Pseudomonas* sp. LK9 for removal of heavy metals (Cd^{2+} and Cu^{2+}) and found that the species have high efficiency to remove heavy metals in wastewater. Sundar et al. (2011) used *Bacillus* species "*Bacillus subtilis* and *Bacillus cereus*" biofilms for Cr removal and found the effectiveness of species. Furthermore, Chen et al. (2005) documented the metals (Cu and Zn) adsorption capacity by *Pseudomonas putida* CZ1. The development and maturation of biofilms are regulated by a communication mechanism known as quorum sensing (QS) (Nadell et al. 2016).

Quorum sensing makes a substantial contribution to the development of biofilm and exopolysaccharide secretion. Quorum sensing bacteria can produce number of chemical signals or autoinducers for cell-to-cell contact. Various signalling molecules among QS bacteria include "autoinducer 1, autoinducer 2, autoinducing peptides, and AHLs" (Yousuf et al. 2022). Among these signalling molecules, AHL is the most common one produced by about 100 *Proteobacteria* species (Huang et al. 2016). Autoinducing peptide (AIPs) serves as a signalling molecule in Gram-positive bacteria, while LUXI and LUXR control QS in Gram-negative bacteria (Todd et al. 2017). The basic regulatory mechanism of quorum sensing (QS) involves the binding of an AI (autoinducer) molecule to a receptor on the surface of a sensory cell. This binding event then initiates either a signal transduction cascade that leads to the transcription of relevant genes or a direct interaction with transcription factors. The QS system causes the bacteria to behave collectively and synchronize their actions in order to adapt to changing environments (Liao and Li 2013). QS system was first observed in luminescent *Vibrio fischeri* and *Vibrio harveyi* (Nealson and Markovitz 1970). Therefore, microbial aggregation and communication is a safe and inexpensive bioremediation technique.

12.5 Factors Affecting Microbial Remediation of Heavy Metals

Various physicochemical and biological parameters including pH, temperature, nutrient availability, chemotaxis, and gene transfer influence the bioremediation of metals (Igiri et al. 2018; Mohapatra et al. 2020).

Temperature: The heavy metals adsorption is notably influenced by temperature. Increased temperature increases mobility and solubility of heavy metals which improves bioavailability and leads to good interaction with biofilm matrix. When temperatures reach suitable levels, microorganisms become more active, leading to increased metabolism and enzyme activity. This enhanced microbial activity accelerates the bioremediation process (Igiri et al. 2018).

pH: The pH of the environment can have a big impact on how biofilms form. It has been demonstrated that this factor impacts microorganism's adherence to surfaces, which is the first stage of biofilm formation. Microbial biofilm has negative charge on surface, and when pH increases, it creates more negative charge and increases the binding capacity for metals also. The optimal pH for the secretion of polysaccharides differs among species, but for the majority of bacteria, it typically falls within a pH range of 6.5–7.0 (Alotaibi and Bukhari 2021). Consequently, the sites' pH can be adjusted to speed up the bioremediation process using the bacterial biofilm (Mohapatra et al. 2020).

Nutrient availability: Availability of nutrients in the surrounding media affects the pace of microbial attachment to surfaces; therefore, an increase in nutrient concentration speeds up the process. Nutrition levels may influence how a biofilm develops and how its cells disperse (Zhao et al. 2017). Furthermore, it has also been determined that variations in the critical nutrient availability have an effect on the physiology of bacteria in developing biofilms (Alotaibi and Bukhari 2021).

Chemotaxis: Chemotaxis is defined as response of organism in reaction to a chemical gradient (nutrient or hazardous chemicals). The chemotaxis property of the bacteria can aid in their growth, survival, and formation of biofilms in any challenging environment. Bacteria equipped with chemotaxis capabilities have the ability to navigate towards xenobiotic compounds in environments with limited nutrients and utilize them as their sole source of carbon and energy. This characteristic proves advantageous for the bioremediation process. Using chemotaxis bacteria, hazardous metal ions can be biodegraded more effectively (Shukla et al. 2014).

Gene transfer: Gene transfer is a widespread phenomenon observed among biofilm bacteria, enabling them to engage in the sharing of genetic material through HGT or lateral gene transfer. The HGT procedure uses "DNA conjugation or transformation" methods that can give the biofilm communities new adaptive and catabolic potential properties. Consequently, the close proximity of biofilm bacteria facilitates autogene translocation, including horizontal gene transfer (HGT), which may contribute to enhancing the effectiveness of bioremediation (Shukla et al. 2014).

12.6 Conclusion and Future Research

The majority of bacteria that thrive in unfavourable, contaminated environments have the ability to create biofilms for increased survival. They have proven as viable candidates for the bioremediation of various toxic substances due to their complex structured-EPS matrix, intercellular gene transfer, quorum sensing, and chemotaxis. Based on the findings of various studies, it is possible to successfully use bacterial biofilms comprising of a single or conjoint species from varied habitats for both in situ and ex situ bioremediation procedures. However, biofilm bioremediation technique has some drawbacks also. When compared to chemical action for contaminant breakdown, bioremediation is comparatively slow. When there is a lot of pollution, not all substances such as plastics and some halogenated aromatic compounds are broken down by biofilms. As biofilm bioremediation is a relatively new area of study, researchers are continually investigating novel strains that can be genetically engineered and possess the ability to break down halogenated aromatic compounds. Quorum sensing between members in biofilm is another area to explore for better outputs in remediation process. Furthermore, application of bacterial–fungal microbial consortia for bioremediation is another topic of interest that needs to be explored.

References

Abadin H, Fay M, Ingerman L, Tencza B, Yu D, Wilbur SB (2012) Toxicological profile for chromium. Agency for Toxic Substances and Disease Registry, Atlanta, GA

Abinandan S, Subashchandrabose SR, Venkateswarlu K, Megharaj M (2018) Microalgae–bacteria biofilms: a sustainable synergistic approach in remediation of acid mine drainage. Appl Microbiol Biotechnol 102(3):1131–1144

Alotaibi GF, Bukhari MA (2021) Factors influencing bacterial biofilm formation and development. Am J Biomed Sci Res 12(6):617–626

Ara A, Usmani JA (2015) Lead toxicity: a review. Interdiscip Toxicol 8(2):55

Baker BJ, Tyson GW, Goosherst L, Banfield JF (2009) Insights into the diversity of eukaryotes in acid mine drainage biofilm communities. Appl Environ Microbiol 75(7):2192–2199

Bates N (2003) Metallic and inorganic mercury poisoning. Emergen Nurse 11(1):25
Bernhoft RA (2012) Mercury toxicity and treatment: a review of the literature. J Environ Public Health 2012:460508
Bernhoft RA (2013) Cadmium toxicity and treatment. Sci World J 18(2):321–336
Bist P, Choudhary S (2022) Impact of heavy metal toxicity on the gut microbiota and its relationship with metabolites and future probiotics strategy: a review. Biol Trace Elem Res 200:1–23
Breton J, Le Clère K, Daniel C, Sauty M, Nakab L, Chassat T et al (2013) Chronic ingestion of cadmium and lead alters the bioavailability of essential and heavy metals, gene expression pathways and genotoxicity in mouse intestine. Arch Toxicol 87(10):1787–1795
Chakraborti D, Rahman MM, Chatterjee A, Das D, Das B, Nayak B et al (2016) Fate of over 480 million inhabitants living in arsenic and fluoride endemic Indian districts: magnitude, health, socio-economic effects and mitigation approaches. J Trace Elem Med Biol 38:33–45
Chen XC, Wang YP, Lin Q, Shi JY, Wu WX, Chen YX (2005) Biosorption of copper (II) and zinc (II) from aqueous solution by pseudomonas putida CZ1. Colloids Surf B: Biointerfaces 46(2):101–107
Chen Q, Cao J, Jia Y, Liu X, Yan Y, Pang G (2012) Modulation of mice fecal microbiota by administration of casein glycomacropeptide. Microbiol Res 3(1):8–12
Choiniere J, Wang L (2016) Exposure to inorganic arsenic can lead to gut microbe perturbations and hepatocellular carcinoma. Acta Pharm Sin B 6(5):426–429
Deng Y, Wang M, Tian T, Lin S, Xu P, Zhou L, Dai Z (2019) The effect of hexavalent chromium on the incidence and mortality of human cancers: a meta-analysis based on published epidemiological cohort studies. Front Oncol 9:24
Dopp E, Hartmann LM, Florea AM, Rettenmeier AW, Hirner AV (2004) Environmental distribution, analysis, and toxicity of organometal (loid) compounds. Crit Rev Toxicol 34(3):301–333
Feng P, Ye Z, Han H, Ling Z, Zhou T, Zhao S, Li X (2020) Tibet plateau probiotic mitigates chromate toxicity in mice by alleviating oxidative stress in gut microbiota. Commun Biol 3(1):1–12
Flemming HC, Wingender J (2010) The biofilm matrix. Nat Rev Microbiol 8(9):623–633
Flemming HC, Wingender J, Szewzyk U, Steinberg P, Rice SA, Kjelleberg S (2016) Biofilms: an emergent form of bacterial life. Nat Rev Microbiol 14(9):563–575
Flora SJ, Flora G, Saxena G (2006) Environmental occurrence, health effects and management of lead poisoning. In: Lead. Elsevier Science BV, Amsterdam, pp 158–228
Flora SJS, Saxena G, Gautam P, Kaur P, Gill KD (2007) Response of lead-induced oxidative stress and alterations in biogenic amines in different rat brain regions to combined administration of DMSA and MiADMSA. Chem Biol Interact 170(3):209–220
Flora G, Gupta D, Tiwari A (2012) Toxicity of lead: a review with recent updates. Interdiscip Toxicol 5(2):47
Garai P, Banerjee P, Mondal P, Saha NC (2021) Effect of heavy metals on fishes: toxicity and bioaccumulation. J Clin Toxicol S:18
Gieg LM, Fowler SJ, Berdugo-Clavijo C (2014) Syntrophic biodegradation of hydrocarbon contaminants. Curr Opin Biotechnol 27:21–29
Has-Schön E, Bogut I, Strelec I (2006) Heavy metal profile in five fish species included in human diet, domiciled in the end flow of river Neretva (Croatia). Arch Environ Contam Toxicol 50(4):545–551
Headley JV, Gandrass J, Kuballa J, Peru KM, Gong Y (1998) Rates of sorption and partitioning of contaminants in river biofilm. Environ Sci Technol 32(24):3968–3973
Huang J, Shi Y, Zeng G, Gu Y, Chen G, Shi L et al (2016) Acyl-homoserine lactone-based quorum sensing and quorum quenching hold promise to determine the performance of biological wastewater treatments: an overview. Chemosphere 157:137–151
Hughes MF (2002) Arsenic toxicity and potential mechanisms of action. Toxicol Lett 133:1–16
Igiri BE, Okoduwa SI, Idoko GO, Akabuogu EP, Adeyi AO, Ejiogu IK (2018) Toxicity and bioremediation of heavy metals contaminated ecosystem from tannery wastewater: a review. J Toxicol 2018:2568038

International Agency for Research on Cancer (1987) Overall evaluation of carcinogenicity: an updating of monographs. In: IARC monographs on the evaluation of carcinogenic risks to humans, vol 1–42. IARC, Lyons, pp 230–232

International Agency for Research on Cancer (1993) Cadmium and cadmium compounds. In: Monographs on evaluation of carcinogenic risks to humans, vol 58. International Agency for Research on Cancer, Lyon, pp 119–237

Järup L, Åkesson A (2009) Current status of cadmium as an environmental health problem. Toxicol Appl Pharmacol 238(3):201–208

Katz SA, Salem H (1993) The toxicology of chromium with respect to its chemical speciation: a review. J Appl Toxicol 13(3):217–224

Kinuthia GK, Ngure V, Beti D, Lugalia R, Wangila A, Kamau L (2020) Levels of heavy metals in wastewater and soil samples from open drainage channels in Nairobi, Kenya: community health implication. Sci Rep 10(1):1–13

Kumar S, Sharma A (2019) Cadmium toxicity: effects on human reproduction and fertility. Rev Environ Health 34(4):327–338

Kumar A, Bisht BS, Joshi VD, Dhewa T (2011) Review on bioremediation of polluted environment: a management tool. Int J Environ Sci 1(6):1079

Liao NQ, Li HM (2013) Conceivable bioremediation techniques based on quorum sensing. In: Applied mechanics and materials, vol 295. Trans Tech Publications Ltd., Durnten, pp 39–44

Lima RA, de Souza SLX, Lima LA, Batista ALX, de Araújo JTC, Sousa FFO et al (2020) Antimicrobial effect of anacardic acid–loaded zein nanoparticles loaded on Streptococcus mutans biofilms. Braz J Microbiol 51(4):1623–1630

Luo S, Li X, Chen L, Chen J, Wan Y, Liu C (2014) Layer-by-layer strategy for adsorption capacity fattening of endophytic bacterial biomass for highly effective removal of heavy metals. Chem Eng J 239:312–321

Mahto KU, Das S (2020) Whole genome characterization and phenanthrene catabolic pathway of a biofilm forming marine bacterium Pseudomonas aeruginosa PFL-P1. Ecotoxicol Environ Saf 206:111087

Mangwani N, Kumari S, Das S (2016) Bacterial biofilms and quorum sensing: fidelity in bioremediation technology. Biotechnol Genet Eng Rev 32(1-2):43–73

Mitra RS (1984) Protein synthesis in Escherichia coli during recovery from exposure to low levels of Cd2+. Appl Environ Microbiol 47(5):1012–1016

Mohapatra RK, Behera SS, Patra JK, Thatoi H, Parhi PK (2020) Potential application of bacterial biofilm for bioremediation of toxic heavy metals and dye-contaminated environments. In: New and future developments in microbial biotechnology and bioengineering: microbial biofilms. Elsevier, pp 267–281

Monalisa M, Kumar PH (2013) Effect of ionic and chelate assisted hexavalent chromium on mung bean seedlings (Vigna radiata L. wilczek. Var k-851) during seedling growth. J Stress Physiol Biochem 9(2):232–241

Nadell CD, Drescher K, Foster KR (2016) Spatial structure, cooperation and competition in biofilms. Nat Rev Microbiol 14(9):589–600

Nealson KH, Markovitz A (1970) Mutant analysis and enzyme subunit complementation in bacterial bioluminescence in photobacterium Fischeri. J Bacteriol 104(1):300–312

Ogura H, Takeuchi T, Morimoto K (1996) A comparison of the 8-hydroxydeoxyguanosine, chromosome aberrations and micronucleus techniques for the assessment of the genotoxicity of mercury compounds in human blood lymphocytes. Mutat Res 340(2-3):175–182

Olojo EAA, Olurin KB, Mbaka G, Oluwemimo AD (2005) Histopathology of the gill and liver tissues of the African catfish Clarias gariepinus exposed to lead. Afr J Biotechnol 4(1):117–122

Pal A, Paul A (2008) Microbial extracellular polymeric substances: central elements in heavy metal bioremediation. Indian J Microbiol 48(1):49–64

Palmeira CM, Madeira VM (1997) Mercuric chloride toxicity in rat liver mitochondria and isolated hepatocytes. Environ Toxicol Pharmacol 3(3):229–235

Paul SDAK (2015) Hexavalent chromate reduction during growth and by immobilized cells of Arthrobacter sp. SUK 1205. Science 34(3):158–168
Prince RC (2000) Bioremediation. In: Kirk-Othmer (ed) Encyclopedia of chemical technology. Wiley, New York
Rahman MM, Asaduzzaman M, Naidu R (2013) Consumption of arsenic and other elements from vegetables and drinking water from an arsenic-contaminated area of Bangladesh. J Hazard Mater 262:1056–1063
Reuben A (2018) Childhood lead exposure and adult neurodegenerative disease. J Alzheimers Dis 64(1):17–42
Rinaldi, M., Micali, A., Marini, H., Adamo, E. B., Puzzolo, D., Pisani, A., .et al (2017). Cadmium, organ toxicity and therapeutic approaches: a review on brain, kidney and testis damage. Curr Med Chem, 24(35), 3879-3893
Saha JC, Dikshit AK, Bandyopadhyay M, Saha KC (1999) A review of arsenic poisoning and its effects on human health. Crit Rev Environ Sci Technol 29(3):281–313
Shukla SK, Mangwani N, Rao TS, Das S (2014) Biofilm-mediated bioremediation of polycyclic aromatic hydrocarbons. In: Microbial biodegradation and bioremediation. Elsevier, Amsterdam, pp 203–232
Singh R, Paul D, Jain RK (2006) Biofilms: implications in bioremediation. Trends Microbiol 14(9):389–397
Sonawdekar S (2012) Bioremediation: a boon to hydrocarbon degradation. Int J Environ Sci 2(4):2408–2424
Stohs SJ, Bagchi D (1995) Oxidative mechanisms in the toxicity of metal ions. Free Radical Biol Med 18:321–336
Sundar K, Sadiq IM, Mukherjee A, Chandrasekaran N (2011) Bioremoval of trivalent chromium using bacillus biofilms through continuous flow reactor. J Hazard Mater 196:44–51
Swem LR, Swem DL, Wingreen NS, Bassler BL (2008) Deducing receptor signaling parameters from in vivo analysis: LuxN/AI-1 quorum sensing in Vibrio harveyi. Cell 134(3):461–473
Tchounwou PB, Yedjou CG, Patlolla AK, Sutton DJ (2012) Heavy metal toxicity and the environment. In: Molecular, clinical and environmental toxicology. Springer, Cham, pp 133–164
Todd DA, Parlet CP, Crosby HA, Malone CL, Heilmann KP, Horswill AR, Cech NB (2017) Signal biosynthesis inhibition with ambuic acid as a strategy to target antibiotic-resistant infections. Antimicrob Agents Chemother 61(8):e00263–e00217
Trasande L, Landrigan PJ, Schechter C (2005) Public health and economic consequences of methyl mercury toxicity to the developing brain. Environ Health Perspect 113(5):590–596
Vaishnav A, Kumari S, Jain S, Varma A, Tuteja N, Choudhary DK (2016) PGPR-mediated expression of salt tolerance gene in soybean through volatiles under sodium nitroprusside. J Basic Microbiol 56(11):1274–1288
Valko M, Izakovic M, Mazur M, Rhodes CJ, Telser J (2004) Role of oxygen radicals in DNA damage and cancer incidence. Mol Cell Biochem 266(1):37–56
Valko M, Rhodes CJB, Moncol J, Izakovic MM, Mazur M (2006) Free radicals, metals and antioxidants in oxidative stress-induced cancer. Chem Biol Interact 160(1):1–40
Vishwakarma GS, Bhattacharjee G, Gohil N, Singh V (2020) Current status, challenges and future of bioremediation. In: Bioremediation of pollutants. Elsevier, Amsterdam, pp 403–415
Wani AL, Ara A, Usmani JA (2015) Lead toxicity: a review. Interdiscip Toxicol 8:55–64
Wani AL, Ansari MO, Ahmad M, Parveen N, Siddique HR, Shadab GG (2019) Influence of zinc levels on the toxic manifestations of lead exposure among the occupationally exposed workers. Environ Sci Pollut Res 26(32):33541–33554
Wilson DN (1988) Cadmium-market trends and influences. In: Cadmium 87. Proceedings of the 6th international cadmium conference, vol 9. Cadmium Association, London, p 16
Wolińska A, Stepniewska Z, Wlosek R (2013) The influence of old leather tannery district on chromium contamination of soils, water and plants. Sci Res 5:6

Yang JL, Wang LC, Chang CY, Liu TY (1999) Singlet oxygen is the major species participating in the induction of DNA strand breakage and 8-hydroxydeoxyguanosine adduct by lead acetate. Environ Mol Mutagen 33(3):194–201

Yoshikawa M, Zhang M, Kurisu F, Toyota K (2017) Bacterial degraders of coexisting dichloromethane, benzene, and toluene, identified by stable-isotope probing. Water Air Soil Pollut 228(11):1–10

Yousuf S, Tyagi A, Singh R (2022) Probiotic supplementation as an emerging alternative to chemical therapeutics in finfish aquaculture: a review. Probiotics Antimicro Prot 15:1151–1168

Zhang X, Das S, Li A, Ma Q, Tan L (2021) Marine microbes for contaminant bioremediation. Front Microbiol 12:762968

Zhao X, Zhao F, Wang J, Zhong N (2017) Biofilm formation and control strategies of foodborne pathogens: food safety perspectives. RSC Adv 7(58):36670–36683

Zhao X, Huang J, Lu J, Sun Y (2019) Study on the influence of soil microbial community on the long-term heavy metal pollution of different land use types and depth layers in mine. Ecotoxicol Environ Saf 170:218–226

Zouboulis AI, Moussas PA, Psaltou SG (2019) Groundwater and soil pollution: bioremediation. Elsevier, Amsterdam

Chapter 13
Rhizospheric Microbes: A Promising Resource for Bioremediation and Agricultural Productivity

Nisha, Rohan Samir Kumar Sachan, Khusbhoo, Inderpal Devgon, Manpreet Kaur, Divya Bhagat, Gaurav Kumar, Abdel Rahman Mohammad Said Al-Tawaha, and Arun Karnwal

Abstract Soil is a foundation of agriculture and home to numerous terrestrial organisms, including prokaryotes. Many biological activities with human interference led to distressed plant growth. The contaminants disposed of by various anthropogenic activities have decreased the soil quality and hence lower agricultural productivity. A means to bioremediate soil contaminants using a natural soil system is challenging. Promising research involving rhizospheric microbes in bioremediation has led to an economical and sustainable approach to increasing agriculture productivity. The rhizosphere has an intricate ecosystem comprising microbial consortiums which perform numerous biological activities concerning plant growth. These microbial consortia also perform a prime duty to bioremediate soil contaminants to improve soil quality and maintain its nutrient quality. Differing actions have been researched for bioremediation strategies of soil contamination by rhizospheric microbes. The review discusses different microbial members of prokaryotes, bacteria, and arbuscular mycorrhizal fungi to bioremediate swept contamination in the soil.

Keywords Arbuscular mycorrhizal fungi · Bioremediation · Ecosystem · Human impact · Microbiome · Phytoremediation · Plant roots · Rhizosphere · Soil contamination

---

Nisha · R. S. K. Sachan · Khusbhoo · I. Devgon · G. Kumar
Department of Microbiology, School of Bioengineering and Biosciences, Phagwara, Punjab, India

M. Kaur · D. Bhagat
Department of Biotechnology, School of Bioengineering and Biosciences, Phagwara, Punjab, India

A. R. Mohammad Said Al-Tawaha
Department of Biological Sciences, Al Hussein Bin Talal University, Ma'an, Jordan
e-mail: Abdel-al-tawaha@ahu.edu.jc

A. Karnwal (✉)
Department of Microbiology, Lovely Professional University, Phagwara, Punjab, India

A. Karnwal, A. R. Mohammad Said Al-Tawaha (eds.), *Microbial Applications for Environmental Sustainability*, https://doi.org/10.1007/978-981-97-0676-1_13

13.1 Introduction

Urban and rural soil contaminants consist of washed-away pesticides; organic and chemical fertilizers; radioactive compounds dumped by industries; asbestos; lead; and many other pollutants (Kumar et al. 2022; Zhang et al. 2021). Collectively, these contaminants sweep into the soil through anthropogenic activities (Dror et al. 2022), affecting the soil quality and indirectly affecting crop productivity. Various approaches were formulated to tackle these contaminants to increase crop productivity; however, no notable findings were established. Henceforth, screening for effective yet sustainable strategies to remediate soil contaminants and increase crop productivity is still in progress. However, promising rhizosphere-based remediation results could be helpful (Singh et al. 2016; Dai et al. 2022; Gonçalves and da Delabona 2022).

The soil ecosystem is believed to have a diversified group of organisms involving prokaryotes and eukaryotes. Depending upon the nature of the microorganisms, these can be beneficial or harmful towards plant growth (Zhang et al. 2017). The beneficial organisms, especially prokaryotes, are proven to improve plant growth by disease suppression (Singh et al. 2022). The close association of soil flora to plants, especially roots, is termed root rhizosphere, a whole complex ecosystem (Olanrewaju and Babalola 2022).

Rhizosphere is considered the most complex and active ecosystem (del Orozco-Mosqueda et al. 2022) that relates to plant health and governs both the abiotic and biotic factors surrounding the plants. Remediation using a biological means is termed bioremediation. In this case, the rhizosphere is an agent for the remediation of soil contaminants. Diversifiable speaking, the rhizosphere consists of a microbial consortium (bacteria and arbuscular mycorrhizal fungi mostly) (Gou et al. 2023) which resides closely with the plant root system. The relationship between plants and the rhizosphere is beneficial as plants (Chamkhi et al. 2022) provide nutrients for microbial consortium growth, whereas the consortium protects harmful-to-plant agents. The usefulness of rhizosphere as a bioremediating agent works in either mode, as shown in Fig. 13.1. (1) Microbial consortiums are directly involved in the biodegradation of contaminants through inbuild secretion of enzymes or metabolic bioactive compounds released externally in contact with contaminants, (2) Plant roots induced activation of microbial enzymes or bioactive compounds when plant roots encounter the contaminants (Afridi et al. 2022).

The following review provides a detailed insight into the kind of contaminants found in the soil and microbiome present as root rhizosphere. The review also focuses on the mode of mechanisms of a different group of microbial consortia associated with the bioremediation of common soil contaminants.

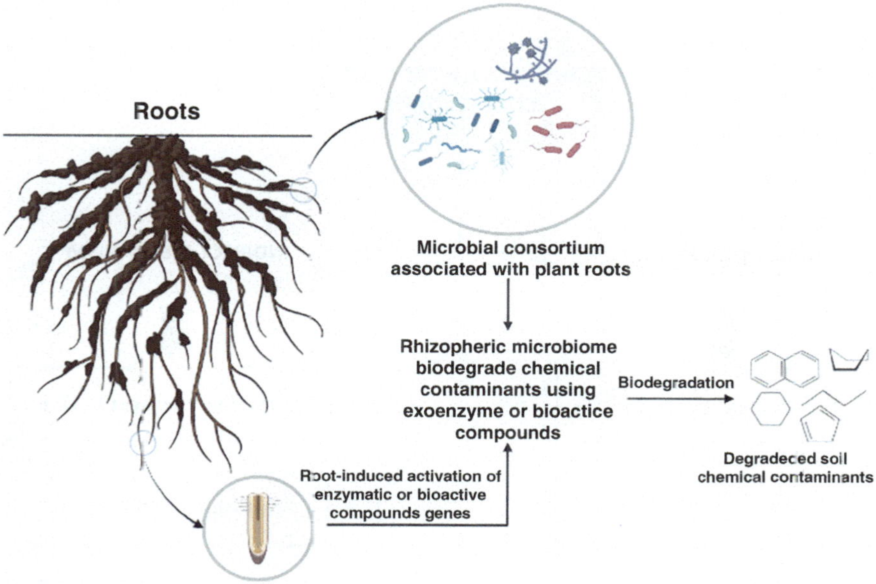

Fig. 13.1 Mode of action of root rhizosphere in bioremediation of soil contaminants

13.2 Plant Growth-Promoting Rhizobacteria (PGPR)

Joe Kloepper and Schroth coined the plant growth-promoting rhizobacteria term in 1978. It was defined as "the soil bacteria that colonize the roots of plants by the following inoculation onto the seed and that enhance plant growth" (Ahemad and Khan 2011). Rhizobacteria act in two mechanisms, i.e., direct and indirect mechanisms, shown in Fig. 13.2.

13.2.1 Direct Mechanism

PGPR includes bio-stimulation of plants against unfavourable environments and stimulates their growth by directly taking minerals present around them. They showcase different functions, like,

1. Nitrogen fixation.
2. Solubilizing of different minerals.
3. Chelated iron siderophores by microorganisms.
4. ACC-deaminase production.

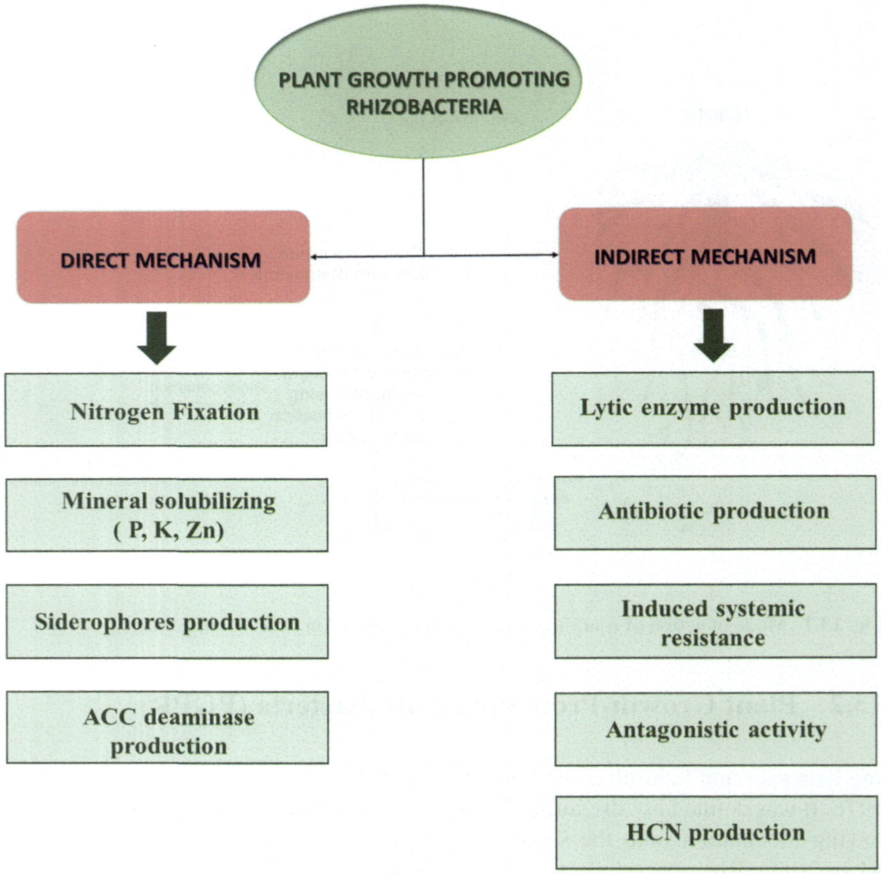

Fig. 13.2 Flow diagram of a mechanism of the PGPR

13.2.2 Indirect Mechanism

The indirect mechanism helps the plant to survive in biotic stress, mainly by plant innate defence activation, disease symptoms reduction post-subsequent infections by pathogens. It produces deleterious products against pests and pathogens (Navarro et al. 2019).

Rhizoremediation of pests includes the biocontrol mechanism of microbes which naturally eradicate the pest and pathogens and help maintain the balance of the plant-rhizosphere zone.

13.3 Rhizosphere and Its Microbiome

The growth and development of a plant are significantly influenced by the soil environment, specifically the region surrounding the roots. The plant creates a microenvironment in which microorganisms play a crucial role. The term "rhizosphere" largely denotes a region distinguished by heightened nutritional activity, enhanced gas exchange, and augmented production of root exudates, all of which contribute to the promotion of plant growth and development. Plants emit carbohydrates, organic acids, and amino acids into the soil throughout their growth, thereby serving as a source of nourishment for soil microorganisms. In reciprocity, the bacteria provide the plants with nutrients in readily utilizable forms and various other benefits. The rhizosphere exhibits a high degree of dynamism in the soil environment, as noted by Berendsen et al. (2018).

13.3.1 Structure of the Rhizosphere

The name "rhizosphere" was initially used by the German scientist Hiltner in 1904, deriving from the Greek word "rhiza" denoting "roots." The rhizosphere is a spatial zone in close proximity to plant roots that harbours diverse microflora populations, which can be influenced by the chemical compounds released by plant sources (Wei et al. 2019). The current definition of the rhizosphere encompasses three distinct zones, namely the endorhizosphere, rhizoplane, and ectorhizosphere. The root, cortex, and endodermis constitute integral components of the endorhizosphere, serving as habitats for endophytic bacteria. The epidermis and mucilage are located in close proximity to the surface of the rhizoplane. The ectorhizosphere is an extension that emerges beyond the soil boundary originating from the rhizosphere. However, it may be more appropriate to use the term "endorhizosphere" when referring to the root portion of the plant, as it specifically distinguishes it from the broader rhizosphere area that encompasses the soil. The concept of the endorhizosphere has contributed to the recognition and understanding of the importance of various microorganisms within the interior region of plant roots. The term used to describe the specific region of soil that is influenced by plant roots is referred to as the ectorhizosphere. The terms "root," "rhizoplane," and "rhizosphere" are currently employed to avoid confusion between the endorhizosphere, rhizoplane, and ectorhizosphere (Esikova et al. 2021).

13.3.2 The Rhizosphere Environment

13.3.2.1 Impact of Physical Change

The various physical attributes of soil, including temperature, humidity, moisture content, and composition, exert a substantial influence on the efficiency with which plants assimilate nutrients. The extension of the rhizosphere is greatly influenced by the physical properties of the soil, as these factors have a direct impact on root development and the movement of ionic and molecular compounds (Compant et al. 2019). Previous studies have shown that the presence of a significant population of arbuscular mycorrhizal (AM) fungus in rhizosphere soils and plant roots can enhance the resilience of adjacent ecosystems in the same rhizosphere area. The soil's physical properties, such as particle size, permeability, and pore size distribution, are notably improved by its substantial water retention capacity.

13.3.2.2 Impact of Chemical Change

The rhizosphere, which represents the interface between plant roots and soil, undergoes a variety of metabolic activities. The aforementioned interactions have a significant impact on various aspects of the soil, such as rhizodeposition, pH, nutrients, and redox potential. Consequently, these factors greatly influence the solubility of nutrients and their subsequent absorption (Hassan et al. 2019b). Plant roots play a crucial role in supporting the overall health and vitality of plants by facilitating the absorption of water and essential nutrients. Additionally, roots contribute to the soil rhizosphere by releasing a variety of organic and inorganic compounds. The secretion of organic and inorganic substances from roots has a significant influence on the availability of nutrients and the composition of rhizosphere microorganisms (Jacoby et al. 2017). These organisms provide defence against the invasion of pathogens, regulate soil moisture levels in the vicinity of plant roots, facilitate the release of different micronutrients, and inhibit the unrestricted growth of competing plant species.

13.3.2.3 Biological Interactions in Rhizosphere

The rhizosphere is characterized by a high level of biological activity, primarily attributed to the presence of nutrient-rich exudates. Certain plant species engage in interactions with specific soil microorganisms. These interactions can have both positive and negative consequences. Beneficial biological interactions in the rhizosphere include mycorrhizae, legume modulators, dinitrogen fixers, nutrient solubilizers, and antibacterial chemical producers. Rhizosphere bacteria have been found to produce several compounds, such as plant hormones, vitamins, antibiotics, and biochemical molecules, which play a crucial role in promoting plant development and mitigating abiotic stress (Gong and Xin 2021). The rhizosphere is an additional crucial site for the introduction of endophytes into plant roots. The critical sites for endophyte colonization within the plant are the proximal regions of root emerging and places of root injury. Subsequently, a complex symbiotic association between plants and endophytes emerges, offering a range of benefits to the plants including enhanced nutrient absorption, synthesis of phytohormones, induction of host resistance, mitigation of abiotic stresses, and production of secondary metabolites (Compant et al. 2019).

13.3.3 Microbial Organisms Associated with the Rhizosphere of Plants

The composition of root exudates released by plants has a direct or indirect impact on the microbial community within the rhizosphere. The root zone is composed of diverse populations of microflora, including bacteria, fungi, and algae, as well as microfauna such as nematodes, protozoa, mites, and insects (Chaparro et al. 2014). The qualitative and statistical analysis of the microbial population and their interactions with the root interface is of utmost importance. According to Fitzpatrick et al. (2020), the involvement in plant health is influenced by the rhizosphere soil, which is characterized by elevated levels of microbial activity.

13.3.3.1 Soil Microflora

Bacteria

As per the findings presented in Bergey's Manual of Systematic Bacteriology, a significant proportion of bacteria found in soil can be classified under the orders Pseudomonadales, Eubacteriales, and Actinomycetales, which belong to the class Schizomycetes. Arthrobacter, Clostridium, Pseudomonas, Enterobacter, Achromobacter, and various other genera have been identified as the predominant bacterial species found in soil ecosystems. The genera of mycobacteria, including Chondrococcus, Micrococcus, Polyangium, Cyptophaga, and Archangiu bacteria, exhibit a broad distribution in soil ecosystems. Arthrobacter, an autochthonous bacteria species found in soil, is widely prevalent and constitutes a significant proportion, ranging from 3% to 70%, of the overall population of soil microflora (Sahu et al. 2016). These organisms can be identified by their tendency to exhibit branching and coccus morphologies. The Pseudomonas genus is widely seen as a prominent kind of Zymogenous bacteria found in soil. It exhibits the ability to degrade a diverse range of chemicals, such as humic acids, hydrocarbons, oils, and numerous synthetic pesticides. The Pseudomonas bacterium demonstrates optimal performance during the process of root exudate secretion. A variety of species produce siderophores that are conjugated with fluorescein and exhibit a high affinity for Fe3+. These siderophores are commonly referred to as diffusible fluorescent pigments. Azotobacter exhibits soil persistence when its application to seeds confers an initial competitive advantage over other bacterial species. Azospirillum, a nitrogen-fixing bacterium, is commonly observed colonizing the roots of grasses in tropical and sub-tropical regions (Rueda et al. 2016). The evaluation of Rhizobium poses challenges because to its high functionality and effectiveness inside the rhizosphere of legumes, while being predominantly suppressed by microbial competition and antagonism in the rhizospheres of other plant species. Actinomycetes, a distinct group of bacteria commonly present in soil, have garnered significant recognition for their remarkable ability to produce antibiotics and effectively impede the proliferation of root-borne illnesses (Soumare et al. 2021).

Fungi

Fungi are a type of eukaryotic organisms that exhibit heterotrophic nutrition and mostly consist of aerobes or a subset of organisms capable of fermentation. The fungal population is predominantly concentrated inside the upper stratum measuring 20–30 cm in depth. At the uppermost strata, the collective biomass of fungi has a comparable magnitude to that of bacteria, and in forest soils, it may even surpass the latter. On average, there exists a population of fungus ranging from 0.001 to 1.0 billion (Hassan et al. 2019a). The rhizosphere harbours both pathogenic and symbiotic fungi, and their prevalence within a particular community is contingent upon a range of soil and plant-related factors. The identification of root exudates has been recognized as a crucial factor in the selection of different rhizosphere fungus (Wen et al. 2021). The genera Aspergillus, Chaetomium, Penicillium, Zygorhynchus, Fusarium, Trichoderma, Rhizopus, Mucor, Verticillium, and others are widely recognized as prominent soil fungi. According to Tedersoo et al. (2020), the fungal partners of endomycorrhizal, such as Glomus, Gigaspora, Sclerocystis, and Acaulospora, as well as ectomycorrhizae, including Gasteromycetes and Agaricales, hold comparable significance in terms of phosphate absorption and pathogen control. Bacteria and fungi contribute to the development and maintenance of soil structure through the process of humus generation. Humus constitutes a crucial component of soil, exerting significant influence on organic molecules, sorption processes, and soil structure. The humus substance plays a crucial part in the development of the soil's crumbly and spongy structure through the formation of mucous capsules and its growth mechanism.

Microfauna

The importance of protozoa feeding on bacteria in the rhizosphere has been demonstrated through plant microcosm studies. There have been reports of enhanced shoot nitrogen and biomass when protozoa and nematode grazers are present. One of the main reasons this happened was that the microbial loop allowed the microfauna to stimulate nitrogen mineralization (Yu et al. 2019a).

Protozoa

Moist soils provide a conducive environment for the development of protozoans. In a gramme of desiccated soil, it is possible for it to sustain populations ranging from a few hundred to several million individuals. Protozoa make up approximately 0.1–0.5 metric tonnes of soil mass each hectare. The majority of organisms within the group are rhizopods, specifically amoebae, and flagellates, although there are also a small number of ciliates. Protozoans exhibit a strong dependence on bacteria as a primary food source, yet they display selectivity in their consumption preferences for specific types of bacteria. Consequently, their influence is discerning in shaping the composition of the bacterial community (Yu et al. 2019b). The Naked Amoebae hold considerable importance as a bacterial grazer group within soil ecosystems, mostly attributed to their substantial biomass, rapid turnover rate, and distinct feeding tactics. These microorganisms generate a biofilm that adheres to both soil and root surfaces, so enabling them to interact with a wide range of soil microbes. According to Hanif et al. (2020), in cases where the water potential in soil is reduced, hence limiting the presence of other protozoa or nematodes, amoeba may have the ability to penetrate bacterial colonies located within soil pore spaces and even within roots that are often inaccessible to other predators. Additionally, it is possible for them to persist in foraging within minuscule water films.

Nematodes

The parasitic nature of nematodes is often influenced by environmental variables. The bulk of plant parasitic forms consist of endoparasites (Heterodera) or ectoparasites such as Tylenchus and Dorylaminus (Mahendra et al. 2020). Nematodes have a high degree of prevalence within the uppermost 30 cm of soil, with their distribution and abundance being influenced by various parameters including soil structure, texture, temperature, moisture, and more variables. The area of soil adjacent to plant roots is commonly referred to as the rhizosphere. The metabolic activity of this zone is influenced by plant roots. In order to optimize the benefits derived from microbial sources in promoting plant growth and health, it is imperative to acknowledge the intricate soil–root connections and ecological interactions and functions within the rhizosphere (Cabal et al. 2021) (Fig. 13.3).

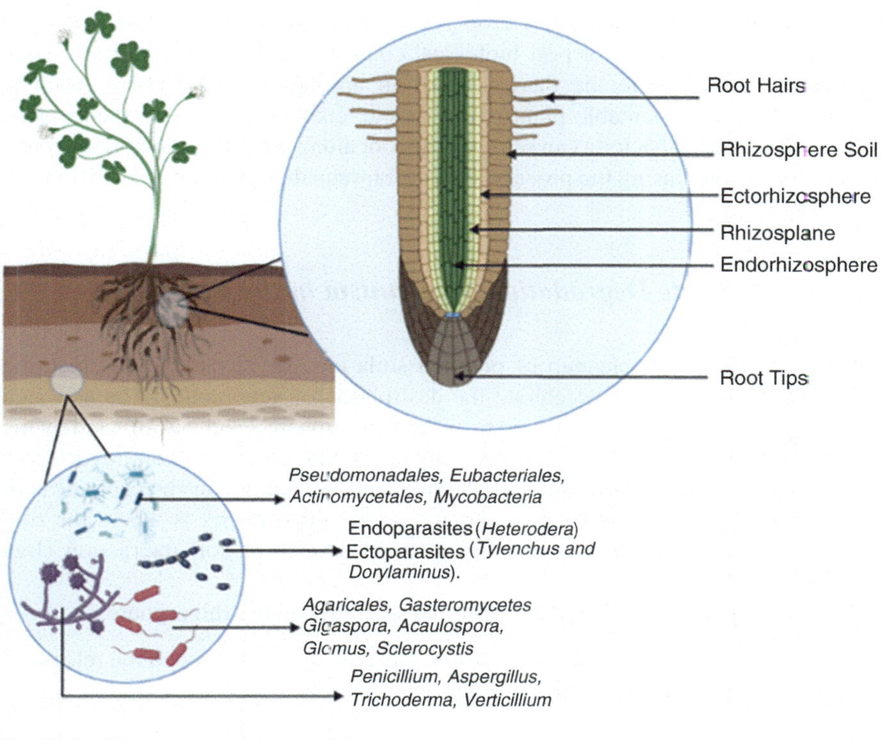

Fig. 13.3 Rhizosphere and its microbiome

13.4 Rhizoremediation of Chemical Pesticides by Microbes

Concerns about environmental pollution are spreading around the globe. Wastes, including complicated inorganic and organic chemicals, were produced due to modern technological breakthroughs, manufacturing, and operations. Environmentalists are now very concerned about how garbage is treated. The amount of harmful environmental pollutants has increased due to pesticide waste produced by the organic or chemical processing in the pesticide industry and their use in commercial, agricultural, and home settings. Pesticide wastes end up in soil and water, polluting the ecosystem. Pesticide contamination of soil, groundwater, and surface water is a severe environmental and biological issue on a global scale (Chi et al. 2018). Rhizospheric microorganisms may break down the bulk of ecological contaminants, and the degradation process ends when the bacterium is starved of food. These microorganisms have access to root exudates, the most significant food supply in the soil. An enrichment technique for microbe isolation has been developed by researchers, which combines the abilities of pollution degradation and superior root colonization (Dangi et al. 2019). Based on the functions of the competent rhizosphere bacterium, the root exudates have named this process "rhizoremediation" rather than "phytoremediation." By damaging the cell membranes and organelles, substituting as a genotoxic substance, disturbing physiological processes like photosynthesis, inactivating protein synthesis, carbohydrate metabolism and respiration, high concentration of metal and their uptake by plants adversely affect plant growth, symbiosis, and ultimately crop yield. *Pseudomonas putida* is a root colonizer that may be useful for pest biological control and rhizoremediation of contaminants. According to the theory, when an appropriate rhizosphere strain is injected alongside a suitable plant (for example, coating bacteria on plant seed), these well-equipped bacteria can settle on the root along with the regular indigenous population, so increasing the process of rhizo-bioremediation (Gan et al. 2018).

13.4.1 Pesticide Degradation Mechanism in the Rhizosphere

Utilizing plants in the cleanup of polluted soils may be advantageous since the chemicals they emit may accelerate the destruction of xenobiotics. The soil area adjoining the plant's roots is referred to as the "rhizosphere," and it is affected by the actions of the live roots. More metabolically active microbial communities are supported in the complex environment of the rhizosphere than in non-rhizospheric soil. The region of soil around the root, known as the rhizosphere, is where the root system impacts microorganisms, creating a dynamic root–soil interaction (Irga et al. 2020).

Three distinct yet connected elements are identified in the rhizosphere:

1. The soil's rhizosphere is the area that is impacted by roots through the release of nutrients that affect microbial activity.

2. Rhizoplane: the surface of the roots that includes soil particles that cling firmly.

A few endophytic microbes (endophytes) can invade root tissue (Irga et al. 2020). The rhizospheric effect is characterized by alterations in the variety of microbial content, its increased numbers and its metabolic activities. The microenvironment is caused by the different physiological, chemical, and biological features of the root-associated soil compared to the other soil. The phrase, rhizoremediation, refers to rhizodegradation and stimulation, highlighting the significance of both the plant and the bacteria in this beneficial relationship (Ma et al. 2015). The rhizosphere's plant–microbial interactions provide a helpful method for cleaning up settings polluted with refractory organic pollutants. Plant roots can take the place of tilling the ground to supply nutrients and promote aeration of the rhizosphere. Different types of grass and leguminous plants have demonstrated that they are ideal for rhizoremediation.

Nutrients are provided in the rhizosphere by the mucigel released by lost root cap cells, starved root cells, or decaying entire roots. Additionally, plants produce a range of organic chemicals derived from photosynthesis. These root exudates include phenolic compounds, amino acids, sugars, proteins, alcohols, organic acids, flavonones, nucleotides, and some enzymes, among other volatile, water-soluble, and insoluble substances. The plant age, the availability of minerals, macro and micronutrients, and the presence of pollutants all affect how quickly a plant exudes (Mishra et al. 2020a). A rhizoremediation process depends on the kind, amount, and intensity of root exudates and the time of exudation. In addition to altering some physiochemical conditions, root exudates drive microbial development and activity in the rhizosphere and help plants acquire nutrients. To manage the metabolic activities of rhizosphere microorganisms, plants may alter the arrangement of root exudates in response to chemical stress in the soil (Mishra et al. 2020a). Root exudates contain certain organic substances that may act as carbon and nitrogen sources for the development and long-term survival of bacteria that may degrade organic contaminants (Irga et al. 2020).

According to Nasr (2019), Rhizomicrobia, which are bacteria residing in the rhizosphere, have the potential to enhance various aspects of plant well-being, such as root growth stimulation, enhanced water and mineral absorption, and inhibition of both pathogenic and non-pathogenic soil microorganisms. Microbial conversion of organic compounds is commonly driven by the imperative to reduce toxicity rather than by energy demands, potentially leading microorganisms to encounter an energy shortfall. Consequently, the activities can be expedited or facilitated by the abundant energy provided by the root exudates. Plants acquire stimulation of soil microbial populations by the release of root exudates, which leads to increased availability of nutrients that are bonded to the soil and the degradation of phytotoxic soil pollutants. These structures facilitate the penetration of roots into deeper layers of soil.

Rhizomicrobia have the ability to enhance the process of humification of organic pollutants, hence expediting the remediation procedures. The catalysis of the polymerization of pollutants onto the humic soil part and root surfaces is primarily

Table 13.1 Different plant species involved in pesticide degradation

Plant species	Pesticide degraded	Process	References
Saccharum officinarum	2,4-Dichlorophenoxyacetic acid	Various 2,4-Dichlorophenoxyacetic acid degrading bacteria present	Salam et al. (2017)
Oryza sativa	Thiocyanate	In the rhizosphere, heterotrophic bacteria have multiplied eight times.	Gao et al. (2022)
Bassia scoparia	Metolachlor, Trifluralin	Degaradation of pesticides	Shi et al. (2022)
Lolium	Chlorpyrifos	Increased mineralization in comparison to soils without a rhizosphere	He et al. (2019)
Pennisetum pedicellatum	Chlorpyrifos Cypermethrin	Chrysobacterium Indologenes	Saravanan et al. (2020a, c)
Zinnia anguistifolia	Mefenoxam	Pseudomonas fluorescens And Chrysobacterium Indologenes	Saravanan et al. (2020a, c)
Pennisetum glaucum	Fenvalerate	Degraders that are enriched explicitly in rhizosphere soil	Nwadinigwe and Obi-Amadi (2014)
Zea mays	Atrazine	An increase in atrazine degradation production	Bhuyan et al. (2022)

facilitated by the release of oxidoreductase enzymes, such as peroxidase, by microorganisms and plant roots (Dangi et al. 2019). According to Salam et al. (2017), bacterial populations frequently exhibit more efficient degradation of contaminants compared to a single species or strain, owing to the presence of partners that effectively exploit various steps of the degradation pathway. The bacterial activity exhibited during the process of rhizoremediation often leads to the degradation of pesticides.

Moreover, it has been observed that different strains exhibit the ability to establish colonies in specific plant root habitats. Notably, the formation of mixed microcolonies and the proximity of various strains were seen exclusively in the immediate region of the contaminant, as reported by Saravanan et al. (2020a, c). However, there is a limited amount of research that addresses the intentional introduction of a microbial community or consortia with the purpose of facilitating the breakdown of xenobiotics, which have the ability to rapidly establish themselves in the root system (Table 13.1).

13.5 Rhizoremediation of Heavy Metals by Microbes

Heavy metals consist of elements with metallic properties in which some are used as essential cofactors for different enzymes, and some are treated as non-essential. The first group includes vital trace elements such as iron, copper, manganese, molybdenum, zinc and cobalt. In contrast, the second group comprises non-essential metals such as lead, mercury, arsenic, cadmium, vanadium, tungsten, and plutonium. Moreover, special attention needs for the non-essential elements as they contain potent toxins during the low level of exposure by entering the cell or tissues. According to The Environmental Protection Agency, lead, mercury, arsenic, and cadmium are the most toxic metals for the environment and constitute a significant threat to life forms (Theron et al. 2012). Rhizoremediation is the process that shows the interaction of the rhizosphere of the plant and the microbes, where both are mutually satisfied. The microorganisms get their essential food from the root exudates of the plant while the plant receives protection and also helps in nitrogen fixation by these microbes (Jeevanantham et al. 2019). Rhizoremediation is mainly categorized into two processes, i.e., an intrinsic process in which the required root exudates can be directly obtained from the plant roots and an engineered process in which the mainly particular microbes are supplied with specific substrates needed. Microbial action, plant–microbe interaction and plant growth-promoting microbes are the mechanisms involved in the rhizoremediation process. The microbial community is vital in removing heavy metals from the rhizosphere. Among them, bacteria and fungi are majorly proven as helpful candidates in the process of rhizoremediation (Table 13.2) Saravanan et al. (2020a, c). It was studied that the inoculum of *Rhizobium* sp., *Enterobacter cloacae, and Pseudomonas* sp. inoculated in the *Vicia faba L.* (Fava bean) plant helped in the reduction of the copper stress under the hydroponic conditions. The pressure due to copper significantly increased the antioxidant activity of the enzymes, while the plant inoculated with strains showed the best growth rate and decreased enzymatic activity levels (Fatnassi et al. 2015). Of the list of toxic heavy metals, cadmium is one of them and is easily absorbed by the plant roots, creating oxidative stress conditions that produce

Table 13.2 Rhizoremediation of heavy metals by different types of microorganisms

Sr. No.	Source of microbes	Microorganisms used	Heavy metals	References
1.	*Festuca arundinacea L.*	*Novosphingobium* sp.	Cu and Pb	Lee et al. (2022)
2.	Soil	*Trichoderma harzianum* and *Bacillus subtilis*	Cd^{2+}, Pb^{2+}, and Zn^{2+}	Henagamage et al. (2022)
3.	*Zea mays* Sturt	*Serratia marcescens, S. Marcescens, Bacillus subtilis* and *Paenibacillus* sp.	Cd	Bhuyan et al. (2022)
4.	Wastewater	Pseudomonas aeruginosa	Pb and Zn	Meliani and Bensoltane (2016)
5.	Soil	*Pseudomonas* sp.	Cr6+	Gupta et al. (2018)
6.	Soil	*Pantoea dispersa*	Cu (II)	Yaashikaa et al. (2020)
7.	Farmland soil	*Purpureocillium* sp.	Cd	Zheng et al. (2021)
8.	*Chlorophytum comosum*	*Acinetobacter beijerinckii* and *Raoultella planticola*	Cr and Ar	Hussain et al. (2023)
9.	Soil	*Staphylococcus arlettae*	Cr	Qadir et al. (2020)
10.	Soil	*Bacillus paramycoides* and *Bacillus subtilis*	Cd	Viji et al. (2022)

ROS. The Arbuscular mycorrhizal fungi were used to eradicate cadmium content in the Solanum lycopersicum plant by increasing their antioxidant enzymes and reinforcing the plant's defence system (Hashem et al. 2016).

13.6 Rhizoremediation of Organic Compounds by Microbes

Compared to bare soil, the soil in rhizospheres frequently has 10–100 times more microorganisms. The microbial communities in rhizospheres create compounds that lower the levels of stress hormones in plants, provide vital nutrients, protect against plant diseases, and degrade pollutants before they harm the plants. Some synthetic organic chemicals of concern as environmental contaminants include pesticides, industrial effluents, petroleum products, dioxins and furans, explosives, and halogenated compounds (McGuinness and Dowling 2009). Rhizoremediation actively remediate soil toxins in the plants' roots, so it is one of the most effective phytoremediation techniques (Ite and Ibok 2019). Plants may be able to digest contaminants in the rhizosphere since they have many kilometres of roots per acre. In the carbon cycle, the assimilation of CO_2 by plants, root exudation, and incorporation into microbial biomass, more significant organic compounds like sugars, organic acids, and more significant organic compounds. Between 10 and 50 per cent of a plant's photosynthate is deposited in the soil (Gerhardt et al. 2009). The general idea of bioremediation is illustrated schematically in Fig. 13.4.

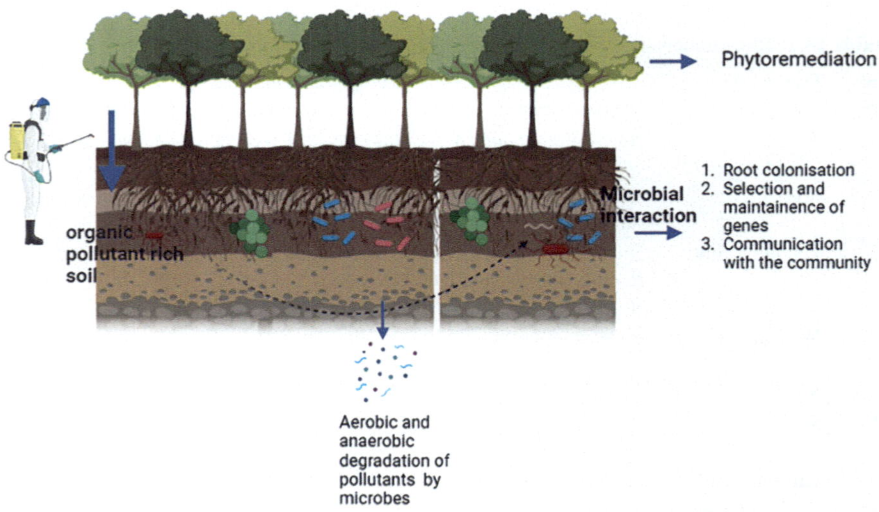

Fig. 13.4 Illustrates the process of rhizoremediation. Pollutants reach the deeper soil. Root exudates attract the microbes resulting in the formation of colonies. Phenolic compounds stimulate the degraders in the rhizospheres' microbes, further speeding up the organic compound's degradation process

Complex aromatic compounds such as flavonoids and coumarins, which share structural similarities with PCBs, PAHs, and PHC, have the potential to enhance microbial colonization of roots. This presents an avenue for the enrichment of analogues and the stimulation of degradative pathways in microorganisms (Holden and Firestone 1997). Rhizoremediation, an integral component of phytoremediation, can occur through natural processes or be initiated by the introduction of pollutant-degrading bacteria or microorganisms that promote plant growth (Megharaj et al. 2011). The requirement to address the existence of pollutants in soils beyond the reach of plant roots involves the implementation of excavation, alternative agronomic methods, or the selection of tree species with greater root depth. This is due to the variability in root depth across herbaceous plants, which is influenced by factors such as plant species, soil composition, and seasonal variations. Rhizoremediation mainly depends on various factors such as:

1. Colonization of large no. of microbes in the rhizosphere.
2. Promotion of primary and secondary catabolic activities.
3. Establishment of ecological communication with the microbial colonies (Kuiper et al. 2004).

The process is influenced by biotic and abiotic factors that include pH, temperature, the types of pollutants present, the condition of the soils, and the local microflora catabolism. Many species have been discovered to be involved in the rhizoremediation of organic substances like PCBs and pesticides, among others (Vyas et al. 2021). *Pseudomonas fluorescence*, *Rhodococcus* species, *Mycobacterium* species, and *Paenibacillus* species are examples of *Pseudomonas*. A few species participate in the breakdown of PAH (Oberai and Khanna 2018).

Organic acids such as citrate, malate, and oxalate are present in the rhizosphere that are essential for several plant processes like nutrient uptake, stress relief, and metal detoxification. They also act as a source of carbon for a variety of microbes. Plants naturally respond to stressful situations by producing more root exudate boosting microbial activity and its biomass in the rhizosphere. Furthermore, root exudates may make the pollutant more accessible for microbial oxidation (Yergeau et al. 2014). Rhizoremediation, or plant-assisted bioremediation, stands out as a possible technique for the efficient microbial breakdown of soil contaminated with petroleum hydrocarbons. The cleaning process may be improved by using the right mix of plant species and bacteria. The types of microorganisms used in the biodegradation of contaminants are listed in Table 13.3.

Table 13.3 Rhizoremediation of organic pollutant

Microorganisms	Organic pollutants	References
Rhodococcus aetherovorans	Phenanthrene	Andreoni et al. (2004)
Panicum bisulcatum	PAH	Lee et al. (2008)
Pseudomonas putida	Phenanthrene	Rentz et al. (2004)
Pseudomonas fluorescens	PAH	Ho et al. (2007)
Burkholderia vietnamiensis BU61	Toluene and trichloroethylene	Filonov et al. (2005)
Bacillus licheniformis, Bacillus mojavensi	PAH	Eskandary et al. (2017)
Rhizobium, pseudomonas, Stenotrophomonas, and *Rhodococcus*	Hydrocarbon	Pawlik et al. (2017)
Pseudomonas	PAH	Shukla et al. (2010)
Mycobacterium vanbaalenii	PAH	Kim et al. (2005)
Pseudomonas putida mt-2	Xylene	Wammer and Peters (2006)
Anabaena oryzae, Chlorella kessleri	Crude oil	Parwani et al. (2021)
Micrococcus luteus	Hydrocarbon	Promsinga et al. (2021)

13.7 Rhizoremediation of Pests by Microbes

Microorganisms are crucial to maintaining a well-functioning ecosystem, since they contribute to various important processes such as ecological balance, stress relief, nutrient recycling, acquisition and mineralization, bioconversion of organic matter, and soil remediation. The rhizosphere of a plant serves as a conducive environment for the proliferation of microorganisms. Plants secrete exudates that provide as a rich source of sugars, amino acids, and metabolites, thereby facilitating the development of a healthy microbial community. The development of microorganisms helps the plant increase its tolerance to changing environments and stabilize its functionality through growth-producing activities. These activities are nitrogen fixation, phosphate solubilizing, potassium solubilizing, siderophores, and zinc solubilizing, which helps create disease suppressing environment around the plant (Singh et al. 2019). The plant–microorganism interaction is symbiotic and/but opposite to it also. The microorganism–microorganism interactions are also established with other rhizosphere interactions. Rhizoremediation, also known as rhizodegradation, degrade many disease-causing pathogens/pests like bacteria fungi and nematodes. It is one of the kinds of microorganism–microorganism interaction. Several factors play a role in the rhizoremediation of pests by microbes.

Pesticides and fungicides are chemically synthesized formulas used to kill the pests and pathogens in/on plants and their rhizosphere. Biogenic controls are formulated to eliminate problems and pathogens and eliminate chemically synthesized procedures' harmful effects. Microorganisms present inside the soil, especially residents of the rhizosphere, are multifunctional in helping the plants in biotic and abiotic stress conditions by different traits and helping them survive in unfavourable conditions. These helping microbes are also known as plant growth-promoting bacteria.

13.7.1 Biocontrol Mechanism of Microbes

13.7.1.1 Antibiosis

Plant growth-promoting rhizobacteria (PGPRs) have the ability to synthesize a diverse array of antibiotics when exposed to adverse environmental conditions. These antibiotics serve as effective biocontrol agents, exhibiting different mechanisms of action, in combating diseases and pests. This phenomenon has been extensively discussed by Glick (2012). Antibiotics are a class of chemical molecules with low molecular weight and heterogeneity, sometimes referred to as oligopeptides. The studies conducted by Kumar et al. (2015) and Maksimov et al. (2011) demonstrate that even at low concentrations, these substances have significant effects against undesired bacteria and other pests. The phenomenon of producing antibiotics to combat phytopathogens is commonly referred to as antibiosis. The presence of a wide range of microorganisms in the rhizosphere leads to the activation of this mechanism, which facilitates competition for nutrient uptake and enhances survival (Nesme and Simonet 2015; Kenawy et al. 2019). In a study conducted by Prasannakumar et al. (2015), it was shown that the PGPR strain Delftia tsuruhatensis WGR-UCM-BT1, which was isolated from Rauwolfia serpentina (L.) Benth, demonstrated noteworthy plant growth-promoting capabilities as well as effective antifungal properties. The bacterial strain under investigation synthesizes a nitrogen-containing heterocyclic chemical, specifically identified as amino(5-(4-methoxyphenyl)-2-methyl-2-(thiophene-2-yl)-2,3-dihydrofuran-3-yl) methanol. This substance, with a molecular formula of C17H19NO3S and abbreviated as AMTM, represents a novel antibiotic within the realm of bacterial production. Phenazine-1-carboxylic acid, oomycin, zwittermicin A, 2,4-diacetylphloroglucinol, kanosamine, pyoluteorin, pyrrolnitrin, and pantocin are among the antibiotics that have been documented and recognized. The regulation of antibiotic synthesis by plant growth-promoting rhizobacteria (PGPR) is governed by a range of intrinsic signalling mechanisms, such as sigma factors, N-acyl homoserine lactones, and gene sequences that are widely conserved. According to Lahiri et al. (2022), antibiotics derived from polyketide, lipopeptide, and heterocyclic nitrogenous chemical families have demonstrated efficacy in combating plant illnesses that hinder plant growth.

13.7.2 Volatile Organic Compound Production

The metabolic activity of bacteria and fungi produces microbial volatile organic compounds (VOCs). According to reports, the VOCs produced by PGPR are molecules that both stimulate plant development and protect plants from infections and pests (Ryu et al. 2003; Lee et al. 2012; Park et al. 2015; Raza et al. 2016; Tahir et al.

2017). Many volatile substances, including terpenes and jasmonates, have been implicated in bacterial signalling and have the potential to promote plant development. Three-hydroxy-2-butanone (acetoin) and 2,3-butanediol, two volatile chemicals produced by the PGPR bacteria *Bacillus subtilis* GB03 and *Bacillus amyloliquefaciens* IN937a, respectively, were used by Ryu et al. in 2003 to stimulate plant development in Arabidopsis thaliana seedlings. Bacterial secreted VOCs have low molecular weight, are lipophilic, have a high vapour pressure, and a low boiling point, resulting in their function in the rhizosphere as short- or long-range signals, influencing competitive microbes and influencing plant health (Mhlongo et al. 2022). This study Mhlongo et al. 2022 used solid-phase micro-extraction gas chromatography-mass spectrometry together with a multivariate data analysis to characterize the released VOCs from four PGPR strains: *Pseudomonas koreensis* N19, *P. fluorescens* N04, *Lysinibacillus sphaericus* T19, and *Paenibacillus alvei* T22 which results in the novel information about the released VOCs blends of the four PGPR strains was discovered using metabolomic profiling and chemometric analysis. *Bacillus amyloliquefaciens* SQR-9 produced the VOCs, a biocontrol strain, and evaluated its result on the growth and virulence traits of tomato wilt pathogen *Ralstonia solanacearum* (Raza et al. 2016).

In their study, Xu et al. (2015) investigated the volatile organic compounds (VOCs) produced by five different bacteria, namely Pseudochrobactrum saccharolyticum, Wautersiella falsenii, Proteus hauseri, Arthrobacter nicotianae, and Achromobacter xylosoxidans. The researchers identified several compounds, including acetophenone, S-methyl thiobutyrate, dimethyl disulfide, ethyl 3,3-dimethylacrylate, nonan-2-one, 1-methoxy-4-methylbenzene, and butyl isovalerate. These compounds were found to exhibit nematicidal activity against Caenorhabditis elegans and Meloidogyne incognita. The substance S-methyl thiobutyrate exhibits superior nematicidal activity compared to the commercial insecticide dimethyl disulfide. Xu et al. were the first to report the nematicidal properties of S-methyl thiobutyrate, as well as other compounds such as ethyl 3,3-dimethylacrylate, 1-methoxy-4-methylbenzene, and butyl isovalerate.

13.7.2.1 Lytic Enzyme Production

Microbial enzymes are vital components for eradicating many pathogens and pests. These are helpful as biocatalysts and help the plant to survive pathogen/attacks by producing hydrolytic enzymes (Mishra et al. 2020b). Lytic enzymes are the most effective and efficient method to deteriorate the pathogens in the rhizosphere's vicinity. In response to attacks by phytopathogens, they exhibit the production of chitinases, cellulases, proteases, and β-glucanases. Numerous microorganisms have been documented for their ability to employ enzymatic mechanisms in the management of pests. *Lysobacter capsici* YS1215 produces lytic enzymes, chitinase and gelatinase, which suppresses the infection of *Meloidogyne incognita*, causing root-knot infection in tomatoes (Lee et al. 2015). In 1998, Spiegel and Chet tested *Trichoderma harzianum*, and *T. lignorum* isolates against the root-knot nematode

Table 13.4 PGPR biocontrol against pathogens

PGPR	Plant causative organisms	Action	References
Bacillus siamensis, Bacillus amyloliquefaciens, Bacillus velezensis, Bacillus methylotrophicus	*Phytophthora capsici; Phythophthora blight*	Lytic enzyme production, antimicrobial production and nutrient competition	Ngo et al. (2020)
PGPR	*P. sojae* Soyabean root rot	Antagonistic activity	
PGPR	*Fusarium oxysporum* f.sp. *ciceris*; chickpea wilt	Enzymes including the phenylalanine ammonia lyase, peroxidase, polyphenol peroxidase,β-1,3 glucanase	Fatima et al. (2022)
Burkholderia cepacia AMMDR1	*Pythium aphanidermatum* and *Aphanomyces euteiches* zoospore	Antibiosis, colonization competition	Heungens and Parke (2000)
Corynebacterium flavescens, Sporosarcina aquimarina Sporosarcina saromensis	*Phytophtora medicaginis*; *Alternaria brassicicola*; *Pseudomonas syringae* pv. Tomato	Antagonistic activity, lytic enzyme production, colonization competition	Booth et al. (2022)
Delftia tsuruhatensis WGR–UOM–BT1		Antibiotic production, antifungal activity	Prasannakumar et al. (2015)
Bacillus subtilis GB03; *Bacillus amyloliquefaciens* IN937a		Antibiotic production, growth enhancer	

Meloidogyne javanica, which show an increase in plant growth and a decrease in egg numbers and a decrease in gall index also. The bacterial strain *Achromobacter xylosoxidans* was isolated from common beetroot for antagonistic activity against *Fusarium solani,* causing root rotting of common beet. The positive result is due to the production of different lytic enzymes (chitinase, protease and pectinase) and other PGPR traits like siderophore production of volatile compounds production (Mohamadpoor et al. 2022). Spiegel and Chet isolated a novel collagenolytic/proteolytic enzyme from Bacillus cereus in 1998 which results in digestion of bovine serum albumin and degrades the nematode cuticle of root-knot, causing *Meloidogyne javanica.* Table 13.4 summarizes PGPR and its mechanisms involved in different disease-causing organisms of plants.

13.8 Conclusion and Future Aspects

The rhizosphere is a multifaceted ecological system that encompasses the region surrounding plant roots, and the microbiome inhabiting the root system remains incompletely understood. The process of rhizosphere-based bioremediation in soil is influenced by both biotic and abiotic variables, which might impede its efficiency. Gaining comprehension and effectively addressing these aspects will yield valuable insights into the operational intricacies of the bioremediation process. It is also necessary to comprehend how plants respond to plant growth-promoting microorganisms or rhizospheric bacteria. Furthermore, the exploration of various interactions among rhizospheric bacteria will contribute to a more comprehensive understanding of bioremediation, allowing for more significant advancements in this field of study. In conclusion, the implementation of specific methodologies such as bioaugmentation, rhizo-engineering, and bio-stimulation has promise for enhancing the efficacy of rhizospheric bioremediation in forthcoming endeavours.

Acknowledgement The author would like to express gratitude and appreciation to individuals who have contributed to the completion of The authors express their sincere gratitude to the Research and Development department for providing the essential resources required for this study.

Conflict of Interest No conflicting interest was found.

References

Afridi MS, Fakhar A, Kumar A, Ali S, Medeiros FHV, Muneer MA, Ali H, Saleem M (2022) Harnessing microbial multitrophic interactions for rhizosphere microbiome engineering. Microbiol Res 265:127199. https://doi.org/10.1016/J.MICRES.2022.127199

Ahemad M, Khan MS (2011) Response of greengram [Vigna radiata (L.) Wilczek] grown in herbicide–amended soil to quizalafop– p–ethyl and clodinafop tolerant plant growth promoting Bradyrhizobium sp. (vigna) MRM6. J Agric Sci Technol 13:1209–1222

Andreoni V, Cavalca L, Rao MA, Nocerino G, Bernasconi S, Dell'Amico E, Gianfreda L (2004) Bacterial communities and enzyme activities of PAHs polluted soils. Chemosphere 57(5):401–441

Berendsen RL, Vismans G, Yu K, Song Y, de Jonge R, Burgman WP et al (2018) Disease-induced assemblage of a plant-beneficial bacterial consortium. ISME J 12:1496–1507. https://doi.org/10.1038/s41396-018-0093-1

Bhuyan B, Kotoky R, Maheshwari DK, Pandey P (2022) Rhizoremediation of cd-contaminated soil using Zea mays Sturt, with heavy metal resistant rhizobacteria that alleviate cd-induced stress in plant. Environ Sustain 5(3):375–387

Booth J, Schenk PM, Mirzaee H (2022) Microbial biopesticides against bacterial, fungal and oomycete pathogens of tomato, cabbage and chickpea. Appl Microbiol 2(1):288–301

Cabal C, De Deurwaerder HPT, Matesanz S (2021) Field methods to study the spatial root density distribution of individual plants. Plant Soil 462:25–43

Chamkhi I, Cheto S, Geistlinger J, Zeroual Y, Kouisni L, Bargaz A et al (2022) Legume-based intercropping systems promote beneficial rhizobacterial community and crop yield under stressing conditions. Ind Crop Prod 183:114958. https://doi.org/10.1016/j.indcrop.2022.114958

Chaparro JM, Badri DV, Vivanco JM (2014) Rhizosphere microbiome assemblage is affected by plant development. ISME J 8:790–803. https://doi.org/10.1038/ismej.2013.196
Chi X, Zhang Y, Wang D et al (2018) The greater roles of indigenous microorganisms in removing nitrobenzene from sediment compared with the exogenous *Phragmites australis* and strain JS45. Front Environ Sci Eng 12:1–11. https://doi.org/10.1007/s11783-018-1016-0
Compant S, Samad A, Faist H, Sessitsch A (2019) A review on the plant microbiome: ecology, functions, and emerging trends in microbial application. J Adv Res 19:29–37. https://doi.org/10.1016/j.jare.2019.03.004
Dai Y, Liu R, Chen J, Li N (2022) Bioremediation of HMW-PAHs-contaminated soils by rhizosphere microbial community of fire Phoenix plants. Chem Eng J 432:134246. https://doi.org/10.1016/J.CEJ.2021.134246
Dangi AK, Sharma B, Hill RT, Shukla P (2019) Bioremediation through microbes: systems biology and metabolic engineering approach. Crit Rev Biotechnol 39:79–98
del Orozco-Mosqueda MC, Fadiji AE, Babalola OO, Glick BR, Santoyo G (2022) Rhizobiome engineering: unveiling complex rhizosphere interactions to enhance plant growth and health. Microbiol Res 263:127137. https://doi.org/10.1016/J.MICRES.2022.127137
Dror I, Yaron B, Berkowitz B (2022) The human impact on all soil-forming factors during the anthropocene. ACS Environ Au 2(1):11–19. https://doi.org/10.1021/ACSENVIRONAU.1C00010/ASSET/IMAGES/LARGE/VG1C00010_0003.JPEG
Esikova TZ, Anokhina TO, Abashina TN, Suzina NE, Solyanikova IP (2021) Characterization of soil bacteria with potential to degrade benzoate and antagonistic to fungal and bacterial phytopathogens. Microorganisms 9(4):755
Eskandary S, Tahmourespour A, Hoodaji M, Abdollahi A (2017) The synergistic use of plant and isolated bacteria to clean up polycyclic aromatic hydrocarbons from contaminated soil. J Environ Health Sci Eng 15:12
Fatima I, Hakim S, Imran A, Ahmad N, Imtiaz M, Ali H, Mubeen F (2022) Exploring biocontrol and growth-promoting potential of multifaceted PGPR isolated from natural suppressive soil against the causal agent of chickpea wilt. Microbiol Res 260:127015
Fatnassi IC, Chiboub M, Saadani O, Jebara M, Jebara SH (2015) Impact of dual inoculation with rhizobium and PGPR on growth and antioxidant status of Vicia faba L. under copper stress. C R Biol 338(4):241–254
Filonov A, Akhmetov L, Puntus I, Esikova T, Gafarov A, Izmalkova TY, Sokolov S, Kosheleva I, Boronin A (2005) The construction and monitoring of genetically tagged, plasmid-containing, naphthalene-degrading strains in soil. Microbiology 74:453–458
Fitzpatrick CR, Salas-González I, Conway JM, Finkel OM, Gilbert S, Russ D et al (2020) The plant microbiome: from ecology to reductionism and beyond. Annu Rev Microbiol 74:81–100. https://doi.org/10.1146/annurev-micro-022620-014327
Gan L, Zhao H, Wang A et al (2018) *Pseudomonas putida* inoculation promotes submerged plant Vallisneria natans growth by carbon conversion in a plant-microbe interaction. Mar Freshw Res 69:851–858. https://doi.org/10.1071/MF17117
Gao JJ, Wang B, Li ZJ, Xu J, Fu XY, Han HJ, Yao QH (2022) Metabolic engineering of Oryza sativa for complete biodegradation of thiocyanate. Sci Total Environ 820:153283
Gerhardt KE, Huang X-D, Glick BR, Greenberg BM (2009) Phytoremediation and rhizoremediation of organic soil contaminants: potential and challenges. Plant Sci 176 20–30
Glick BR (2012) Plant growth-promoting bacteria: mechanisms and applications. Scientifica 2012:1
Gonçalves CR, da Delabona PS (2022) Strategies for bioremediation of pesticides: challenges and perspectives of the Brazilian scenario for global application—a review. Environ Adv 8:100220. https://doi.org/10.1016/J.ENVADV.2022.100220
Gong T, Xin X-F (2021) Phyllosphere microbiota: community dynamics and its interaction with plant hosts. J Integr Plant Biol 63:297–304. https://doi.org/10.1111/jipb.13060
Gou X, Ni H, Sadowsky MJ, Chang X, Liu W, Wei X (2023) Arbuscular mycorrhizal fungi alleviate erosion-induced soil nutrient losses in experimental agro-ecosystems. *CATENA* 220:106687. https://doi.org/10.1016/J.CATENA.2022.106687

Gupta P, Rani R, Chandra A, Kumar V (2018) Potential applications of pseudomonas sp.(strain CPSB21) to ameliorate Cr6+ stress and phytoremediation of tannery effluent contaminated agricultural soils. Sci Rep 8(1):1–10

Hanif MK, Malik KA, Hameed S, Saddique MJ, Fatima K, Naqqash T et al (2020) Growth stimulatory effect of AHL producing Serratia spp. from potato on homologous and non-homologous host plants. Microbiol Res 238:126506. https://doi.org/10.1016/j.micres.2020.126506

Hashem A, Abd Allah EF, Alqarawi AA, Al Huqail AA, Egamberdieva D, Wirth S (2016) Alleviation of cadmium stress in Solanum lycopersicum L. by arbuscular mycorrhizal fungi via induction of acquired systemic tolerance. Saudi J Biol Sci 23(2):272–281

Hassan MK, Mcinroy JA, Jones J, Shantharaj D, Liles MR, Kloepper JW (2019a) Pectin-rich amendment enhances soybean growth promotion and nodulation mediated by bacillus velezensis strains. Plan Theory 8:120. https://doi.org/10.3390/plants8050120

Hassan MK, Mcinroy JA, Kloepper JW (2019b) The interactions of rhizodeposits with plant growth-promoting rhizobacteria in the rhizosphere: a review. Agriculture 9:142. https://doi.org/10.3390/agriculture9070142

He S, Guo H, He Z et al (2019) Interaction of Lolium perenne and Hyphomicrobium sp. GHH enhances the removal of 17α-ethinyestradiol (EE2) from soil. J Soils Sedim 19:1297–1305. https://doi.org/10.1007/s11368-018-2116-y

Henagamage AP, Peries CM, Seneviratne G (2022) Fungal-bacterial biofilm mediated heavy metal rhizo-remediation. World J Microbiol Biotechnol 38(5):1–13

Heungens K, Parke J (2000) Zoospore homing and infection events: effects of the biocontrol bacterium Burkholderia cepacia AMMDR1 on two oomycete pathogens of pea (Pisum sativum L.). Appl Environ Microbiol 66(12):5192–5200

Hiltner L (1904) UÈ ber neuere Erfahrungen und Probleme auf dem Gebiet der Bodenback teriologie und unter besonderer BeruÈcksichtigung der GruÈnduÈngung und Brache. ArbeitenDeutscher Landwirtschafts Gesellschaft 98:59–78

Ho CH, Applegate B, Banks MK (2007) Impact of microbial/plant interactions on the transformation of polycyclic aromatic hydrocarbons in rhizosphere of Festuca arundinacea. Int J Phytoremediation 9(2):107–114

Holden PA, Firestone MK (1997) Soil microorganisms in soil cleanup: How can we improve our understanding? J Environ Qual 26:32–40

Hussain A, Shah M, Hamayun M, Iqbal A, Qadir M, Alataway A, Lee IJ (2023) Phytohormones producing rhizobacteria alleviate heavy metals stress in soybean through multilayered response. Microbiol Res 266:127237

Irga PJ, Shagol CC, Kim KJ, Pettit T, Torpy FR (2020) Plant–microbe interaction within phytosystems used for air treatment. In: From biofiltration to promising options in gaseous fluxes biotreatment. Elsevier. https://doi.org/10.1016/b978-0-12-819064-7.00012-1

Ite AE, Ibok UJ (2019) Role of plants and microbes in bioremediation of petroleum hydrocarbons contaminated soils. Int J Environ Bioremediat Biodegrad 7(1):1–19

Jacoby R, Peukert M, Succurro A, Koprivova A, Kopriva S (2017) The role of soil microorganisms in plant mineral nutrition-current knowledge and future directions. Front Plant Sci 8:1617. https://doi.org/10.3389/fpls.2017.01617

Jeevanantham S, Saravanan A, Hemavathy RV, Kumar PS, Yaashikaa PR, Yuvaraj D (2019) Removal of toxic pollutants from water environment by phytoremediation: a survey on application and future prospects. Environ Technol Innov 13:264–276

Kenawy A, Dailin DJ, Abo-Zaid GA, Malek RA, Ambehabati KK, Zakaria KHN, Enshasy HAE (2019) Biosynthesis of antibiotics by PGPR and their roles in biocontrol of plant diseases. In: Plant growth promoting rhizobacteria for sustainable stress management. Springer, Singapore, pp 1–35

Kim YH, Freeman JP, Moody JD, Engesser KH, Cerniglia CE (2005) Effects of pH on the degradation of phenanthrene and pyrene by mycobacterium vanbaalenii. Appl Microbiol Biotechnol 67:275–285

Kuiper I, Lagendijk EL, Bloemberg GV, Lugtenberg BJJ (2004) Rhizoremediation: a beneficial plant-microbe interaction. Mol Plant-Microbe Interact 17(1):6–15

Kumar AJAY, Vandana RS, Singh MONIKA, Pandey KD (2015) Plant growth promoting rhizobacteria (PGPR). A promising approach for disease management. In: Microbes and environmental management. Studium Press, New Delhi, pp 195–209

Kumar M, Bolan N, Jasemizad T, Padhye LP, Sridharan S, Singh L, Bolan S, O'Connor J, Zhao H, Shaheen SM, Song H, Siddique KHM, Wang H, Kirkham MB, Rinklebe J (2022) Mobilization of contaminants: potential for soil remediation and unintended consequences. Sci Total Environ 839:156373. https://doi.org/10.1016/J.SCITOTENV.2022.156373

Lahiri D, Nag M, Sayyed RZ, Gafur A, Ansari MJ, Ray RR (2022) PGPR in biofilm formation and antibiotic production. In: Antifungal metabolites of rhizobacteria for sustainable agriculture. Springer, Cham, pp 65–82

Lee SH, Lee WS, Lee CH, Kim JG (2008) Degradation of phenanthrene and pyrene in rhizosphere of grasses and legumes. J Hazard Mater 153(1–2):892–898

Lee IM, Shiroma EJ, Lobelo F, Puska P, Blair SN, Katzmarzyk PT (2012) Effect of physical inactivity on major non-communicable diseases worldwide: an analysis of burden of disease and life expectancy. Lancet 380(9838):219–229

Lee YS, Nguyen XH, Naing KW, Park YS, Kim KY (2015) Role of lytic enzymes secreted by Lysobacter capsici YS1215 in the control of root-knot nematode of tomato plants Indian J Microbiol 55(1):74–80

Lee SY, Lee YY, Cho KS (2022) Effect of Novosphingobium sp. CuT1 inoculation on the rhizoremediation of heavy metal-and diesel-contaminated soil planted with tall fescue. Environ Sci Pollut Res 1:1–14

Ma Y, Oliveira RS, Wu L et al (2015) Inoculation with metal-mobilizing plant-growth-promoting Rhizobacterium bacillus sp. SC2b and its role in rhizoremediation. J Toxicol Environ Health A 78:931–944. https://doi.org/10.1080/15287394.2015.1051205

Mahendra R, Chauhan N, Sharma JB, Rana K, Bakshi M (2020) Ex vitro establishment of tissue cultured plants in fruit crops-a review. Int J Curr Microbiol App Sci 9:3321–3329

Maksimov IV, Abizgil RR, Pusenkova LI (2011) Plant growth promoting rhizobacteria as alternative to chemical crop protectors from pathogens (review). Appl Biochem Microbiol 47(4):373–385. https://doi.org/10.1134/S0003683811040090

McGuinness M, Dowling D (2009) Plant-associated bacterial degradation of toxic organic compounds in soil. Int J Environ Res Public Health 6(8):2226–2247

Megharaj M, Ramakrishnan B, Venkateswarlu K, Sethunathan N, Naidu R (2011) Bioremediation approaches for organic pollutants: a critical perspective. Environ Int 37(8):1362–1375

Meliani A, Bensoltane A (2016) Biofilm-mediated heavy metals bioremediation in PGPR pseudomonas. J Bioremed Biodegr 7(370):2

Mhlongo MI, Piater LA, Dubery IA (2022) Profiling of Volatile Organic Compounds from Four Plant Growth-Promoting Rhizobacteria by SPME-GC-MS: A Metabolomics Study. Meta 12(8):763. https://doi.org/10.3390/metabo12080763

Mishra A, Mishra SP, Arshi A et al (2020a) Plant-microbe interactions for bioremediation and phytoremediation of environmental pollutants and agro-ecosystem development. In: Bioremediation of industrial waste for environmental safety: volume II: biological agents and methods for industrial waste management. https://doi.org/10.1007/978-981-13-3426-9_17

Mishra P, Mishra J, Dwivedi SK, Arora NK (2020b) Microbial enzymes in biocontrol of phytopathogens. In: Microbial enzymes: roles and applications in industries. Springer, Singapore, pp 259–285

Mohamadpoor M, Amini J, Ashengroph M, Azizi A (2022) Evaluation of biocontrol potential of Achromobacter xylosoxidans strain CTA8689 against common bean root rot. Physiol Mol Plant Pathol 117:101769

Nasr M (2019) Environmental perspectives of plant-microbe nexus for soil and water remediation. In: Kumar V, Prasad R, Kumar M, Choudhary D (eds) Microbiome in plant health and disease. Springer, Singapore. https://doi.org/10.1007/978-981-13-8495-0_18

Navarro P, Arias Á, Arias-González L, Laserna-Mendieta EJ, Ruiz-Ponce M, Lucendo AJ (2019) Systematic review with meta-analysis: the growing incidence and prevalence of eosinophilic oesophagitis in children and adults in population-based studies. Aliment Pharmacol Ther 49(9):1116–1125. https://doi.org/10.1111/apt.15231

Nesme J, Simonet P (2015) The soil resistome: a critical review on antibiotic resistance origins, ecology and dissemination potential in telluric bacteria. Environ Microbiol 17(4):913–930

Ngo VA, Wang SL, Nguyen VB, Doan CT, Tran TN, Tran DM, Nguyen AD (2020) Phytophthora antagonism of endophytic bacteria isolated from roots of black pepper (Piper nigrum L.). Agronomy 10(2):286

Nwadinigwe AO, Obi-Amadi A (2014) Crude oil degrading potential of Pennisetum glaucum (L.) R. Br Afr J Biotechnol 13(34):3489–3495

Oberai M, Khanna V (2018) Rhizoremediation–plant microbe interactions in the removal of pollutants. Int J Curr Microbiol App Sci 7(1):2280–2287

Olanrewaju OS, Babalola OO (2022) The rhizosphere microbial complex in plant health: a review of interaction dynamics. J Integr Agric 21(8):2168–2182. https://doi.org/10.1016/S2095-3119(21)63817-0

Park KH, Kim DH, Kim SK, Yi YH, Jeong JH, Chae J, Hwang J, Roh H (2015) The relationships between empathy, stress and social support among medical students. Int J Med Educ 6:103–108. https://doi.org/10.5116/ijme.55e6.0d44

Parwani L, Bhatt M, Singh J (2021) Potential biotechnological applications of cyanobacterial exopolysaccharides. Braz Arch Biol Technol 64:1

Pawlik M, Cania B, Thijs S, Vangronsveld J, Piotrowska-Seget Z (2017) Hydrocarbon degradation potential and plant growth-promoting activity of culturable endophytic bacteria of Lotus corniculatus and Oenothera biennis from a long-term polluted site. Environ Sci Pollut Res Int 24(24):19640–19652

Prasannakumar SP, Gowtham HG, Hariprasad P, Shivaprasad K, Niranjana SR (2015) Delftia tsuruhatensis WGR–UOM–BT1, a novel rhizobacterium with PGPR properties from Rauwolfia serpentina (L.) Benth. Ex Kurz also suppresses fungal phytopathogens by producing a new antibiotic—AMTM. Lett Appl Microbiol 61(5):460–468

Promsinga S, Pokethitiyooka P, Kruatrachuea M, Ounjaia P, Meetama M, Onparna N, Kumsopac A (2021) Rhizoremediation of fuel oil by Vetiveria zizanioides in association with Kocuria sp. no. MU1 and Micrococcus luteus WN01. J Sci Soc Thail 47:96–105

Qadir M, Hussain A, Hamayun M, Shah M, Iqbal A, Murad W (2020) Phytohormones producing rhizobacterium alleviates chromium toxicity in Helianthus annuus L. by reducing chromate uptake and strengthening antioxidant system. Chemosphere 258:127386

Raza W, Ling N, Yang L, Huang Q, Shen Q (2016) Response of tomato wilt pathogen Ralstonia solanacearum to the volatile organic compounds produced by a biocontrol strain bacillus amyloliquefaciens SQR-9. Sci Rep 6(1):1–13

Rentz JA, Alvarez PJ, Schnoor JL (2004) Repression of pseudomonas putida phenanthrene-degrading activity by plant root extracts and exudates. Environ Microbiol 6(6):574–583

Rueda D, Valencia G, Soria N, Rueda BB, Manjunatha B, Kundapur RR, Selvanayagam M (2016) Effect of Azospirillum spp. and Azotobacter spp. on the growth and yield of strawberry (Fragaria vesca) in hydroponic system under different nitrogen levels. J Appl Pharm Sci 6(01):048–054

Ryu C-M, Farag MA, Hu C-H, Reddy MS, Wei HX, Paré PW, Kloepper JW (2003) Bacterial volatiles promote growth in Arabidopsis. Proc Natl Acad Sci USA 100:4927–4932

Sahu PK, Sharma L, Gupta A, Renu (2016) Rhizospheric and endophytic beneficial microorganisms:treasure for biological control of plant pathogens. In: Santra S, Mallick A (eds) Recent biotechnological applications in India, ENVIS centre on environmental biotechnology. University of Kalyani, West Bengal, pp 50–63

Salam JA, Hatha MAA, Das N (2017) Microbial-enhanced lindane removal by sugarcane (Saccharum officinarum) in doped soil-applications in phytoremediation and bioaugmentation. J Environ Manag 193:394–399. https://doi.org/10.1016/j.jenvman.2017.02.006

Saravanan A, Jeevanantham S, Narayanan VA, Kumar PS, Yaashikaa PR, Muthu CM (2020a) Rhizoremediation–A promising tool for the removal of soil contaminants: a review. J Environ Chem Eng 8(2):103543

Saravanan A, Jeevanantham S, Narayanan VA et al (2020c) Rhizoremediation-A promising tool for the removal of soil contaminants: a review, vol 8. Elsevier, Amsterdam, p 103543

Shi R, Liang L, Liu W, Zeb A (2022) Kochia scoparia L., a newfound candidate halophyte, for phytoremediation of cadmium-contaminated saline soils. Environ Sci Pollut Res 1:1–10

Shukla KP, Singh NK, Sharma S (2010) Bioremediation: developments, current practices and perspectives. Genet Eng Biotechnol J 3:1–20

Singh N, Srivastava S, Rathaur S, Singh N (2016) Assessing the bioremediation potential of arsenic tolerant bacterial strains in rice rhizosphere interface. J Environ Sci 48:112–119. https://doi.org/10.1016/J.JES.2015.12.034

Singh RJ, Deshwal JS, Sharma NK, Ghosh BN, Bhattacharyya R (2019) Effects of conservation tillage based agro-geo-textiles on resource conservation in sloping croplands of Indian Himalayan Region. Soil Tillage Res 191:37–47. https://doi.org/10.1016/j.still.2019.03.012

Singh K, Chandra R, Purchase D (2022) Unraveling the secrets of rhizobacteria signaling in rhizosphere. Rhizosphere 21:100484. https://doi.org/10.1016/J.RHISPH.2022.100484

Soumare A, Diédhiou AG, Arora NK, Tawfeeq Al-Ani LK, Ngom M, Fall S, Sy MO (2021) Potential role and utilization of plant growth promoting microbes in plant tissue culture. Front Microbiol 12:649878

Spiegel Y, Chet I (1998) Evaluation of Trichoderma spp. as a biocontrol agent against soilborne fungi and plant-parasitic nematodes in Israel. Integr Pest Manag Rev 3(3):169–175

Tahir HA, Gu Q, Wu H, Raza W, Hanif A, Wu L, Colman MV, Gao X (2017) Plant growth promotion by volatile organic compounds produced by Bacillus subtilis SYST2. Front Microbiol 8:171

Tedersoo L, Bahram M, Zobel M (2020) How mycorrhizal associations drive plant population and community biology. Science 367:eaba1223. https://doi.org/10.1126/science.aba1223

Theron AJ, Tintinger GR, Anderson R (2012) Harmful interactions of non-essential heavy metals with cells of the innate immune system, p s3

Viji AS, Antony BT, Wagh MS, Osborne WJ (2022) Bioremoval of cadmium by co-cultivated bacterial strains, Bacillus paramycoides and Bacillus subtilis, in a pilot-scale phyto-and rhizoremediation approach. Int J Environ Sci Technol 19(8):7565–7574

Vyas P, Singh S, Dhaka V, Dhanjal DS, Singh J (2021) Rhizoremediation for efficient management of inorganic and organic contaminants in soil. In: Microbe mediated remediation of environmental contaminants. Woodhead Publishing, pp 95–106

Wammer KH, Peters CA (2006) A molecular modeling analysis of polycyclic aromatic hydrocarbon biodegradation by naphthalene dioxygenase. Environ Toxicol Chem 25:912–920

Wei Z, Gu Y, Friman V-P, Kowalchuk GA, Xu Y, Shen Q et al (2019) Initial soil microbiome composition and functioning predetermine future plant health. Sci Adv 5:eaaw0759. https://doi.org/10.1126/sciadv.aaw0759

Wen T, Zhao M, Yuan J, Kowalchuk GA, Shen Q (2021) Root exudates mediate plant defense against foliar pathogens by recruiting beneficial microbes. Soil Ecol Lett 3:42–51. https://doi.org/10.1007/s42832-020-0057-z

Xu P, Ma W, Han H, Jia S, Hou B (2015) Isolation of a Naphthalene-Degrading Strain from Activated Sludge and Bioaugmentation with it in a MBR Treating Coal Gasification Wastewater. Bull Environ Contam Toxicol 94:358–364. https://doi.org/10.1007/s00128-014-1366-7

Yaashikaa PR, Senthil Kumar P, Varjani S, Saravanan A (2020) Rhizoremediation of Cu (II) ions from contaminated soil using plant growth promoting bacteria: an outlook on pyrolysis conditions on plant residues for methylene orange dye biosorption. Bioengineered 11(1):175–187

Yergeau E, Sanschagrin S, Maynard C, St-Arnaud M, Greer CW (2014) Microbial expression profiles in the rhizosphere of willows depend on soil contamination. ISME J 8(2):344–358

Yu K, Liu Y, Tichelaar R, Savant N, Lagendijk E, van Kuijk SJL et al (2019a) Rhizosphere-associated pseudomonas suppress local root immune responses by gluconic acid-mediated lowering of environmental pH. Curr Biol 29:3913–3920. https://doi.org/10.1016/j.cub.2019.09.015

Yu K, Pieterse CMJ, Bakker PAHM, Berendsen RL (2019b) Beneficial microbes going underground of root immunity. Plant Cell Environ 42:2860–2870. https://doi.org/10.1111/pce.13632
Zhang R, Vivanco JM, Shen Q (2017) The unseen rhizosphere root–soil–microbe interactions for crop production. Curr Opin Microbiol 37:8–14. https://doi.org/10.1016/J.MIB.2017.03.008
Zhang LJ, Qian L, Ding LY, Wang L, Wong MH, Tao HC (2021) Ecological and toxicological assessments of anthropogenic contaminants based on environmental metabolomics. Environ Sci Ecotechnol 5:100081. https://doi.org/10.1016/J.ESE.2021.100081
Zheng X, Yang S, Chen L, Kimotho RN, Chen M, Chen J, Li X (2021) A newly-isolated cd-loving Purpureocillium sp. strain YZ1 substantially alleviates cd toxicity to wheat. Plant Soil 464(1):289–302

Chapter 14
Microbial Nanotechnology: Nanoparticles for Bioremediation of Toxic Pollutants

Vipul Kumar, Shweta Meshram, Akhilesh Chandrapati, Amit Mandlik, Adesh Kumar, and Geetika

Abstract The interest in nanotechnology has increased recently among both the scientific and business worlds. In terms of preserving the environment, generating energy, reducing pollution, and cleaning up after it, nanotechnology has a lot more to offer. For commercial purposes, removing contaminants from industrial pollutants is a significant challenge These pollutants impart serious risk to the ecosystem and nanotechnology can reduce the amount of money spent by businesses to counteract these pollutants by producing environmentally friendly nanomaterials. Nanomaterials are becoming more and more common due to their enhanced physical, chemical, and mechanical properties. Green biotechnology is now receiving increasingly more attention as owing to the use of microbes in the creation of nanoparticles, a new field of nanotechnology has been developed for efficient and sustainable production. This succinct review's goal is to cover the various aspects of microbial nanotechnology-based industrial wastewater bioremediation. Enzyme activity and reusability have increased thanks to the use of nanotechnology in conjunction with enzymes. This chapter gives insight into the possibility and changes of using microbial nanotechnology in bioremediation of toxic pollution.

Keywords Nanotechnology in environmental preservation · Industrial pollution remediation with nanomaterials · Microbial nanotechnology for sustainable wastewater treatment · Enzyme enhancement through nanotechnology · Green biotechnology and nanoparticles · Business opportunities in nanotechnology for pollution control

V. Kumar · S. Meshram (✉) · A. Chandrapati · A. Mandlik · A. Kumar (✉) · Geetika
Department of Plant Pathology, Lovely Professional University, Phagwara, Punjab, India
e-mail: shweta.26662@lpu.co.in; adesh.19078@lpu.co.in

A. Karnwal, A. R. Mohammad Said Al-Tawaha (eds.), *Microbial Applications for Environmental Sustainability*, https://doi.org/10.1007/978-981-97-0676-1_14

14.1 Introduction

With the development of environmentally friendly nanomaterials, industry can spend less money on mitigating harmful pollutants thanks to nanotechnology. Because of their improved physical, chemical, and mechanical qualities, nanomaterials are attracting more attention. Green biotechnology, a developing area of nanotechnology for sustainable manufacturing and cost reduction, is given an even larger push by the use of microorganisms in the manufacture of nanoparticles. The goal of this brief review is to address the various facets of microbial nanotechnology-integrated industrial wastewater bioremediation. Enzymes now have increased activity and reusability thanks to the usage of nanotechnology (Shukla 2020). Nanomaterials are being employed in more and more new products and gadgets, and this has a significant impact on several industries, including biomedicine and sensoristics. In contrast to conventional chemical and physical techniques, the biosynthesis of nanomaterials by microorganisms is gaining attention as a novel, interesting method for the development of "greener" nanomanufacturing (Fang et al. 2019). In order to produce sensoristic devices and for therapeutic/diagnostic purposes, bacteria, yeast, molds, and microalgae are used in this review to provide light on the microbial production of nanomaterials. The last 10 years' worth of literature were focused on, with a particular emphasis on scientific studies that discussed the properties, characterization, and applications of biosynthesis.

A description of the understanding, difficulties, and possibilities of microbial-mediated biosynthesis was also provided. The primary microorganisms employed in nanobiosynthesis, mostly for biomedical applications, are bacteria and microalgae (Srivastava and Mukhopadhyay 2014). Bacterial nanocellulose, exopolysaccharides, bacterial nanowires, and biomineralized nanoscale materials are only a few examples of the novel nanostructures that some bacteria and microalgae have demonstrated the ability to synthesize (magnetosomes, frustules, and coccoliths). Extracellular synthesis, a characteristic of yeasts and molds, is helpful for the potential reuse of cell cultures and for speeding up the purification of nanomaterials. Because of the inherent unpredictability of microbiological systems, it is necessary to standardize processes more thoroughly in order to produce nanomaterials with more consistent and repeatable chemical-physical properties (Grasso et al. 2019). The field of nanotechnology, which is centered on the exact design, production, development, and use of materials at a nanoscale (10^{-9} nm), is developing at an unheard-of rate. Separately, the fields of microbiology and nanotechnology have produced cutting-edge solutions for maintaining the proper ecological and environmental balance and improving human well-being (Sharon et al. 2010). Although it is urgently necessary to develop interdisciplinary research practices combining nanotechnology and microbiology, it is necessary due to instances where the frequent or inappropriate use of drugs has resulted in the emergence of multidrug resistance in microorganisms and the delivery of nanoparticles is affecting the food cycle (Kaittanis et al. 2010).

14.2 Nanoparticles in Nature

The word "Nano" is derived from the Greek word "nanos," meaning extremely small. The dimensions of nanoparticles is about 0.1–100 nm, which is nearly a billionth of a meter in nanoscale (1 nm = 10^{-9} m). It is a microscopic particle with a size of less than 100 nm and having a large surface area in comparison to its volume, so that it consists of both enhanced chemical and biological reactivity. These nanoparticles can be designed to possess new properties for certain applications which are not exhibited at micro- or macroscale by particles of the same material. The structures of nanoparticles vary from particle to particle in shape and size, i.e., spheres, rods, tubes, and wires, etc.

Nanoparticles are grouped in three different classes: natural, engineered, and incidental. In nanoparticle naturally found material are volcanic ash, magnetosomes, and ocean spray, in nanoparticle magnetosomes are found in magnetotactic microorganisms, mineral composites, and industrial processes such as coal combustion, welding fumes, and diesel exhaust, among others are anthropogenic nanoparticle. In nanoparticle there are various categories such as micelles, dendrimers, liposomes, hybrid, and compact polymeric NPs but Engineered nanoparticle are sub classified by various type of basic material likewise metal, semiconductor, metal oxides, nanoclay, nanotubes, and quantum dot the incidence of nanoparticle by naturally and the shape of is having irregular or regular. Engineered nanoparticles are regular shapes, tubes, sphere, ring, etc.

14.3 Nanoparticles Role in Bioremediation

Nanoparticles play a vital role in the purification of major environmental contaminants, i.e., chlorinated organic solvents, poly-chlorinated hydrocarbons, organochlorine pesticides, toxic metal compounds by the usage of microorganisms and their products leads to bioremediation (Vázquez-Núñez et al. 2020). In recent years, nanotechnology is becoming an emerging field by having achievements of eco-friendly remediation in purifying environmental soil, water, and toxic pollutants (Rizwan et al. 2014).

14.4 Hydrophobic Compounds and Their Degradation Using Nanoparticles

Nanoparticles are used in recent years for increasing the availability of hydrophobic organic compounds to improve its bioremediation. For improving the availability of polynuclear aromatic hydrocarbons (PAHs) in aqueous and soil solutions, the

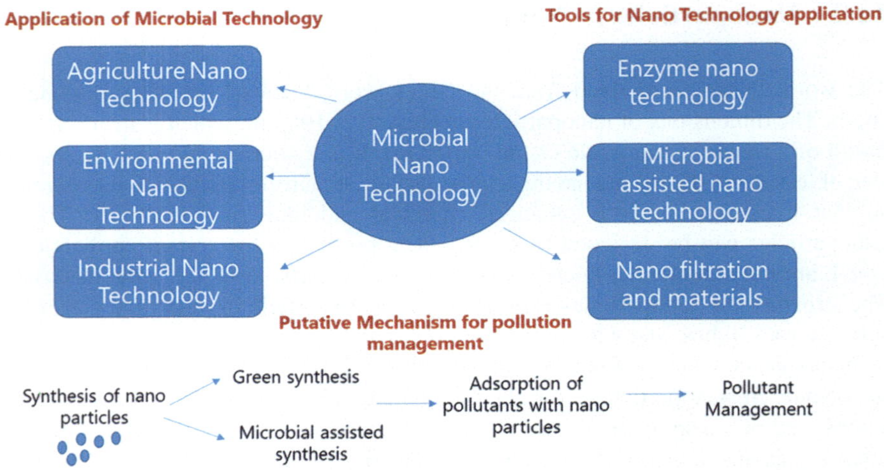

Fig. 14.1 Putative representation of microbial nanotechnology assisted bioremediation

application of polymeric nanoparticles made up of polyethylene glycol modified urethane acrylate (PMUA) is used. Because of the hydrophobicity of PMUA's interior regions, PAH's have a stronger affinity for PMUA's outer hydrophilic surfaces, since it increases the amount of PAHs released into the aqueous phase, promotes particle mobility, and speeds up mineralization (Tungittiplakorn et al. 2004, 2005). After microbial PAH breakdown, the liberated PAHs are treated using a pump and treat technique, allowing for the recycling and collection of polymeric nanoparticles. Later these nanoparticles can be used for immobilization of bacterial cells capable of degrading selective toxic compounds. By applying electro-magnetic field to microorganisms' cells, the nanoparticles that are present in the cells as a coated material are able to target specific sites of the receptor wall which separates them from the whole solution and helps in recycling of cells for treating similar compounds (Fig. 14.1).

14.5 Agriculture and Food Nanotechnology

Without the usage of agrochemicals such as pesticides, fertilizers, etc., contemporary agriculture cannot be sustained in terms of productivity and efficiency. Every agrochemical does, however, have certain potential drawbacks, such as the possibility of contaminating water supplies or leaving residues on foods that are harmful to human and environmental health. As a result, careful management and control of inputs may be able to lessen these risks. The creation of a high-tech agricultural system using engineered smart nanotools could be a great way to revolutionize agricultural methods and, in doing so, lessen or even completely remove the environmental impact of modern agriculture while also improving crop quality and quantity (Sekhon 2014; Kah 2015; Prasad et al. 2017). Specific applications like nanofertilizers and nanopesticides to track products and nutrient levels to increase productivity without decontaminating soils, waters, and protection against a variety of insect pests and microbial diseases are among the significant interests of using nanotechnology in agriculture (Vaghasiya et al. 2022). Nanoparticles indeed can assist in the biodegradation process but on the other hand it also helps in reduction of pesticide abuse in the ecosystem by managing the plant diseases (Tables 14.1 and 14.2).

Table 14.1 Plant pathogen managed by microbial assisted nanotechnology

Type of nanoparticles	Pathogens	Remarks	References
Silver nanoparticle (Ag)	*Biploaris sorokinniana, Botrytis cinerea, Colletotrichum gloeosporioides, Fusarium culmorum, Phythium ultimum, Phoma, Megnaporthe grisea, Sclerotinia sclerotiorum*	Broad-spectrum active agent against phytopathogen	Kumar et al. (2010), Kashyap et al. (2013)
Silica nanoparticle	Powdery mildew disease in cucurbits	Completely active powdery mildew	Park et al. (2006)
Ag (Si–Ag NP)	Powdery mildew, e.g., *Alternaria alternata*	Effective against foliar pathogen	Park et al. (2006), Kumar et al. (2010)
Si + validamycin	Gram-positive and gram-negative bacteria and fungi. E.g., *Xanthomonas oryzae, Xanthomonas campestris fusarium*	Controlled delivery of pesticide	Giannousi et al. (2014)
Cu-NP	*Phytophthora infestans, Xanthomonas oryzae,* and several plant pathogens	Effective against broad range of pathogen	Esteban-Tejeda et al. (2009)
Cu-chitosan NPs	*A. alternata, M. phaseolina, and R. solani*	Effectively inhibit growth and spore germination	Saharan et al. (2013)
Nanosulphur + nanoformulation of hexaconazole	*R. solani, Erysiphe cichoracearum*	Effective nanopesticide	Rani et al. (2016)
Zn NP	*Penicillium expansum, B. cinerea, Aspergillus flavus*	Nanofertilizer + Nanopesticide	Jayaseelan et al. (2012)

Table 14.2 Sources of nanomaterials and their application in different ways

Nanomaterials available in nature	Examples of nanomaterials	Application of nanomaterials	References
Carbon-based nanomaterials	Fullerene	Medical industry as photosensitizers and biological antioxidants	Bakry et al. (2007)
Metal based nanomaterials	Gold, zinc	Displays for television, computer screens	Arivalagan et al. (2011)
Gaseous based nanomaterials	Nitrogen dioxide	Sensors of gases	Arivalagan et al. (2011)
Metal oxide-based nanomaterials	Silicon dioxide, zinc oxide, titanium oxide	Construction of buildings, energy conductivity in power plants, sunscreen for body	Findik (2021)
Silica based nanomaterials	Carbon nanotubes	Semiconductors in electronic devices	Popov (2004)
Biological based nanomaterials	Nanocellulose	Wastewater treatment	Trache et al. (2020)
Ceramic based nanomaterials	Zirconia	Dental biomaterial	Della Bona et al. (2015)
Semiconductor based nanomaterials	Cadmium sulphide	Making batteries and solar cells	Vanalakar et al. (2018), Koli et al. (2020)
Polymeric based nanomaterials	Polylactic acid	Food and beverage packaging	Lagarón (2011)
Organic based nanomaterials	Cyclodextrins	Complexing agents for aqueous solubilization of poorly water-soluble drugs in pharmaceutical industries	Tiwari et al. (2010)

14.6 Application of Nanotechnology/Nanomaterial as an Alternative to Replace Convention Materials for Reduction of Pollutants

In terms of rehabilitating the environment, nanomaterials have generated a lot of interest. Nanomaterials outperform more traditional methods for environmental cleanup due to their enormous surface area and strong reactivity. For particular applications, they can be altered to include new functionalities. Nanoscale materials can be very reactive due to the high surface-area-to-volume ratio and the presence of a greater number of reactive sites. These qualities enable greater contact with pollutants, which causes a rapid decrease in contaminant concentration (Adeleye et al. 2016; Roy et al. 2021; Ahmed et al. 2021). An overview of several nanomaterials that may be used to clean up environmental pollution has been discussed (Table 14.2).

For some waste kinds, bioremediation offers a viable cleanup technique, but as is to be expected, it won't work for all. For instance, bioremediation would not be a practical option at locations where chemical concentrations are high and the majority of microorganisms are poisonous. These include salt and heavy metals. The development of science and technology has also raised the standard of living, which has a direct or indirect impact on the growth of toxic waste. As a result, using current technologies to remove toxins from the environment does not effectively and efficiently clean it up. Therefore, nanomaterials can be utilized in bioremediation, which will not only have a less damaging effect on microorganisms but will also increase the microbial activity of the specific waste and toxic material, lowering the overall time consumption and cost (Li and Li 2011; Roy et al. 2021; Gong et al. 2009).

14.7 Nanoparticles in Remediation of Water and Soil

Herbicides, persistent organic pollutants (POPs), heavy metals, and pesticide residues all lead to soil contamination, adding to the problem. Because of biomagnification, contaminated soil with these pollutants increases the likelihood that the food chain will become poisoned (Bakshi and Abhilash 2020). Engineered nanomaterials have been used to immobilize soil contaminants, including carbon-based nanomaterials like carbon nanotubes, metal oxides (such as ferric oxide and titanium oxide), and different nanocomposites (Arif et al. 2020). With improved results of 88.34%, iron nanoparticles are effective at removing total petroleum hydrocarbons (TPHs) from water (Murgueitio et al. 2018). Numerous NM-based whole form alterations have been looked at in order to find a suitable material that combines low cost, high efficiency, improved stability, least harmful environmental impact, and maximum reusability. Nanohydroxyapatite particles efficiently immobilized metal concentrations in pore water by lowering the interchangeable percentage of metals in contaminated sediments and soils (Wang et al. 2014) (Table 14.3).

Commercial and industrial industries' wastewater discharge, untreated domestic sewage, and the influx of synthetic toxins into aquatic resources are all adverse for the ecosystem and for people. The main water pollutants in this regard are heavy metal ions, organic compounds, and oils, which can render any water stream unsafe for eating (Lu et al. 2016). In comparison to activated carbon, nanomaterials as adsorbent systems can effectively remove pollutants and have a large surface area for reactions at a relatively small weight. a broad range of nanomaterial-based adsorbents with different forms and specifications (Ahsan et al. 2020; Homaeigohar 2020). Various remediation techniques, including lysis, filtration, adsorption, and oxidation, use ceramic, nanowire, and polymeric membranes, carbon nanotubes, and submicron particles. Due to their ability to cause cancer, dye pollution are fatal. For thousands of years, dyes have been utilized in textiles, artwork, pigments, and other things. Industrial-scale production of synthetic dyes currently produces about 0.1 million distinct varieties. On an annual basis, about 1.6 million tonnes of dyes are used. Aprrox. 15% of the total is wasted during use (Singh et al. 2012; Tan et al. 2015).

Table 14.3 Application of nanotechnology in remediation of pollutants using microorganisms

Nanotechnology applied	Modification	Associated microorganisms	Ionic adsorption or release	Process	Remarks	References
Oxides of Ni and Mg as nanoparticles	Silica embedded	–	Maximum uptake of 41.36, 13.76, 7.23 (ions per nm^2) for Cr 3+, cu 2+, and Zn 2+	Endothermic, and physical adsorption of copper 2^+ and 3^+ ion spontaneously	Regeneration and reusability proved sustainability Density 3.58 g/cm (metric) 0.129 lb./in^3 (Imperical) | 0.129 lb./in^3	Hassan et al. (2018), Abuhatab et al. (2020) https://www.mdpi.com/2073-4360/12/12/3066/htm
Electrospun nanofibrous webs	Bacterial encapsulation	*Pseudomonas aeruginosa*	At various concentrations, methylene blue is removed ranging 57%–70%	Biological dye removal	Potent bacterial cell	Hasan et al. (2008), Sarioglu et al. (2017)
Cobalt ion/ oxide-based nanoparticles	Heating of chemical	–	Murexide dye is degraded by cobalt and cobalt oxide nanoparticles by 43.6 and 39.4%, respectively. At maximum Langmuir adsorption capacity 238.10 mg/g	Irradiation and large surface	Cost-efficient, quicker, simpler, and greener sensors, piezoelectric devices, fuel cells, coatings for the passivation of surfaces against corrosion, and as catalysts	Adekunle et al. (2020) https://www.researchgate.net
Zirconia nanoparticles	Synthesis from microbial cell free culture supernatant	*Pseudomonas aeruginosa*	Tetracycline adsorption-526.32 mg/g	Chemical absorption along with strong electrostatic ionic bond among zwitterions	Green nanoparticle production and long-term bioremediation	Debnath et al. (2020)

Electrospun cyclodextrin fibers	Bacterial encapsulation	*Lysinibacillus* sp. *NOSK*	Removal efficiency of Ni(II) = 70 ± 0.2%, Cr(VI) = 58 ± 1.4% and reactive black 5 = 82 ± 0.8	Bacterial bioremediation	Cyclodextrin provides extra carbon source for growth of bacteria	San Keskin et al. (2018), Zhang and Hu (2018)
Silica nanoparticles	Synthesized from actinomycetes	*Actinomycetes*	80% of industrial wastewater decolorized	Photocatalytic degradation	Sustainable and affordable	Fang et al. (2019)
Mesoporous organosilica nanoparticles (MONs)	Incorporation of ferrocene	–	Methylene blue adsorption–222 mg/g	The presence of pores between MOFs and platforms, high surface area, hydrophobic and/or -interactions, and a variety of morphological characteristics in mixed nanocomposites	Excellent interaction of nanocomposite	Yang et al. (2019)
Enzyme immobilized nanoparticles	Laccase immobilization	*P. ostreatus*	90% of bisphenol-A and 10% of carbamazepine degraded	Oxidation by immobilized laccase	Cost-effective and reusable enzyme, more stable and robust than their free counterparts due to their fixed conformation	Ji et al. (2017) https://www.ncbi.nlm.nih.gov/pmc/articles/PMC3787205/

14.8 Nanoparticles in Remediation of Water and Soil

Filters are useful when dealing with air pollution. A filter has a narrow porous structure that lets gas pass through while keeping particles within where they belong. Three methods are used by the membrane of the filter to remove toxins: surface contact with the filter structure, inertial force provided when the gas is moving, and electric charge produced by particle and filter construction. In comparison to conventional gas separation methods, carbon nanotubes can capture gases 100 times faster. They are therefore appropriate for large-scale applications (Zhu et al. 2017). In several ways, nanotechnology can aid in air purification. One approach is to utilize nano-catalysts with a larger surface area for gaseous exchange/reactions. Catalysts quicken chemical reactions that change dangerous fumes from vehicles and factories into harmless gases. Because it is ineffectual to remove air pollution to the ground, the filtered chemicals presented a challenge for disposal (Jain and Singh 2012).

14.9 Conclusion

Research on microbial-based nano-synthesis is being sparked by an improved understanding of biosynthetic pathways and the potential of genetic engineering for future scaling-up and potential industrial exploitation of these promising "nanofactories." The future strategy for environmental and biotechnology lawsuits has been changed by nanoparticles. Industrialization and resource consumption have increased along with human development in nature. While not regretting the negative impacts of previous, prevailing procedures such as adsorption, advanced oxidation, coagulation, and Fenton, etc. or their combinations, there is an urgent need to remove these toxic pollutants with the least negative effects to the Earth. Nanoparticles, which may be active while operating on both small- and large-scale levels, may provide a more sustainable, all-encompassing approach to the solution of this serious problem. Diversified nanoparticles are useful for removing certain pollutants from their natural environment, such as sewage spills and landmines.

Acknowledgements The authors are highly grateful to the School of agriculture, Lovely Professional University, for providing necessary research facilities.

References

Abuhatab S, El-Qanni A, Al-Qalaq H, Hmoudah M, Al-Zerei W (2020) Effective adsorptive removal of Zn2+, Cu2+, and Cr3+ heavy metals from aqueous solutions using silica-based embedded with NiO and MgO nanoparticles. J Environ Manag 268:110–713

Adekunle AS, Oyekunle JA, Durosinmi LM, Oluwafemi OS, Olayanju DS, Akinola AS, Ajayeoba TA (2020) Potential of cobalt and cobalt oxide nanoparticles as nano catalyst towards dyes degradation in wastewater. Nano-Struct Nano-Objects 21:100–405

Adeleye AS, Conway JR, Garner K, Huang Y, Su Y, Keller AA (2016) Engineered nanomaterials for water treatment and remediation: costs, benefits, and applicability. Chem Eng J 286:640–662
Ahmed HM, Roy A, Wahab M, Ahmed M, Othman-Qadir G, Elesawy BH, Emran TB (2021) Applications of nanomaterials in agrifood and pharmaceutical industry. J Nanomater 2021:1
Ahsan MA, Jabbari V, Imam MA, Castro E, Kim H, Curry ML, Noveron JC (2020) Nanoscale nickel metal organic framework decorated over graphene oxide and carbon nanotubes for water remediation. Sci Total Environ 698:134–214
Arif Z, Sethy NK, Kumari L, Mishra PK, Upadhyay SN (2020) Recent advances in functionalized polymer-based composite photocatalysts for wastewater treatment. In: Nano-materials as photocatalysts for degradation of environmental pollutants, pp 39–64
Arivalagan K, Ravichandran S, Rangasamy K, Karthikeyan E (2011) Nanomaterials and its potential applications. Int J Chem Technol Res 3(2):534–538
Bakry R, Vallant RM, Najam-ul-Haq M, Rainer M, Szabo Z, Huck CW, Bonn GK (2007) Medicinal applications of fullerenes. Int J Nanomedicine 2(4):639
Bakshi M, Abhilash PC (2020) Nanotechnology for soil remediation: revitalizing the tarnished resource. In: Nano-materials as photocatalysts for degradation of environmental pollutants. Elsevier, pp 345–370
Debnath B, Majumdar M, Bhowmik M, Bhowmik KL, Debnath A, Roy DN (2020) The effective adsorption of tetracycline onto zirconia nanoparticles synthesized by novel microbial green technology. J Environ Manag 261:110–235
Della Bona A, Pecho OE, Alessandretti R (2015) Zirconia as a dental biomaterial. Materials (Basel) 8(8):4978–4991
Esteban-Tejeda L, Malpartida F, Esteban-Cubillo A, Pecharromán C, Moya JS (2009) Antibacterial and antifungal activity of a soda-lime glass containing copper nanoparticles. Nanotechnology 20(50):505–701
Fang X, Wang Y, Wang Z, Jiang Z, Dong M (2019) Microorganism assisted synthesized nanoparticles for catalytic applications. Energies 12(1):190
Findik F (2021) Nanomaterials and their applications. Period Eng Nat Sci 9(3):62–75
Giannousi K, Sarafidis G, Mourdikoudis S, Pantazaki A, Dendrinou-Samara C (2014) Selective synthesis of Cu2O and cu/Cu2O NPs: antifungal activity to yeast Saccharomyces cerevisiae and DNA interaction. Inorg Chem 53(18):9657–9666
Gong JL, Wang B, Zeng GM, Yang CP, Niu CG, Niu QY, Liang Y (2009) Removal of cationic dyes from aqueous solution using magnetic multi-wall carbon nanotube nanocomposite as adsorbent. J Hazard Mater 164(2–3):1517–1522
Grasso G, Zane D, Dragone R (2019) Microbial nanotechnology: challenges and prospects for green biocatalytic synthesis of nanoscale materials for sensoristic and biomedical applications. Nano 10(1):11
Hasan M, Ahmad AL, Hameed BH (2008) Adsorption of reactive dye onto cross-linked chitosan/oil palm ash composite beads. Chem Eng J 136(2–3):164–172
Hassan SED, Salem SS, Fouda A, Awad MA, El-Gamal MS, Abdo AM (2018) New approach for antimicrobial activity and bio-control of various pathogens by biosynthesized copper nanoparticles using endophytic actinomycetes. J Radiat Res Appl Sci 11(3):262–270
Homaeigohar S (2020) The nanosized dye adsorbents for water treatment. Nano 10(2):295
Jain CK, Singh RD (2012) Technological options for the removal of arsenic with special reference to South East Asia. J Environ Manag 107:1–18
Jayaseelan C, Rahuman AA, Kirthi AV, Marimuthu S, Santhoshkumar T, Bagavan A, Rao KB (2012) Novel microbial route to synthesize ZnO nanoparticles using Aeromonas hydrophila and their activity against pathogenic bacteria and fungi. Spectrochim Acta A Mol Biomol Spectrosc 90:78–84
Ji C, Nguyen LN, Hou J, Hai FI, Chen V (2017) Direct immobilization of laccase on titania nanoparticles from crude enzyme extracts of *P. ostreatus* culture for micro-pollutant degradation. Sep Purif Technol 178:215–223

Kah M (2015) Nano pesticides and nano fertilizers: emerging contaminants or opportunities for risk mitigation. Front Chem 3:64

Kaittanis C, Santra S, Perez JM (2010) Emerging nanotechnology-based strategies for the identification of microbial pathogenesis. Adv Drug Deliv Rev 62(4–5):408–423

Kashyap PL, Srivastava AK, Tiwari SP, Kumar S (2018) Microbes for climate resilient agriculture. John Wiley & Sons

Koli PB, Kapadnis KH, Deshpande UG, More BP, Tupe UJ (2020) Sol-gel fabricated transition metal Cr3+, Co2+ doped lanthanum ferric oxide (LFO-LaFeO3) thin flm sensors for the detection of toxic, fammable gases: a comparative study. Mater Sci Res India 17:70–83

Kumar P, Robins A, Vardoulakis S, Britter R (2010) A review of the characteristics of nanoparticles in the urban atmosphere and the prospects for developing regulatory controls. Atmospheric Environment 44(39):5035–5052

Lagarón JM (2011) Multifunctional and nano-reinforced polymers for food packaging. Woodhead Publishing, pp 1–28

Li YY, Li B (2011) Study on fungi-bacteria consortium bioremediation of petroleum contaminated mangrove sediments amended with mixed biosurfactants. In: Advanced materials research, vol 183. Trans Tech Publications Ltd, pp 1163–1167

Lu H, Wang J, Stoller M, Wang T, Bao Y, Hao H (2016) An overview of nanomaterials for water and wastewater treatment. Adv Mater Sci Eng 2016:1

Murgueitio E, Cumbal L, Abril M, Izquierdo A, Debut A, Tinoco O (2018) Green synthesis of iron nanoparticles: application on the removal of petroleum oil from contaminated water and soils. J Nanotechnol 2018:2018

Park HJ, Kim SH, Kim HJ, Choi SH (2006) A new composition of nanosized silica-silver for control of various plant diseases. Plant Pathol J 22(3):295–302

Popov VN (2004) Carbon nanotubes: properties and application. Mater Sci Eng R Rep 43(3):61–102

Prasad R, Bhattacharyya A, Nguyen QD (2017) Nanotechnology in sustainable agriculture: recent developments, challenges, and perspectives. Front Microbiol 8:1014

Rani PU, Laxmi KP, Vadlapudi V, Sreedhar B (2016) Phyto fabrication of silver nanoparticles using the mangrove associate, Hibiscus tiliaceus plant and its biological activity against certain insect and microbial pests. J Biopest 9(2):167

Rizwan M, Singh M, Mitra CK, Morve RK (2014) Eco-friendly application of nanomaterials: nano-bioremediation. J Nanopart 2014:1

Roy A, Sharma A, Yadav S, Jule LT, Krishnaraj R (2021) Nanomaterials for remediation of environmental pollutants. Bioinorg Chem Appl 2021:1

Saharan V, Mehrotra A, Khatik R, Rawal P, Sharma SS, Pal A (2013) Synthesis of chitosan-based nanoparticles and their in vitro evaluation against phytopathogenic fungi. Int J Biol Macromol 62:677–683

San Keskin NO, Celebioglu A, Sarioglu OF, Uyar T, Tekinay T (2018) Encapsulation of living bacteria in electro spun cyclodextrin ultrathin fibers for bioremediation of heavy metals and reactive dye from wastewater. Colloids Surf B Biointerfaces 161:169–176

Sarioglu OF, San Keskin NO, Celebioglu A, Tekinay T, Uyar T (2017) Bacteria encapsulated electro spun nanofibrous webs for remediation of methylene blue dye in water. Colloids Surf B Biointerfaces 152:245–251

Sekhon BS (2014) Nanotechnology in agri-food production: an overview. Nanotechnol Sci Appl 7:31–53

Sharon M, Choudhary AK, Kumar R (2010) Nanotechnology in agricultural diseases and food safety. J Phytology 2(4):1

Shukla P (2020) Microbial nanotechnology for bioremediation of industrial wastewater. Front Microbiol 11:590–631

Singh PKTR, Vats S, Kumar D, Tyagi S (2012) Nanomaterials use in wastewater treatment. In: Proceedings of the international conference on nanotechnology and chemical engineering, p 21

Srivastava N, Mukhopadhyay M (2014) Biosynthesis of SnO2 nanoparticles using bacterium Erwinia herbicola and their photocatalytic activity for degradation of dyes. Ind Eng Chem Res 53(36):13971–13979

Tan KB, Vakili M, Horri BA, Poh PE, Abdullah AZ, Salamatinia B (2015) Adsorption of dyes by nanomaterials: recent developments and adsorption mechanisms. Sep Purif Technol 150:229–242

Tiwari G, Tiwari R, Rai AK (2010) Cyclodextrins in delivery systems: applications. J Pharm Bioallied Sci 2(2):72

Trache D, Tarchoun AF, Derradji M, Hamidon TS, Masruchin N, Brosse N, Hussin MH (2020) Nanocellulose: from fundamentals to advanced applications. Front Chem 8:392

Tungittiplakorn W, Lion LW, Cohen C, Kim JY (2004) Engineered polymeric nanoparticles for soil remediation. Environmental science & technology 38(5):1605–1610

Tungittiplakorn W, Cohen C, Lion LW (2005) Engineered polymeric nanoparticles for bioremediation of hydrophobic contaminants. Environmental Science & Technology 39(5):1354–1358

Vaghasiya TP, Kumar A, Nakum K (2022) A review on wide range application of nanoparticles in agriculture and its implications in plant disease management. J Nanoworld 8(2):55–65

Vanalakar SA, Patil PS, Kim JH (2018) Recent advances in synthesis of Cu2FeSnS4 materials for solar cell applications: a review. Sol Energy Mater Sol Cells 182:204–219

Vázquez-Núñez E, Molina-Guerrero CE, Peña-Castro JM, Fernández-Luqueño F, de la Rosa-Álvarez MG (2020) Use of nanotechnology for the bioremediation of contaminants: a review. PRO 8(7):826

Wang Y, Fang Z, Liang B, Tsang EP (2014) Remediation of hexavalent chromium contaminated soil by stabilized nanoscale zero-valent iron prepared from steel pickling waste liquor. Chem Eng J 247:283–290

Yang S, Chen S, Fan J, Shang T, Huang D, Li G (2019) Novel mesoporous organosilica nanoparticles with ferrocene group for efficient removal of contaminants from wastewater. J Colloid Interface Sci 554:565–571

Zhang H, Hu X (2018) Biosynthesis of Pd and Au as nanoparticles by a marine Bacterium bacillus sp. GP and their enhanced catalytic performance using metal oxides for 4-nitrophenol reduction. Enzym Microb Technol 113:59–66

Zhu M, Han J, Wang F, Shao W, Xiong R, Zhang Q, Huang C (2017) Electro spun nanofibers membranes for effective air filtration. Macromol Mater Eng 302(1):160–353

Chapter 15
Phytoremediation: A Clean and Green Approach for Heavy Metal Remediation

Akhil Sharma, Shivika Sharma, Sonu Sharma, Amit Kumar, and Vikas Sharma

Abstract The buildup of heavy metals in soil has intensified the outcome of a variety of human (industrial) activities and geological cycles. Heavy metals cannot biodegrade, therefore, they remain in the surroundings, represent a danger of infiltrating the supply chain of food via agricultural plants, and may eventually accumulate within individuals due to biomagnification. Heavy metals are harmful to humans, plants, and ecosystems because they are toxic. Heavy metal ions that are still present in the soil may be accumulated by plant tissues, the biosphere, and by the various tropical levels present in the food chain. For the biosorption of metal-contaminated soil, several biological, chemical, and physical remediation approaches (both in situ and ex situ) are acknowledged. The importance of removing soil pollution cannot be emphasized. Phytoremediation allows for the environmentally beneficial replanting of soil which is contaminated with heavy metals. Increasing the efficacy of remediation with the help of plants necessitates a better knowledge of the processes behind the accumulation of heavy metals and tolerance in plants. In this review, we explain the heavy metal absorption, detoxification, and translocation, mechanisms in plants. We focus on techniques that employ microbial assistance, genetic engineering, and chelate assistance to boost the efficacy of photo extraction and phytostabilization. The limitations of effective phytoremediation and potential solutions are also covered in this research.

A. Sharma
Department of Biotechnology, Lovely Professional University, Phagwara, Punjab, India

S. Sharma · V. Sharma (✉)
Department of Molecular Biology and Genetic Engineering, School of Bioengineering and Biosciences, Lovely Professional University, Phagwara, Punjab, India
e-mail: vikas.25269@lpu.co.in

S. Sharma
Department of Biotechnology Maharishi Markandeshwar (Deemed to be University), Ambala, India

A. Kumar
Kehloor Biosciences and Research Centre (KBRC), Bilaspur, India

A. Karnwal, A. R. Mohammad Said Al-Tawaha (eds.), *Microbial Applications for Environmental Sustainability*, https://doi.org/10.1007/978-981-97-0676-1_15

Keywords Heavy metals · Soil pollution · Phytoremediation · Biomagnification · Metal ions · Detoxification mechanisms · Plant tolerance · Microbial assistance

15.1 Introduction

Although heavy metals have a broad variety of industrial purposes, their enormous production and utilization have detrimental implications on both human health and the vast tracts of land used to raise food for the globe (Banuelos and Ajwa 1999; Hou et al. 2020; Järup 2003). The above issue becomes more pertinent and urgent as the modern business depends more on cutting-edge technology for viability, many of which involve possibly hazardous metals. In battery cells, for example, lead (Pb), nickel (Ni), and cadmium are employed, as solar cells, and electric-car batteries (Cd) (Hou et al. 2020). The demand for additional productive soils to cultivate crops grows even more urgently as the human population expands. Because of the metalliferous ore mining and smelting, power lines, downwash community garbage, fertilizers, sludge, and insecticides are released into the environment, many portions of land have become too toxic to produce crops (Banuelos and Ajwa 1999; Hou et al. 2020). It is simple for toxic substances (HMs) to accumulate in the ecosystem. For instance, when the quantity of HMs surpasses the established limitations, the planet's biota as a whole is affected, causing biomagnification in the food supply chain. As a result, it is crucial to get rid of these metal contaminants if we want to lessen the harm to all living forms and the environment. HMs have been removed from the environment utilizing several techniques/processes, including reverse osmosis (Huang et al. 2017) and chemical precipitation (Levchuk et al. 2018), ion exchange (Burakov et al. 2018), extraction of solvent, and adsorption (Liu et al. 2020). These methods are typically not reliable and entail extensive maintenance costs and functions. Toxic metal pollution may be eliminated using one of the ecologically friendly methods known as phytoremediation (Fig. 15.1), which is also a less expensive and more effective method of cleaning up HM-contaminated locations (Lone et al. 2008). Because it is less expensive than other clean-up techniques, phytoremediation is extensively used all around the world (Yao et al. 2012; He et al. 2012). Because no modification of the soil's structure is necessary, this technology has a limited environmental impact (Pusz et al. 2021). Reverse osmosis (Huang et al. 2017), chemical precipitation (Levchuk et al. 2018), ion exchange (Burakov et al. 2018), extraction of solvent and uptake (Liu et al. 2020) are just a few of the methods/processes that have been utilized to remove HMs from the environment and eliminate metal toxicity from polluted locations (Macnair 2003). After that, the region may be utilized once again for farming or other agricultural pursuits. The bioaccumulation factor, a crucial component, affects plants' capacity to eliminate heavy metals (BCF). It offers a measurement of the plant's capacity for metal

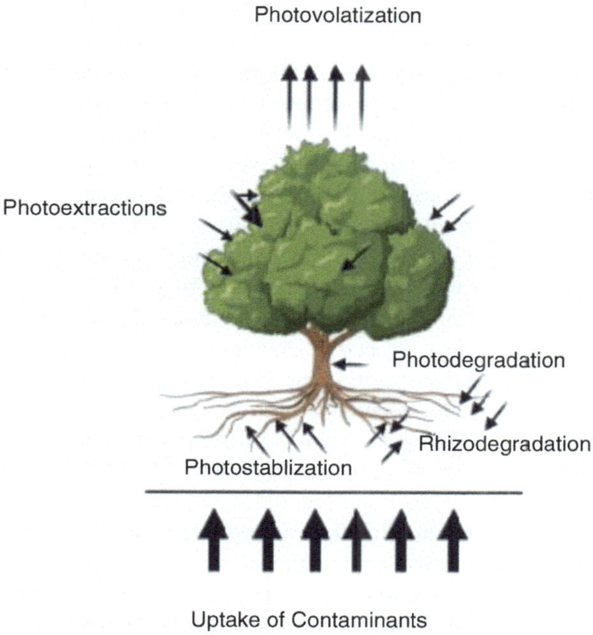

Fig. 15.1 Remediation of heavy metals in soil and water with the help of plants

agglomeration in response to metal concentrations in the substrate. The Compound varies depending on the kind of media and kind of plants used. High levels of absorption into the symplasm of the root cells and decreased root vacuolar transport are characteristics of hyperaccumulators, which often develop roots in places with high metal ion concentrations (Baker and Whiting 2002). Hyperaccumulators have a range of traits, including a shoot-root metal concentration quotient larger than one, a Bio Concentration Factor greater than one, and outstanding metal tolerance, thanks in large part to excellent detoxification (Nedjimi 2020; Zhao et al. 2000). High accumulating HM developing of several hyper accumulators has been investigated. An effort has been made to give a thorough overview of the many facets of bioremediation. Additionally, a thorough explanation of the efforts made by various organisms to remove contaminants from the atmosphere, notably HMs, has been provided.

For marine life to survive and for the ecosystem to remain healthy, water and sediment quality is essential (Ahmed et al. 2021; Hossain et al. 2021). Although, these probable regions of our environment are persistently contaminated by both anthropogenic and natural activity (Rahman et al. 2019; Ahmed et al. 2019, 2020). These toxins either actively or passively enter our systems through the food chain. Non-biodegradable harmful by-products provide a persistent environmental hazard (Ahmadpour et al. 2012). Nowadays, it is typical to discover harmful compounds, which is quite detrimental to the ecosystem (Ebrahimi et al. 2018). Ni, Co, Cu, Zn, and Mn are some of these metals that are crucial for some species' physiological processes, but too much of these can be harmful to the organisms (Bhattacharya et al. 2016). Some elements are so poisonous that they might slow down the capability of a plant to release water. Plant chloroplasts can be damaged by toxic elements, which lowers the increasing soil levels of metallic elements and can negatively influence the process of photosynthesis. When the concentration of Cd, for example, surpasses the required quantity, finally limits the death of cells and the growth of plants. Reactive oxygen species (ROS), Cadmium poisoning causes biomolecules in the cellular environment to be damaged (Cui and Du 2011). It is crucial to get rid of these metal contaminants if we want to lessen the harm to all living forms and the environment. Many procedures/techniques have been implemented to remove HMs from the environment, comprising solvent extraction, ion exchange, adsorption, and chemical precipitation caused by reverse osmosis. These methods, however, often are not viable and require a lot of upkeep costs and functions. One of the environmentally friendly approaches to dealing with hazardous metal contamination is phytoremediation, which provides a rapid and affordable substitute to traditional methods of decontaminating areas polluted with heavy metals (Liu et al. 2020). Because it is less expensive than other cleanup techniques, phytoremediation is extensively used across the world (Yao et al. 2012). Considering that the soil structure does not need to be altered, this technology has a minimum environmental impact. Physical and chemical remediation techniques are now employed widely to clear polluted soils. Although these techniques are effective and constant heavily they are commercialized, exorbitantly, costly, time-consuming, harm the earth and frequently raise ecological and environmental problems (He et al. 2012). The world's most important issue, environmental contamination, requires top emphasis. The use of phytoremediation as a method of environmental cleanup has long been acknowledged. The word "phytoremediation" refers to the removal procedure or detoxification of poisons from the surrounding with the help of plants. Plants play two roles in phytoremediation, the first is the encouragement of microbial degradation-friendly circumstances, notably by bacteria that colonies plant roots, the additional component being the actual plant root, which provides a manageable economical way to reach pollutants contained in subterranean soils and water. The waste products from the roots and decomposing biomass are sources of plant nutrients that support a microbial population in the rhizosphere.

15.2 Impact of Heavy Metal on Human Health

Many HMs and metalloids have negative effects on human health, making them harmful even in small quantities (Arora et al. 2008). Due to their extremely toxic effects, an abundance of many different heavy metals can present serious threats to both human health and the surroundings. Among the most hazardous metals are cadmium (Cd), mercury (Hg), chromium (Cr), arsenic (As), copper (Cu), zinc (Zn)

Table 15.1 Heavy metals impact on the health of human

Heavy metals	The detrimental effect of heavy metal	Reference
Hg	Symptoms include autoimmune problems, balance issues, anxiety, and damage to the kidneys, lungs, and brain	Verma et al. (2021)
Cu	Elevated levels have been linked to intestinal inflammation, renal damage, and liver cirrhosis	Wuana and Okieimen (2011)
Pb	Infants have balance problems, learning disabilities, declining IQ, renal failure, and cardiovascular illness	Mishra et al. (2010)
Ni	Hepatic, nephrotoxic, and neurotoxic. Genotoxic and a hematotoxin that results in hair loss	Das et al. (2008)
Cr	Loss of hairs	Salem et al. (2000)

and lead (Pb), tin (Sn) (Ghosh 2010). The effects of hazardous HMs on one's well-being might vary according to the substance in question, its quantity, and its level of oxidation (Table 15.1).

15.3 Heavy Metals Impact on the Environment

Organic and artificial (geological) processes take a substantial impact on metal ions in their environment. The constant and random discharge of industrial wastes heavy in metals is the main cause of HM contamination. An urgent, well-defined strategy for its reduction is of utmost importance since the expulsion from metal-based sectors, particularly textile companies, is a severe impact on the surrounding, particularly on soil and water (Marella et al. 2020). Similar to how excessive pesticide use and fertilizers on farmland have greatly increased the permissible levels of metal ions in the soil, this is mostly due to the growing global population (Zarcinas et al. 2004). Using wastewater as an irrigation supply and conveyance has also been linked to the accumulation of significant amounts of metal ions within the soil's subsurface. Underground and surface contaminants are produced during tasks like road maintenance and ice removal, which harms the ecosystem (García-Delgado et al. 2007).

15.4 Phytoremediation

In phytoremediation, plants and the rhizobacteria with which they are associated are used to remove environmental pollutants including organic compounds, radioactive and heavy metals (HMs) chemicals from dirt, subterranean water, or commercial effluent (phyto, which means "plant," and medium, which means "restoring balance"). When compared to alternative processes such as burning, soil excavation, soil washing, and the probable of scrap metal, bioremediation is inviolable, economically beneficial, affordable, and spirited (Sabir et al. 2014). The most significant factors influencing HM absorption by plant age, climatic conditions, bacterial colonization, metal size, and transportation to various areas are all factors that affect plants. According to several experts, some species of trees, such as Population and Salix, etc., have been demonstrated to satisfy all of these characteristics (Soudek et al. 2004). These trees produce a lot of energy, have deep roots, transpire a lot, and are simple to propagate. Meanwhile, numerous investigations have shown a striking clonal variation in their capacity to collect or absorb HMs (Dos Santos et al. 2007). The capability of a live plant to operate as a solar-powered compressor that can draw and store materials from the surroundings is used in phytoremediation (Raskin et al. 1997).

15.5 Phytoremediation of Heavy Metals in Soil and Water

To clean up contaminated major sources, such as soil and water, phytoremediation uses plants that accumulate metals (Muthusaravanan et al. 2018). Plants used in phytoremediation have the physiological properties to adapt to varied settings and the potential to absorb significant content of metals through their roots (Prieto et al. 2018). According to estimates, certain water plants including *E. stagnina, L. stolonifera, E. crassipes L. stolonifera, and P. australis* have the ability to collect Ni, Cd, and Pb. These plants might be used to identify wetlands contaminated by HMs and undertake phytoremediation on them (Eid et al. 2020). The findings revealed that Echinochloa stagnina may store a substantial amount of Pb in its organs, whereas Phragmites australis can accumulate the most Cd and Ni. Apart from Ni in Echinochloa stagnina and Cd in *Ludwigia stolonifera*, which are suited for phytoremediation of heavy metals, these investigated species may also phytostabilize these evaluated HMs. Thus, phytoextraction is a practical and productive method for removing metal ions from wastewater as well as soil (Eid et al. 2020).

15.6 Occurrence of Heavy Metals in Aquatic System and Soil

Due to its toxicity to living things via biomagnification, HM pollution has attracted the attention of the entire globe (Ekmekyapar et al. 2012). Metal ions are often categorized as greater fundamental density elements or metalloids. All of them were divided based on their weight, atomic number, and density (Kumar et al. 2017; Ali and Khan 2018). A new protocol was suggested by Ali and co-workers (2019), who gave the instances of the Different Heavy metals such as Chromium, Cadmium,

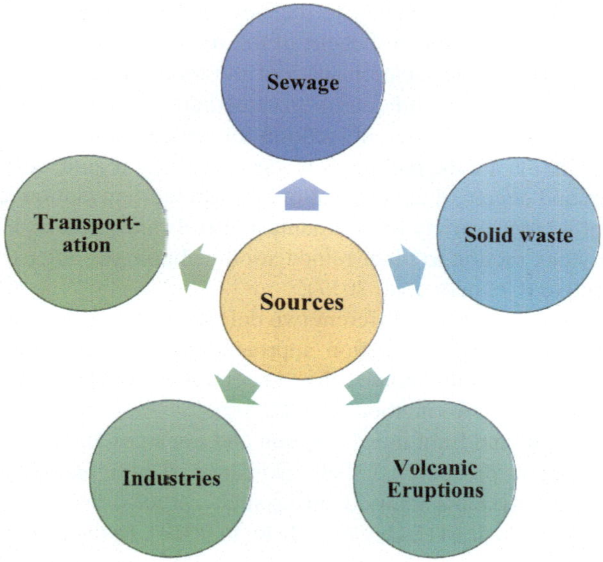

Fig. 15.2 Sources of heavy metal phytoremediation in soil and water

Copper, Cobalt, Mercury, Lead, and Zinc. They claimed that HM should have an atomic density of greater than 41 g/cm^3. Heavy Metals in the environment come from different sources mainly from air pollutants, which are widely acknowledged as the main cause of soil and water contamination. While industrial effluents include incomplete fossil combustion, mineral extraction, landfilling, metal refining, the production of electrical goods, dyes, agrochemicals, combat activities, and vehicle emissions, alternative sources include things like soil erosion, rock disintegration, and volcanoes. However, industrial activities cause more contamination of the environment than do natural causes (Duruibe et al. 2007) (Fig. 15.2).

15.7 Phytoremediation Procedure

Conventional and recombinant plants are both used in the phytoextraction method to clean up contaminated habitats (Beharti 2014). The use of chelators for the recovery, absorption, and decomposition of environmental contaminants was first introduced in 1983 (Sarwar et al. 2017). The procedure makes use of a wide range of phytotechnologies to remove environmental waste using both naturally occurring and genetically altered plant species. It is possible to complete the phytoextraction process utilizing both in situ and ex situ methods. Because it reduces the growth of pollutants in soil, water, and airborne waste, in situ treatment is more commonly utilized, thereby reducing the danger to the surrounding environment (Raskin and Ensley 2000). Another significant benefit of the in situ method is the treatment of several contaminants at one location without the requirement for a disposal site. Similar to this, the ex situ technique of bioremediation entails removing polluted soil and then moving it to a different location for remediation. The key parameters for ex situ remediation are the polluted site's geographical region, cost of treatment, pollutant types, and degree of contamination. Ex situ bioremediation techniques are utilized to treat a larger variety of pollutants and soils and are simpler to regulate. Comparing phytoextraction to other remediation technologies after treatment, it is more cost-effective (Cristaldi et al. 2017).

Avoidance and tolerance are different two defensive tactics that may be used for the application of the phytoextraction approach for the metal ions remediation (Thakur et al. 2016). Plants balance the concentration of HMs below their fatal standard limits via these two methods (Hall 2002).

In order to control and limit the absorption and transit of HMs into the plant tissues, plants adopt a procedure called avoidance (Dalvi and Bhalerao 2013). Numerous defensive systems are used during such a procedure (root sorption, metal precipitation, and exclusion) (Dalvi and Bhalerao 2013). In the root system, a variety of root exudates serve as HM ligands to create HM complexes, which limit the bioavailability and lethality of HMs (Dalvi and Bhalerao 2013). The accessibility of HMs from the soil to the roots is further limited by the exclusion boundaries that exist between both the shoot and root system. Since the pectin groups (carboxylic groups) in the cell wall function as cation exchangers to restrict the passage of HMs in the cells, HM incorporation in the cell wall of plants offers another avoidance too (Ernst et al. 1992).

Once an HM ion has entered the cytoplasm, plants use the endurance approach to deal with its cytotoxicity. This is done through the mechanisms of inactivation, metal chelation, and HM compartmentalization. As a result of sequestration, different organic and inorganic ligands in the cytoplasm lower the amount of HMs (Gupta et al. 2013). Following chelation, the HM binding compounds are moved from the cytosol to non-toxic storage areas such the vacuole, leaf petioles, leaf sheaths, and trichomes (Eapen and D'souza 2005). The aforementioned methods are occasionally insufficient to clean up polluted locations if there is a significant buildup of HMs because this buildup can cause the cytosol to produce reactive oxygen species (ROS), which can lead to osmotic damage (DalCorso et al. 2019). The plant cells use flavonoids, glutathione, ascorbate, carotenoids, and tocopherols as non-enzymatic antioxidants as well as enzymatic antioxidants superoxide dismutase (SOD), catalase (CAT), peroxidase (POD), and glutathione peroxidase (GR) to activate ROS scavenging (Jozefczak et al. 2012). Therefore, the cellular antioxidant mechanism is absolutely essential and necessary with regard to HM stress (Fig. 15.3).

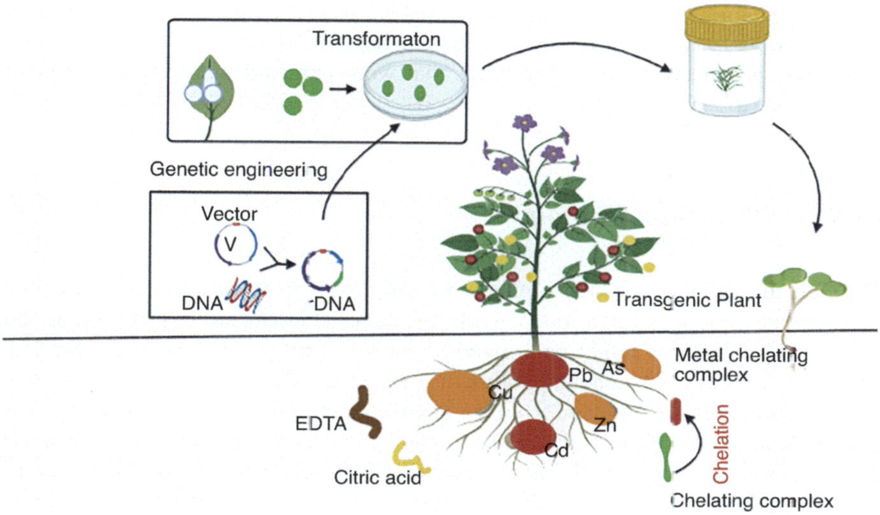

Fig. 15.3 Process of phytoremediation

15.8 Methods of Phytoremediation

15.8.1 Phytoextraction

Phytoextraction, often referred to as phytosequestration, phytoaccumulation, or photoabsorption is the process by which plant roots absorb and transmit poisons into shoots. The roots and shoots are picked to get rid of soil contaminants. According to a review of the literature, a variety of uptake and translocation metallophytes have a lot of promise for use in the remediation of HM-contaminated soils (Jakovljević et al. 2016). The growth and operation of uptake and translocation metallophytes are unaffected by the accumulation of HMs in their upper sections at concentrations between 100 and 500 times greater than those of other plants (Sheoran et al. 2016). But since the mechanism of heavy metal accumulation by the hyperaccumulator metallophytes is still poorly understood, it may be investigated and further extended to comprehend the basic mechanism of accumulation of heavy metals (Khalid et al. 2017). A. spinosus L. was shown to be a Zn and Mn shoot accumulator, while Chandra and Kumar 2017 reported that 12 eligible native plants growing on sludge were evaluated for their potential to collect Fe. All plants were discovered to be leaf accumulators for Fe, Mn, and Zn, with the exception of A. spinosus L. and Ricinus communis. Lajayer et al. research's from Asgari Lajayer et al. 2019 has shown that ornamental plants have the capacity to phytoextract heavy metals (HMs) from contaminated surroundings. This approach seems to be more environmentally sustainable when compared to conventional cleansing methods. The phytoremediation abilities of specific vegetables were examined by Ugya and co-workers in 2019, as well as potential health hazards associated with consuming vegetables produced in cultivated fields that were irrigated with refinery effluent (Ugya et al. 2019).

15.8.2 Phytovolatilization

Through the foliar mechanism, plants are employed to transform environmental contaminants into less dangerous volatile substances that are subsequently released into the sky. This idea might be utilized to get rid of toxic substances like Se, Hg, and As, as well as environmental pollutants (Yan et al. 2020). In phytovolatilization, HM contaminants are removed from the environment or harvested and dispersed as gaseous chemicals (Yan et al. 2020). Typical plants used for phytovolatilization include Nicotiana tabacum, Triticum aestivum, Crinum americanum, Bacopa monnieri, Arabidopsis thaliana, and Trifolium repens (Singh et al. 2018). Either a direct or indirect approach can be taken. Volatile organic compounds (VOCs) are directly vaporized by the leaves and stems, while root system interactions with the soil cause indirect volatilization (Limmer and Burken 2016). Organic pollutants, such as acetone, phenol, and chlorinated benzene, are degraded through phytovolatilization (BTEX) (Herath and Vithanage 2015). Previous studies have shown that Astragalus racemosus may phytovolatilize Se into dimethyl diselenide whereas Arabidopsis thaliana may phytovolatilize Hg2+ into HgO, increasing Hg instability (Awa and Hadibarata 2020). According to Sakakibara and co-workers in 2010 Pteris vittate may convert arsenic into forms that are volatile.

15.8.3 Photostablization

In order to immobilize heavy metals (HM) in polluted soils and reduce their bioavailability in the ecosystem, phytostabilization is used (Yan et al. 2020). The procedure mostly depends on the use of certain plants to maintain pollutants in contaminated areas (Singh and Prasad 2011). In the past 10 years, there has been a great deal of research on the stability of HM through adsorption, binding, or co-precipitation with soil amendments (biosolids, manures, and composts) (Karami et al. 2011). Luo et al. (2019) examined the viability of phytostabilizing zinc smelting slag by combining four trees and shrubs with organic amendments. The findings demonstrated that intensive vegetation planting boosted nutrient retention and reduced heavy metal solubility in zinc smelting waste slag (Cu, Zn, and Cd). In a different investigation, (Galal et al. 2017) looked into Vossia cuspidate's capacity to absorb HM from polluted sources of water when used as a phytoremediator (emergent macrophytes). In the various plant tissues of V. cuspidate, the results showed significant seasonal change in the amounts of the examined HMs, with under regions collecting more of them than above-ground shoots.

15.8.4 Rhizofiltration

Rhizofiltration is a process that uses plant roots to filter out wastewater contaminants. HMs may be absorbed by leguminous plants, which changes the pH of the root system. The settlement of metal pollutants on the root surface is greatly influenced by environmental parameters in the bottom microbiome, including pH in the rhizosphere, root exudates, and root turnover. The plant may be simply gathered and disposed of in a safe spot once it has absorbed all the metal contaminants (Zhu et al. 1999). In this mechanism, there is a mutual association between the plant and the associated microorganisms. Microbes clean the metal element while the plants stimulate microbial population. Rhodococcus spp., Mycobacterium spp., and Pseudomonas aeruginosa are the most often utilized microorganisms in rhizoremediation (Verma and Rawat 2021). Typically, wild-type organisms are chosen for this procedure, therefore modified microorganisms are not necessary. Simply said, rhizoremediation is remediation focused on roots, microorganisms, and rhizospheric soil. The plants used in the rhizofiltration process must, though, have the ability to produce a broad root system, must collect HMs in elevated doses, must be simple to handle and harvest, and must have a reduced preservation budget (Kushwaha et al. 2018). A rhizofiltration device was planted with two species (Phragmites australis and Kyllinga nemoralis) and evaluated for its effectiveness in removing HMs from industrial effluents in a wastewater treatment facility in KwaZulu-Natal (a province of South Africa). Different amounts of HMs were deposited in the standard and planted portions of the rhizofilter. Additionally, it was discovered that the rhizomes of K nemoralis had increased its Cd levels by 33% and 21%, showing that the system is effective in removing HMs from sewage water (Odinga et al. 2019).

15.8.5 Rhizodegradation

Rhizodegradation is the process by which the organic pollutants in the soil are biodegraded in conjunction with rhizospheric microorganisms that release certain enzymes that either destroy or convert very polluted organic contaminants into less harmful forms. As these microorganisms collect the vital components (nutrients) from the plant's root exudate, rhizodegradation is accelerated. This increases plant effectiveness and speeds up the absorption and removal of contaminants by the plant (Li et al. 2016). The dissolving of the contaminant at the source is one of the key aspects of rhizodegradation, which also emphasizes the full mineralization of the organic pollutant after compound transfer to the plant or environment (Fiorentino et al. 2018). Rhizodegradation has various downsides, such as the fact that it is a sluggish, time-consuming process that only works up to a specific depth, generally between 20 and 25 cm. Select species of plants and soil types can affect rhizodegradation (Kaimi et al. 2006).

15.8.6 Phytodegradation

When organic contaminants are subjected to degradation with the help of plants, they are either metabolized by enzymes that are part of the plant's metabolism or stored by the plant through various metabolic activities (Sharma and Pandey 2014). The dehalogenase, peroxidase, nitroreductase, nitrilase, and phosphatase enzymes degrade pollutants (Deng and Cao 2017). It includes the direct absorption of pollutants into plant tissue through the root system and mostly relies on the soil's transpiration rate, physiochemical composition, and other soil factors. Phytodegradation is a method for decontaminating areas that have been polluted by organic pollutants like pesticides and chlorinated solvents (Wang et al. 2017). Additionally, it may be used to recover both surface and ground fluids. This method is restricted in that the soil must be 3 ft. deep and the groundwater must be no more than 10 ft. below the surface. To increase plant absorption by fusing dirt particles with pollutants, chelating agents are necessary (Miller 1996).

15.8.7 Phytodesalination

The most widely used biological treatment for such remediation is phytodesalination, a freshly developed and developing technique that uses halophytic plants to repair the saline soils (Ali et al. 2013). There is relatively little information on this procedure in the literature when compared to the other phytoremediation methods. Halophytes are thought to be inherently well-adapted to metal ions, in contrast to glycophytic plants (Manousaki and Kalogerakis 2011). The plant's ability to phytodesalinate relies on its species in addition to the soil's salinity, sodicity, and porosity as well as other environmental variables, mainly rainfall (Hussain et al. 2018). According to a research, two halophytic plants, Sesavium partulacastrum and Suaeda maritime can take around 504 kg and 474 kg of NaCl, respectively, from 1 hectare of salty soil over the course of 4 months (Ravindran et al. 2007). The

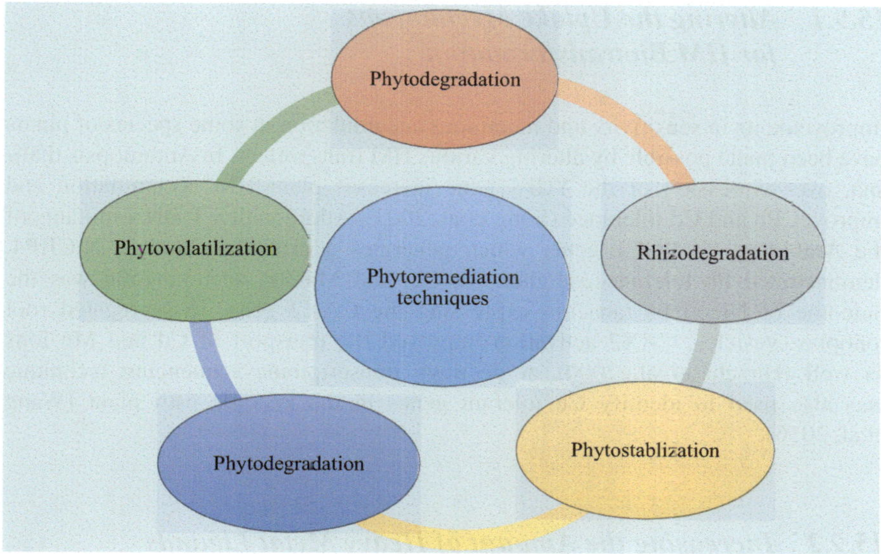

Fig. 15.4 Different techniques used in phytoremediation

rehabilitation of soil impacted by sodium (Na+) and chloride (Cl) ions has been reported to exhibit encouraging outcomes in desalination tests of halophytic plants. This phytoremediation method is promising for soils impacted by salt, but it is not suited for soils contaminated with HMs and polycyclic aromatic hydrocarbons (PAHs) (Zorrig et al. 2012) (Fig. 15.4).

15.9 Bioengineering Strategies That Might Be Used for Phytoremediation

Using genetically altered techniques to increase tolerance of plants and harmful metal accumulation, phytoremediation holds a lot of potential. Additionally, several novel research studies have been conducted using omics technology to understand the genetic components and underlying processes of plant HM tolerance. With the use of biotechnological techniques, HMs such as Cd, Hg, Pb As, Cu, and Se are now being phytoremediated. Three main microbiological techniques are used to develop plants for HM phytoextraction. These techniques include improving the HM ligand production, altering the HM transporter genes and their uptake systems, and converting the metals ion into less dangerous and gaseous forms (Mosa et al. 2016).

15.9.1 Altering the Uptake Mechanisms for HM Biotransformation

Improvements in sensitivity and metal ions accumulation in some species of plants have been made possible by altering various HM transporters. In Arabidopsis thaliana, over-expression of the YCF1 gene increases plant HM accumulation and improves Pb and Cd tolerance (Song et al. 2003). Additionally, T-DNA mutants of the Arabidopsis CNGC1 gene, which generates a protein related to NtCBP4, demonstrated Pb tolerance. Higher Ca, Cd, and Mn ion accumulation was the outcome of Nicotiana tabacum expressing the CAX2 gene. In segregated root tonoplast vesicles, CAX2 activation improved the transport of Cd and Mn ions as well (Hirschi et al. 2000). A de novo transcriptome sequencing technique was also used to identify Cu tolerant genes in the Paeonia ostii plant (Wang et al. 2016).

15.9.2 Increasing the Amount of Heavy Metal Ligands

Cysteine-rich peptides such metallothioneins (MTs), phytochelatins, and GSH are used as metal-binding ligands for HM purification or buildup. Increased Cd tolerance was seen in transgenic plants that had the *E. coli* GSH synthetase (GS) gene overexpressed in the Brassica juncea (Liang Zhu et al. 1999). Additionally, species of plants can be changed to increase their resistance to Pb, Cd, and As. The double transgenic plants accumulated three times more as in above-ground biomass when two genes of bacteria were co-expressed in Arabidopsis and grown in media containing sodium arsenate (125 M) (Dhankher et al. 2002). The potential to improve HM and metalloid tolerance by gene modification to boost the production of metal chelation compounds is substantial (Mosa et al. 2016).

15.9.3 Changing Dangerous Heavy Metals Into Less Volatile Forms

Despite being a necessary element, selenium (Se) is hazardous at greater doses and is thus regarded as a worldwide contaminant (Zwolak and Zaporowska 2012). Selenium is taken up by plants, who then transform it into volatile forms like dimethyl selenide (DMSe). Se tolerance, bioaccumulation, and volatilization have been improved using microbiological Se phytoremediation (Pilon-Smits and LeDuc 2009) (Fig. 15.5).

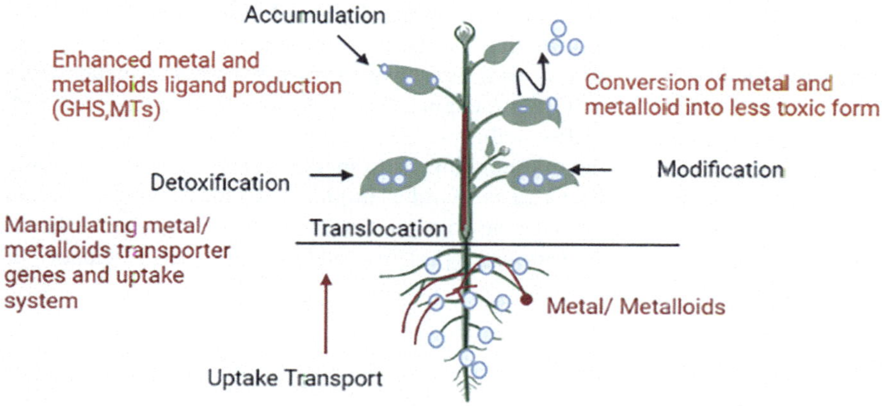

Fig. 15.5 Bioengineered strategies used for phytoremediation

15.10 Conclusion

Among the biggest threats to water, soil, and people is heavy metals (HMs). An option for removing these toxins from the ecosystem appears to be the adoption of inexpensive and ecologically friendly methods. As a result, phytoextraction is one of the most well-liked and successful plant-based techniques for cleaning up polluted regions while inflicting the least amount of harm to water bodies or soil structure. In addition to the advantages that bioremediation offers as an ecosystem functioning, investigation and implementations revealed the need to increase and speed it due to the problems connected with the lengthy removal of HMs as well as the dangers that HM toxicity poses to plant integrity. The study investigates the causes of HM pollutants, the impact of HM contamination on the ecosystem, the detrimental consequences of HM pollution on public health in this setting, and possible heavy metal remediation techniques. The entire study helps us determine the best method for a particular metal-bearing effluent based on time requirements and effectiveness, as well as identify frequently used plants for bioremediation. Additionally, numerous phytoremediation techniques, biotechnological approaches, plants employed in diverse ways, and factors effecting phytoremediation have been examined as potential and strategies for boosting bioremediation, providing views for the creation of new methods. In a word, phytoremediation refers to the use of metal-accumulating plants as an alternative way for HM cleaning. It is cost-efficient, efficient, ecologically beneficial, and environmental friendly. To enhance the procedure, new plant species that are appropriate for removing metals ions from contaminated soils must be introduced. Lowering the pH of contaminated soil and increasing the availability of heavy metals to plants can be accomplished by adopting metal bioavailability-enhancing techniques, including the use of bio-augmented acidified manure. Furthermore, bioremediation can be a useful strategy for enhancing and enhancing quality of the soil as well as a way to recover essential metals from highly contaminated places. Genetic engineering and molecular science research can offer substantial and creative additions to this effort. This method may be used with chelating chemicals and microorganisms to increase HM solubility, enabling higher soil cleansing and heavy metal buildup in plants. To make bioremediation more feasible for environmental conservation, further study is required to better understand the effects of various types of catalysts on the efficiency of bioremediation.

References

Ahmadpour P, Ahmadpour F, Mahmud TMM, Abdu A, Soleimani M, Tayefeh FH (2012) Phytoremediation of heavy metals: a green technology. Afr J Biotechnol 11(76):14036–14043

Ahmed AS, Sultana S, Habib A, Ullah H, Musa N, Hossain MB, Sarker MSI (2019) Bioaccumulation of heavy metals in some commercially important fishes from a tropical river estuary suggests higher potential health risk in children than adults. PLoS One 14(10):e0219336

Ahmed ASS, Hossain MB, Semme SA, Babu SMOF, Hossain K, Moniruzzaman M (2020) Accumulation of trace elements in selected fish and shellfish species from the largest natural carp fish breeding basin in Asia: a probabilistic human health risk implication. Environ Sci Pollut Res 27(30):37852–37865

Ahmed AS, Hossain MB, Babu SOF, Rahman MM, Sarker MSI (2021) Human health risk assessment of heavy metals in water from the subtropical river, Gomti, Bangladesh. Environ Nanotechnol Monit Manag 15:100416

Ali H, Khan E (2018) What are heavy metals? Long-standing controversy over the scientific use of the term 'heavy metals'–proposal of a comprehensive definition. Toxicol Environ Chem 100(1):6–19

Ali H, Khan E, Sajad MA (2013) Phytoremediation of heavy metals—concepts and applications. Chemosphere 91(7):869–881

Ali H, Khan E, Ilahi I (2019) Environmental chemistry and ecotoxicology of hazardous heavy metals: environmental persistence, toxicity, and bioaccumulation. J Chem 2019:1

Arora M, Kiran B, Rani S, Rani A, Kaur B, Mittal N (2008) Heavy metal accumulation in vegetables irrigated with water from different sources. Food Chem 111(4):811–815

Asgari Lajayer B, Khadem Moghadam N, Maghsoodi MR, Ghorbanpour M, Kariman K (2019) Phytoextraction of heavy metals from contaminated soil, water and atmosphere using ornamental plants: mechanisms and efficiency improvement strategies. Environ Sci Pollut Res 26(9):8468–8484

Awa SH, Hadibarata T (2020) Removal of heavy metals in contaminated soil by phytoremediation mechanism: a review. Water Air Soil Pollut 231(2):1–15

Baker AJ, Whiting SN (2002) In search of the holy grail: a further step in understanding metal hyperaccumulation? New Phytol 155:1–4

Banuelos GS, Ajwa HA (1999) Trace elements in soils and plants: an overview. J Environ Sci Health A 34(4):951–974

Beharti A (2014) Phytoremediation: as a degradation of heavy metals. Int J Eng Technol Res 2:137–139

Bhattacharya PT, Misra SR, Hussain M (2016) Nutritional aspects of essential trace elements in oral health and disease: an extensive review. Scientifica 2016:1

Burakov AE, Galunin EV, Burakova IV, Kucherova AE, Agarwal S, Tkachev AG, Gupta VK (2018) Adsorption of heavy metals on conventional and nanostructured materials for wastewater treatment purposes: a review. Ecotoxicol Environ Saf 148:702–712

Chandra R, Kumar V (2017) Phytoextraction of heavy metals by potential native plants and their microscopic observation of root growing on stabilised distillery sludge as a prospective tool for in situ phytoremediation of industrial waste. Environ Sci Pollut Res 24(3):2605–2619

Cristaldi A, Conti GO, Jho EH, Zuccarello P, Grasso A, Copat C, Ferrante M (2017) Phytoremediation of contaminated soils by heavy metals and PAHs. A brief review. Environ Technol Innov 8:309–326

Cui Y, Du X (2011) Soil heavy-metal speciation and wheat phytotoxicity in the vicinity of an abandoned lead–zinc mine in Shangyu City, Eastern China. Environ Earth Sci 62(2) 257–264

DalCorso G, Fasani E, Manara A, Visioli G, Furini A (2019) Heavy metal pollutions: state of the art and innovation in phytoremediation. Int J Mol Sci 20(14):3412

Dalvi AA, Bhalerao SA (2013) Response of plants towards heavy metal toxicity: an overview of avoidance, tolerance and uptake mechanism. Ann Plant Sci 2(9):362–368

Das KK, Das SN, Dhundasi SA (2008) Nickel, its adverse health effects & oxidative stress. Indian J Med Res 128(4):412

Deng Z, Cao L (2017) Fungal endophytes and their interactions with plants in phytoremediation: a review. Chemosphere 168:1100–1106

Dhankher OP, Li Y, Rosen BP, Shi J, Salt D, Senecoff JF, Meagher RB (2002) Engineering tolerance and hyperaccumulation of arsenic in plants by combining arsenate reductase and γ-glutamylcysteine synthetase expression. Nat Biotechnol 20(11):1140–1145

Dos Santos U, Wieshammer G, Vega R, Wenzel WW (2007) Hydroponic screening for metal resistance and accumulation of cadmium and zinc in twenty clones of willows and poplars. Environ Pollut 148(1):155–165

Duruibe, Ogwuegbu, Egwurugwu (2007) Heavy metal pollution and human biotoxic effects. Int J Phys Sci 2(5):112–118

Eapen S, D'souza SF (2005) Prospects of genetic engineering of plants for phytoremediation of toxic metals. Biotechnol Adv 23(2):97–114

Ebrahimi A, Hashemi H, Eslami H, Fallahzadeh RA, Khosravi R, Askari R, Ghahramani E (2018) Kinetics of biogas production and chemical oxygen demand removal from compost leachate in an anaerobic migrating blanket reactor. J Environ Manag 206:707–714

Eid EM, Galal TM, Sewelam NA, Talha NI, Abdallah SM (2020) Phytoremediation of heavy metals by four aquatic macrophytes and their potential use as contamination indicators: a comparative assessment. Environ Sci Pollut Res 27:12138–12151

Ekmekyapar F, Sabudak T, Seren G (2012) Assessment of heavy metal contamination in soil and wheat (Triticum aestivum L.) plant around the Çorlu–Çerkezkoy highway in Thrace region. Global NEST J 14(4):496–504

Ernst WHO, Verkleij JAC, Schat H (1992) Metal tolerance in plants. Acta Bot Neerl 41(3):229–248

Fiorentino N, Mori M, Cenvinzo V, Duri LG, Gioia L, Visconti D, Fagnano M (2018) Assisted phytoremediation for restoring soil fertility in contaminated and degraded land. Ital J Agron 13(1S):34–44

Galal TM, Gharib FA, Ghazi SM, Mansour KH (2017) Phytostabilization of heavy metals by the emergent macrophyte Vossia cuspidata (Roxb.) Griff.: a phytoremediation approach. Int J Phytoremed 19(11):992–999

García-Delgado M, Rodríguez-Cruz MS, Lorenzo LF, Arienzo M, Sánchez-Martín MJ (2007) Seasonal and time variability of heavy metal content and of its chemical forms in sewage sludges from different wastewater treatment plants. Sci Total Environ 382(1):82–92

Ghosh S (2010) Wetland macrophytes as toxic metal accumulators. Int J Environ Sci 1(4):523

Gupta DK, Vandenhove H, Inouhe M (2013) Role of phytochelatins in heavy metal stress and detoxification mechanisms in plants. In: Heavy metal stress in plants. Springer, Berlin, Heidelberg, pp 73–94

Hall JÁ (2002) Cellular mechanisms for heavy metal detoxification and tolerance. J Exp Bot 53(366):1–11

He S, He Z, Yang X, Baligar VC (2012) Mechanisms of nickel uptake and hyperaccumulation by plants and implications for soil remediation. Adv Agron 117:117–189

Herath I, Vithanage M (2015) Phytoremediation in constructed wetlands. In: Phytoremediation. Springer, Cham, pp 243–263

Hirschi KD, Korenkov VD, Wilganowski NL, Wagner GJ (2000) Expression of arabidopsis CAX2 in tobacco. Altered metal accumulation and increased manganese tolerance. Plant Physiol 124(1):125–134

Hossain MB, Semme SA, Ahmed ASS, Hossain M, Porag GS, Parvin A, Sekar S (2021) Contamination levels and ecological risk of heavy metals in sediments from the tidal river Halda, Bangladesh. Arab J Geosci 14(3):1–12

Hou D, O'Connor D, Igalavithana AD, Alessi DS, Luo J, Tsang DC, Ok YS (2020) Metal contamination and bioremediation of agricultural soils for food safety and sustainability. Nat Rev Earth Environ 1(7):366–381

Huang H, Zhang D, Zhao Z, Zhang P, Gao F (2017) Comparison investigation on phosphate recovery from sludge anaerobic supernatant using the electrocoagulation process and chemical precipitation. J Clean Prod 141:429–438

Hussain I, Puschenreiter M, Gerhard S, Schöftner P, Yousaf S, Wang A, Reichenauer TG (2018) Rhizoremediation of petroleum hydrocarbon-contaminated soils: improvement opportunities and field applications. Environ Exp Bot 147:202–219

Jakovljević T, Radojčić-Redovniković I, Laslo A (2016) Phytoremediation of heavy metals: applications and experiences in croatia abstract. Zašt Mater 57(3):496–501

Järup L (2003) Hazards of heavy metal contamination. Br Med Bull 68(1):167–182

Jozefczak M, Remans T, Vangronsveld J, Cuypers A (2012) Glutathione is a key player in metal-induced oxidative stress defenses. Int J Mol Sci 13(3):3145–3175

Kaimi E, Mukaidani T, Miyoshi S, Tamaki M (2006) Ryegrass enhancement of biodegradation in diesel-contaminated soil. Environ Exp Bot 55(1–2):110–119

Karami N, Clemente R, Moreno-Jiménez E, Lepp NW, Beesley L (2011) Efficiency of green waste compost and biochar soil amendments for reducing lead and copper mobility and uptake to ryegrass. J Hazard Mater 191(1–3):41–48

Khalid S, Shahid M, Niazi NK, Murtaza B, Bibi I, Dumat C (2017) A comparison of technologies for remediation of heavy metal contaminated soils. J Geochem Explor 182:247–268

Kumar SS, Kadier A, Malyan SK, Ahmad A, Bishnoi NR (2017) Phytoremediation and rhizoremediation: uptake, mobilization and sequestration of heavy metals by plants. In: Plant-microbe interactions in agro-ecological perspectives, pp 367–394

Kushwaha A, Hans N, Kumar S, Rani R (2018) A critical review on speciation, mobilization and toxicity of lead in soil-microbe-plant system and bioremediation strategies. Ecotoxicol Environ Saf 147:1035–1045

Levchuk I, Màrquez JJR, Sillanpää M (2018) Removal of natural organic matter (NOM) from water by ion exchange–a review. Chemosphere 192:90–104

Li Y, Zhang J, Zhu G, Liu Y, Wu B, Ng WJ, Tan SK (2016) Phytoextraction, phytotransformation and rhizodegradation of ibuprofen associated with Typha angustifolia in a horizontal subsurface flow constructed wetland. Water Res 102:294–304

Liang Zhu Y, Pilon-Smits EA, Jouanin L, Terry N (1999) Overexpression of glutathione synthetase in Indian mustard enhances cadmium accumulation and tolerance. Plant Physiol 119(1):73–80

Limmer M, Burken J (2016) Phytovolatilization of organic contaminants. Environ Sci Technol 50(13):6632–6643

Liu S, Yang B, Liang Y, Xiao Y, Fang J (2020) Prospect of phytoremediation combined with other approaches for remediation of heavy metal-polluted soils. Environ Sci Pollut Res 27(14):16069–16085

Lone MI, He ZL, Stoffella PJ, Yang XE (2008) Phytoremediation of heavy metal polluted soils and water: progresses and perspectives. J Zhejiang Univ Sci B 9(3):210–220

Luo Y, Wu Y, Qiu J, Wang H, Yang L (2019) Suitability of four woody plant species for the phytostabilization of a zinc smelting slag site after 5 years of assisted revegetation. J Soils Sediments 19(2):702–715
Macnair MR (2003) The hyperaccumulation of metals by plants. Adv Bot Res 40:63–105
Manousaki E, Kalogerakis N (2011) Halophytes present new opportunities in phytoremediation of heavy metals and saline soils. Ind Eng Chem Res 50(2):656–660
Marella TK, Saxena A, Tiwari A (2020) Diatom mediated heavy metal remediation: a review. Bioresour Technol 305:123068
Miller RR (1996) Phytoremediation—technology overview. Groundwater Remediation Technologies Analysis Center, Pittsburgh, PA
Mishra S, Dwivedi SP, Singh RB (2010) A review on epigenetic effect of heavy metal carcinogens on human health. Open Nutraceuticals J 3(1):188
Mosa KA, Saadoun I, Kumar K, Helmy M, Dhankher OP (2016) Potential biotechnological strategies for the cleanup of heavy metals and metalloids. Front Plant Sci 7:303
Muthusaravanan S, Sivarajasekar N, Vivek JS, Paramasivan T, Naushad M, Prakashmaran J, Al-Duaij OK (2018) Phytoremediation of heavy metals: mechanisms, methods and enhancements. Environ Chem Lett 16(4):1339–1359
Nedjimi B (2020) Germination characteristics of Peganum harmala L.(Nitrariaceae) subjected to heavy metals: implications for the use in polluted dryland restoration. Int J Environ Sci Technol 17(4):2113–2122
Odinga CA, Kumar A, Mthembu MS, Bux F, Swalaha FM (2019) Rhizofiltration system consisting of Phragmites australis and Kyllinga nemoralis: evaluation of efficient removal of metals and pathogenic microorganisms. Desalin Water Treat 169:120–132
Pilon-Smits EA, LeDuc DL (2009) Phytoremediation of selenium using transgenic plants. Curr Opin Biotechnol 20(2):207–212
Prieto MJ, Acevedo SOA, Prieto GF (2018) Phytoremediation of soils contaminated with heavy metals. Biodivers Int J 2(4):362–376
Pusz A, Wiśniewska M, Rogalski D (2021) Assessment of the accumulation ability of Festuca rubra L. and Alyssum saxatile L. tested on soils contaminated with Zn, Cd, Ni, Pb, Cr, and Cu. Resources 10(5):46
Rahman MS, Hossain MB, Babu SOF, Rahman M, Ahmed AS, Jolly YN, Akter S (2019) Source of metal contamination in sediment, their ecological risk, and phytoremediation ability of the studied mangrove plants in ship breaking area, Bangladesh. Mar Pollut Bull 141:137–146
Raskin I, Ensley BD (2000) Phytoremediation of toxic metals. John Wiley & Sons, Hoboken, NJ
Raskin I, Smith RD, Salt DE (1997) Phytoremediation of metals: using plants to remove pollutants from the environment. Curr Opin Biotechnol 8(2):221–226
Ravindran KC, Venkatesan K, Balakrishnan V, Chellappan KP, Balasubramanian T (2007) Restoration of saline land by halophytes for Indian soils. Soil Biol Biochem 39(10):2661–2664
Sabir M, Waraich EA, Hakeem KR, Öztürk M, Ahmad HR, Shahid M (2014) Phytoremediation: mechanisms and adaptations. Soil Remediat Plants Prospect Challeng 85:85–105
Salem HM, Eweida EA, Farag A (2000) Heavy metals in drinking water and their environmental impact on human health. In: Int conference on the environ hazards mitigation. Cairo Univ, Giza, pp 542–556
Sarwar N, Imran M, Shaheen MR, Ishaque W, Kamran MA, Matloob A, Hussain S (2017) Phytoremediation strategies for soils contaminated with heavy metals: modifications and future perspectives. Chemosphere 171:710–721
Sharma P, Pandey S (2014) Status of phytoremediation in world scenario. Int J Environ Bioremediat Biodegrad 2(4):178–191
Sheoran V, Sheoran AS, Poonia P (2016) Factors affecting phytoextraction: a review. Pedosphere 26(2):148–166
Singh A, Prasad SM (2011) Reduction of heavy metal load in food chain: technology assessment. Rev Environ Sci Biotechnol 10:199–214

Singh R, Ahirwar NK, Tiwari J, Pathak J (2018) Review on sources and effect of heavy metal in soil: its bioremediation. Int J Res Appl Nat Soc Sci 2018:1–22

Song WY, Ju Sohn E, Martinoia E, Jik Lee Y, Yang YY, Jasinski M, Lee Y (2003) Engineering tolerance and accumulation of lead and cadmium in transgenic plants. Nat Biotechnol 21(8):914–919

Soudek P, Tykva R, Vaněk T (2004) Laboratory analyses of 137Cs uptake by sunflower, reed and poplar. Chemosphere 55(7):1081–1087

Thakur S, Singh L, Wahid ZA, Siddiqui MF, Atnaw SM, Din MFM (2016) Plant-driven removal of heavy metals from soil: uptake, translocation, tolerance mechanism, challenges, and future perspectives. Environ Monit Assess 188(4):1–11

Ugya AY, Ahmad AM, Adamu HI, Giwa SM, Imam TS (2019) Phytoextraction of heavy metals and risk associated with vegetables grown from soil irrigated with refinery wastewater. J Appl Biol Biotechnol 7(2):1–9

Verma P, Rawat S (2021) Rhizoremediation of heavy metal-and xenobiotic-contaminated soil: an eco-friendly approach. In: Removal of emerging contaminants through microbial processes. Springer, Singapore, pp 95–113

Verma A, Roy A, Bharadvaja N (2021) Remediation of heavy metals using nanophytoremediation. In: Advanced oxidation processes for effluent treatment plants. Elsevier, pp 273–296

Wang Y, Dong C, Xue Z, Jin Q, Xu Y (2016) De novo transcriptome sequencing and discovery of genes related to copper tolerance in Paeonia ostii. Gene 576(1):126–135

Wang L, Ji B, Hu Y, Liu R, Sun W (2017) A review on in situ phytoremediation of mine tailings. Chemosphere 184:594–600

Wuana RA, Okieimen FE (2011) Heavy metals in contaminated soils: a review of sources, chemistry, risks and best available strategies for remediation. International scholarly research notices

Yan A, Wang Y, Tan SN, Mohd Yusof ML, Ghosh S, Chen Z (2020) Phytoremediation: a promising approach for revegetation of heavy metal-polluted land. Front Plant Sci 11:359

Yao Z, Li J, Xie H, Yu C (2012) Review on remediation technologies of soil contaminated by heavy metals. Procedia Environ Sci 16:722–729

Zarcinas BA, Ishak CF, McLaughlin MJ, Cozens G (2004) Heavy metals in soils and crops in Southeast Asia. Environ Geochem Health 26(3):343–357

Zhao FJ, Lombi E, Breedon TMSP (2000) Zinc hyperaccumulation and cellular distribution in Arabidopsis halleri. Plant Cell Environ 23(5):507–514

Zhu YL, Zayed AM, Qian JH, De Souza M, Terry N (1999) Phytoaccumulation of trace elements by wetland plants: II. Water hyacinth, Vol 28, no 1. American Society of Agronomy, Crop Science Society of America, and Soil Science Society of America, pp 339–344

Zorrig W, Rabhi M, Ferchichi S, Smaoui A, Abdelly C (2012) Phytodesalination: a solution for salt-affected soils in arid and semi-arid regions. J Arid Land Stud 22(1):299–302

Zwolak I, Zaporowska H (2012) Selenium interactions and toxicity: a review. Cell Biol Toxicol 28(1):31–46

Chapter 16
Second-Generation Biofuels: Concepts, Applications, and Challenges

Aarushi Gautam, Manu Pant, Gaurav Pant, and Gaurav Kumar

Abstract For the last two decades, the escalating demands and subsequent depletion of fossil fuels have led to the search for suitable alternatives. A potential solution to the limitation of fossil fuels is "Biofuel." Biofuels are referred to as fuels that are derived from biological substances like crops, microorganisms, wood, waste matter, etc. First-generation biofuels were the first category of biofuels discovered which incorporated the use of biomass from crops like corn, Jatropha, cooking oils, etc. First-generation biofuels were less successful due to the issues of stock feeding, poor crop quality, and feeding world hunger because they relied on food crops. The production of second-generation biofuel was a need-based alternative that relied on non-food crops, forest waste, and crop waste for fuel production. Second-generation biofuels have a more sustainable approach and are more favored as compared to first-generation biofuels. The present review highlights the shortfalls of first-generation biofuels, production, uses, and advantages of second-generation biofuels.

Keywords Fossil fuels · Biofuel · First-generation biofuels · Second-generation biofuels · Biomass · Sustainability

A. Gautam · M. Pant (✉)
Department of Biotechnology, Graphic Era (Deemed to be University), Dehradun, Uttarakhand, India
e-mail: manupant@geu.ac.in

G. Pant
Department of Microbiology, Graphic Era (Deemed to be Univeristy), Dehradun, Uttarakhand, India

G. Kumar (✉)
Department of Microbiology, School of Bioengineering and Biosciences, Lovely Professional University, Phagwara, Punjab, India

A. Karnwal, A. R. Mohammad Said Al-Tawaha (eds.), *Microbial Applications for Environmental Sustainability*, https://doi.org/10.1007/978-981-97-0676-1_16

16.1 Introduction

Rising petroleum prices and increased awareness regarding climate change, which is also linked to the use of fossil fuels, have emphasized the need for effective alternatives to conventional fossil fuels (Nagle and Lemke 1990; Sawayama et al. 1995; Gavrilescu and Chisti 2005). Biofuels have emerged as a viable answer to the current crisis. Biofuels are a broad category of liquid or gaseous fuels obtained from biological materials (known as biomass/feedstock) that may be used as transportation fuel or as an energy source for industrial or domestic level power generation (Bessou et al. 2010; Singh and Dhar 2011; Gasparatos et al. 2013).

The biofuels are classified as first-, second-, or third-generation biofuels based on the biomass and manufacturing process (Demirbas 2011). More recently, fourth-generation biofuels have also been conceptualized (Scaife et al. 2015; Hays and Ducat 2014) (Table 16.1). First-generation biofuels refer to those made from plant material (mostly cereals, oilseeds, and sugar crops) via processes of fermentation, transesterification, and enzymatic hydrolysis. These include biobutanol, biodiesel, and bioethanol. They were best described by OECD (2022) as "liquid, solid, or gaseous fuel" produced by conversion of biomass like sugarcane, corn, charcoal, woodchips or waste matter decompose. The common first-generation biofuel sources are oils from palm, canola, beans, corn, jatropha, and even waste cooking oil (Barnwal and Sharma 2005; Felizardo et al. 2006; Kulkarni and Dalai 2006).

However, these biofuels face challenges such as rising global feeding demands, which limit the continuous supply of food/feed crops for biofuel production, lower crop potential, delayed fruiting in species like Jatropha and Pongamia, and the need for a large production area (Pimentel and Patzek 2005; Fargione et al. 2008; Searchinger et al. 2008). Competition for food crop feedstocks remains to be the most pertinent issue which is linked to the rise of prices in food commodities. First-generation biofuels are associated with unsustainable production practices, e.g., deforestation, use of water, land management, etc. As a result, there is a lot of promise from second-generation biofuels. These are considered to have more sustainable production, and greater land use prospects, including rural areas, where they utilize high-yielding, non-food energy crops or crop and forest wastes. As compared to the first-generation biofuels, second-generation biofuels are a better source of energy and power and provide a large spectrum of feedstock to be used as raw material (Carriquiry Miguel et al. 2011). Since these biofuels are based on lignocellulosic biomass, they minimize competition in food production and have far lesser ecological impacts (Dufosse et al. 2017).

The present review provides information on the uses, sustainability, benefits of secondary biofuels over first-generation biofuels, and associated challenges.

Table 16.1 Production of various biofuels

Generation of biofuel	About	Biofuel	Feedstock biomass	Production process
First	Fuels that have been derived from sources like starch, sugar, animal fats, and vegetable oil. The oil is obtained using conventional production techniques Generally used biomass: Crops like wheat, barley, corn, potato, sugarcane, sugar beet, rapeseed, soybeans, sunflower, trees like coconut, palm, and animal fats	Biobutanol	Sugar crops	Fermentation and saccharification
		Biodiesel	Oil-based crops	Transesterification
		Bioethanol	Sugar and crops	Fermentation
		Bioethanol	Lignocellulosic materials	Advanced enzymatic hydrolysis and fermentation
		Biomethanol	Lignocellulosic materials	Gasification, synthesis, and catalytic cracking
Second	Also known as advanced biofuels. Can be derived from various types of biomass (non-food), including plant and animal materials Generally used biomass: Non-food crops like *Jatropha, Miscanthus,* cassava, lignocellulosic materials like straw, wood, and grass	Dimethyl ether	Lignocellulosic materials	Gasification, synthesis, and catalytic cracking
		Fischer-Tropsch diesel	Lignocellulosic materials	Gasification, synthesis, and catalytic cracking
		Biogas	Lignocellulosic materials	Synthesis and anaerobic digestion
		Biohydrogen	Lignocellulosic materials	Synthesis, gasification, fermentation
Third	Aquatic autotroph (e.g., algae) based carbon for biofuel generation	Vegetable oil, biodiesel	Algae	Transesterification
Fourth	Not a very common category. Biomass captures atmospheric carbon dioxide in the form of plant parts which are subsequently converted into fuels	Synthetic biofuels		Carbon dioxide is captured by precombustion, oxyfuel, or post-combustion processes

16.2 Second-Generation Biofuels and Their Method of Production

The first-generation biofuels rely heavily on food crops as raw materials. These crops are used as a source of starch which is fermented to generate biofuel (Deenanath et al. 2012). Over time, these first-generation biofuels were associated with several limitations which emphasized the need to look for suitable alternatives, ultimately giving way to second-generation biofuels. These limitations are described as under:

16.2.1 Food Security

This happens to be the strongest opponent of first-generation biofuels since they require food crops as the raw material which also serves as food for humans. It has been largely claimed that using food crops to produce biofuel has resulted in inflation in food prices, although a section of experts call it to be a negligible effect. Overall, this has led to a heated debate on "food versus fuel."

16.2.2 Soil Erosion

Extensive cultivation of food crops has also been related to an escalation in soil erosion. Annual feedstock reportedly causes more soil erosion than perennial feedstock (Gasparatos et al. 2013). In the particular instance of sugarcane, the rate of soil erosion was reported to be approximately 5.2 times greater than that of soil development (de Oliveira et al. 2005). According to an Argentinian study, soybean cultivation on degraded grassland prompted more soil erosion than switchgrass (van Dam et al. 2008). Oil palm has been shown to increase soil infertility as well as the decline of aquatic and terrestrial habitats (Obidzinski et al. 2012). All these crops are increasingly used in the first-generation biofuel production.

16.2.3 Loss of Biodiversity

The large-scale production of feedstock in monocultures has posed a threat to biodiversity. A classic example has been an expansion of oil palm plantations which was also used for biofuel generation. The plantation was done by removing the predominant forest area in Malaysia (55–59%) and Indonesia (approximately 56%) as reported by Koh and Wilcove (2008), resulting in a substantial loss of floral and faunal diversity in these regions (Koh et al. 2011).

16.2.4 Release of Atmospheric Pollutants

Biofuel generation has also been reported to release atmospheric pollutants. According to a study carried out in The United of America (Millet et al. 2012), corn-ethanol (a type of first-generation biofuel produced and used in the United States) is a significant source of air pollutants. Tessum et al. (2012) also stated that the majority of the emissions were more pronounced in the Midwestern "corn belt," where a significant proportion of the crop is grown.

16.2.5 Water Utilization and Pollution

First-generation biofuels utilize more water as compared to other energy carriers throughout their life cycle. In parts of Southeast Asia, palm oil industries have become a major reason for water pollution. This is owing to runoff palm oil mill effluent waste and harmful pesticides and fertilizer runoff in natural water bodies (Muyibi et al. 2008; Obidzinski et al. 2012). In a study conducted by Donner and Kucharik (2008), corn-based biofuels have been identified in the United States as a potential source of increased nitrogen loading in the Mississippi River, amplifying the lack of oxygen conditions in the Gulf of Mexico.

These shortfalls associated with first-generation biofuels led to advancements toward second-generation biofuels. The feedstock for second-generation biofuel is the by-product of non-edible food crops, waste cooking oil, waste from food crops, agricultural waste, wood chips, etc. These non-edible compounds diminish the "food-versus-fuel" competition.

Lignocellulosic material is the raw material for second-gen biofuels. The Lignocellulosic crops can be classified into three types: annual crops, perennial herbaceous crops, and SRC (Short Rotation Coppice). Among the different types, the most versatile crops have been identified as a sterile hybrid of Asian origin (*Miscanthus x Giganteus*), and switchgrass (*Panicum virgatum*) (Dufosse et al. 2017). These crops are low maintenance, have an efficient nutrient cycling system and dense root system, emit low greenhouse gases, and can be cultivated on any type of land ranging from degraded, low-productive, or marginal (Cadoux et al. 2012; Strullu et al. 2011; De Cara et al. 2012).

16.3 Lignocellulosic Feedstock

It incorporates cellulose and hemicelluloses, which are converted into sugars. There are three types of lignocellulosic feedstocks:

(a) agricultural by-products,
(b) forest waste,
(c) Woody and herbaceous energy crops.

The cell wall lyses to form sugars from cellulose and this sugar is then converted to ethanol or fuels like diesel or butanol (Gnansounou and Dauriat 2010).

16.3.1 Pretreatment of Lignocellulosic Biomass

Pretreatment of biomass is the most crucial step in biofuel production. Pretreatment consists of two steps:

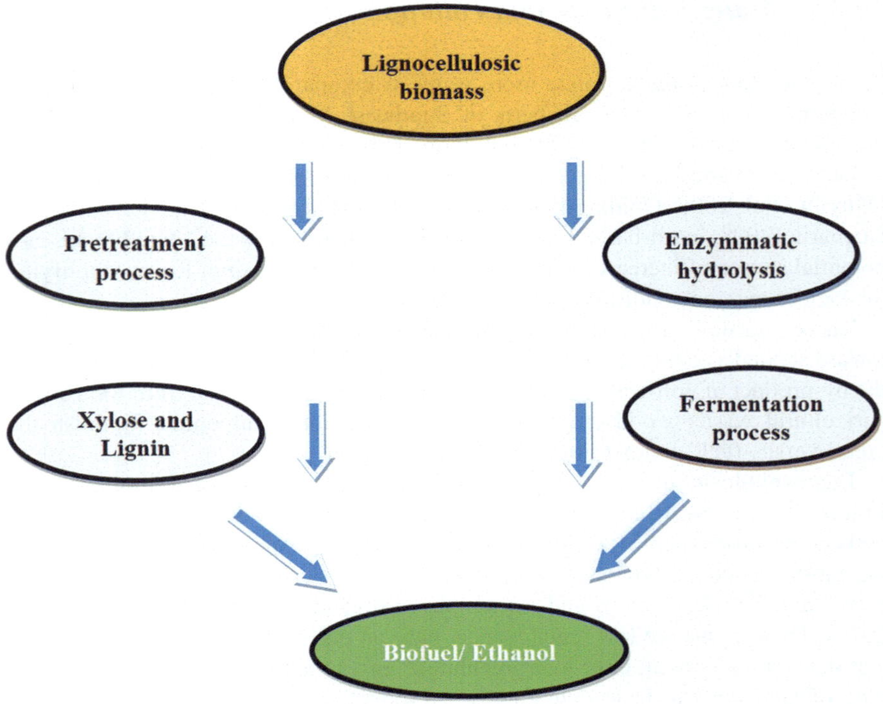

Fig. 16.1 Lignocellulosic biomass conversion (Source: Naik et al. 2010)

(a) Chemical and Physical pretreatment: Such methods have been used to improve the quality of the substrate during the physical process by incorporating procedures such as hot water, steam, ultrasonics, pressure, heat, and so on. Base or acid process conditions, oxidation, and ozonization are used during the chemical treatment (Wagner et al. 2018).
(b) Biological precure: This procedure dissolves lignin coatings and disrupts the cellulose structure, facilitating microbial and enzymatic digestion (Fig. 16.1).

16.4 Production Pathways

Biochemical and thermochemical processing has been recognized as the two pathways for biomass-based production of liquid biofuel. Thermochemical processing involves heating biomass in varying oxygen concentrations to thermally decay and chemically transform the biomass (Naik et al. 2010). While biochemical processing majorly focuses on peptide conversion, the thermochemical method targets all the organic components of biomass; hence making it a better choice (Gomez et al. 2008).

16.4.1 Biochemical Processes

Biomass feedstocks contain organic compounds which can be broken down into food/feed, alcohols, biofuel, biogas, and other chemicals with the help of several enzymes and microorganisms. This process is non-polluting because a chemical reaction takes place at a low temperature and has a low conversion rate thus, requires low energy and few chemicals. At the industrial level, two biochemical processes occur: Sugars in biomass crops are fermented to alcohol and biomass, and their wastes are digested anaerobically to form methane (Biogas), and the residue is used as fertilizer (Demirel 2017).

16.4.2 Chemical Processes

This happens to be the most widely used method wherein biodiesel (Fatty acid methyl ester) is generated by transesterification of triglycerides of fatty acids in the presence of ethanol or methanol and with a catalyst like some acid or salts. Bioethanol is also produced by a chemical process of inorganic acid-based hydrolysis of hemicellulose and cellulose and converted into sugars, which in turn are converted to bioethanol (Demirel 2017).

16.4.3 Thermochemical Process

The feedstock and wastes are converted thermochemically (at a high temperature of 450–1200 °C) through the processes of pyrolysis, gasification, or combustion, (Demirel 2017). In the process of gasification via the conventional method, biomass reacts with a limited amount of oxygen alongside carbon dioxide and steam to produce syngas between the temperature range of 750–1100 °C. Syngas is composed of H_2 and CO, CO_2, CH_4, etc. (Huber and Corma 2007) and is formed through the following reaction.

$$C_nH_m + H_2O = n\,CO + (m/2 + n)H_2$$

Pyrolysis uses rapid heating under anaerobic conditions for the breakdown of biomass into a hydrocarbon mixture. The mixture along with condensed hot gases is converted to biofuels (Demirel 2017).

16.4.4 Hydrothermal Liquefaction

This process is also utilized for producing bio-oil by biomass under controlled reaction rates and reaction mechanisms in the presence of pressure and catalysts (Das and Veziroğlu 2001).

Some of the catalysts used in the process are nickel, alkali, metals, and ruthenium heterogeneous catalysts. The process liquefies biomass to bio-oil using subcritical or supercritical water. The reaction is carried out in a pressurized vessel at a temperature between 200 and 400 °C. High temperature and pressure convert the cellulosic, hemicellulosic, and lignin content into a mixture that is subjected to depolymerization to develop the bio-oil (Demirbas 2011).

16.5 Approaches to Enhance Biofuel Production

Improvement in fuel production can be done by changing the cell wall composition. Various approaches to improve production are discussed below:

16.5.1 Approaches Based Upon the Microbial Community

Monitoring natural microbial communities that efficiently degrade cellulose, recognizing their structure and function, analyzing fermentation reaction mechanism, assessing the process of their effectual breakdown and transition, and then recreating the design of similar approaches and thereafter enhancing their activities during the transition will require new ideas for establishing a new cellulose deterioration system (Zhu et al. 2016; Wang et al. 2016; Liang et al. 2017). Lipase-producing or lipolytic bacteria are also a possible source of biodiesel, as lipases are involved in the transesterification reaction (Al Makishah et al. 2019). A strain (Micrococcus luteus) was isolated by Al Makishah et al. (2019) with unique transesterification activity.

16.5.2 Approaches Based on Nanotechnology

In terms of achieving cost-effective and convenient methods of vastly increased biofuel production, nanotechnology, and its nanoparticles also resurfaced as a viable solution for the biofuel industry (Nizami and Rehan 2018). Several international researchers looked into and addressed the use of nanostructured materials in improved biofuel manufacturing (Nizami and Rehan 2018; Sekoai et al. 2019; Badawy et al. 2021; Elumalai et al. 2021). Nizami and Rehan (2018) discussed the application of nanotechnology and its enormous potential for establishing a premium and reliable biofuel marketplace. Correspondingly, Sekoai et al. (2019) investigated the incorporation of nanomaterials through biofuel procedures (such as the production of biogas, biodiesel, and bioethanol) to improve return.

16.5.3 Synthetic and System Biology

Recent advances in metagenomics, transcriptomics, and proteomics can overcome microbial pure breeding and help to provide a better understanding of bacterial diversity concerning cellulase genes. The study of enzyme expression levels and mechanisms paved the way for further research into novel cellulase gene retrieval, degradation mechanisms, heterologous protein expression, and purification (Liu et al. 2013; Ilmberger and Streit 2017). Several institutions, including the Energy Department of the US, collaborated to create a metagene library that extracted complete DNA and complete RNA from the intestines of the termite. Sequencing data helped in discovering that there is a diverse set of hemicellulose and cellulose and hemicellulose hydrolysis-related genes, hence improving awareness and understanding of cellulase diversity (Nimchua et al. 2012; Rashamuse et al. 2017).

16.5.4 Metabolic Engineering Methods

Simple genetic or metabolic engineering is difficult to achieve the desired result because of the complex nature of the mechanism of sugar utilization and those of stress tolerance, which incorporate various levels of genes, proteins, regulatory factors, and stress behavior patterns (Fu et al. 2016). Our understanding of microbial physiology and metabolic processes has grown in tandem with the interactions between metabolic pathways and their regulatory networks. With appropriate computer simulations of complex metabolic networks can be utilized to standardize growth as well as generate a product under specific conditions (King et al. 2015).

16.5.5 Integration of Diverse Approaches

Scientists have been able to conduct in-depth investigations into cell metabolism and ensure that strains are designed rationally thanks to a new approach for sophisticated automation of gene expression profiling, microbiome records, metabolomics, and proteomics. New artificial biology tools now have the capability of incorporating a broad array of genetic mutations into a microbial host. Patel and Shah (2021) recently discussed an extensive lignocellulosic biorefinery that will produce biofuel in addition to more than 200 valuable products. In 2020, Sarkar and

Sarkar started a multi-stage system for economically efficient biofuel generation for purified biofuel production, ensuring lesser carbon and energy emission. Developed biorefineries may indeed be beneficial in terms of lowering second-generation biofuel costs, whereas enhanced biorefineries will improve processing by utilizing the entire biomass. Another efficient technique would be the use of second-generation lignocellulosic biomass from the timber and pulp industries (Rai et al. 2022). In addition to the approaches described above, Rosson et al. (2021) proposed the waste consists of animal fats that were proposed as bioliquids. Also, it opens up new avenues for sustainable power. Although waste shark liver oil (WSLO) is a non-edible feedstock, Al Hatrooshi et al. (2020) used it to consider making biodiesel as inexpensively as possible.

16.6 Types of Second-Generation Biofuels

The common examples of second-generation biofuels and their production methods are as follows.

16.6.1 Biohydrogen

This is a fuel with no emissions and little NOx generation during the combustion process. It is produced via fast pyrolysis and gasification (steam reforming) of second-generation biomass feedstock which includes agricultural waste, sewage, sludge, and municipal organic waste.

16.6.2 Bioethanol

Bioethanol is produced using critical elements of lignocellulosic biomass (cellulose, hemicelluloses, and lignin). The raw materials that are commonly used include wheat straws, corn cobs, stover, sorghum stalks, and alfalfa stems. The biochemical production strategy is summarized in Fig. 16.2 while the process of production of bioethanol is summarized in Fig. 16.3 (Huber and Corma 2007).

The biomass is transformed into syngas components (CO, H_2, CO_2, CH_4, and so on), tars, and solid char, which are then converted back into CO and H_2. Cooled biogas is transmitted to an amine unit for CO_2 and H_2S removal. The syngas is then converted into methanol, ethane, and other liquid fuels via the Fischer-Tropsch process. In a separate section, the derived alcohol is guided into a flash separator, whereas the syngas is repurposed for gas cleanup (Unruh et al. 2010). Thereafter a distillation column is used for the separation of alcohol mixtures.

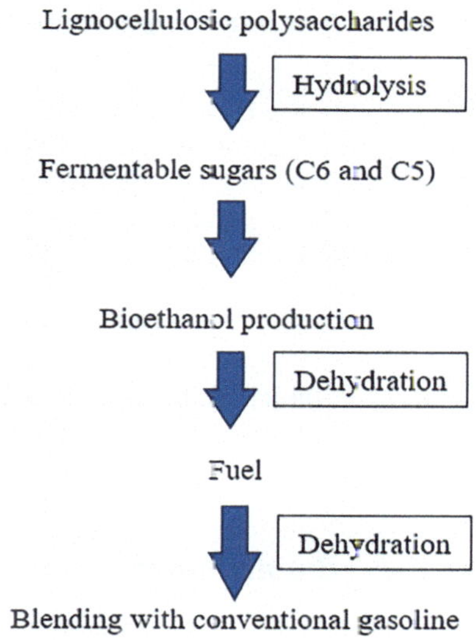

Fig. 16.2 The biochemical production strategy of bioethanol

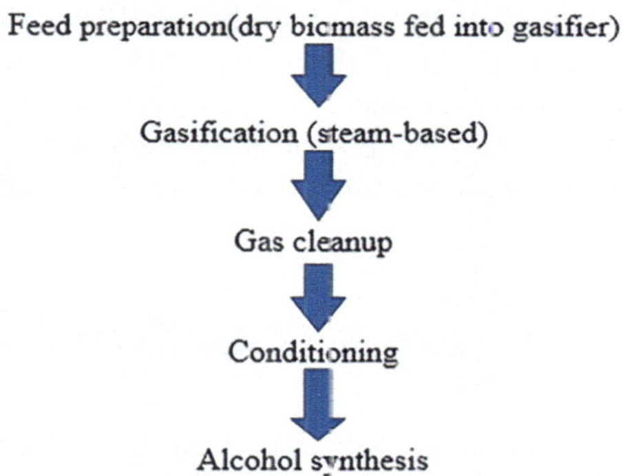

Fig. 16.3 Thermochemical process of bioethanol production

A novel method for bioethanol production has been discovered by using *Ciona intestinalis* tunic which is a marine filter feeder used as a biostock for the production of bioethanol. The tunic was enriched with new organosolv pretreatment which increased the cellulose content to 58.5% w/w, i.e., 3.4% higher as compared to untreated biomass. Pretreated biomass saccharified easily and the glucose was further used for bioethanol production. It was observed that fermentation yields rose up to 84.5% and the total ethanol concentration was 34.9 g/L. In the most favorable conditions, the maximum ethanol concentration obtained was 38.7 g/L with a total yield of 78.3% as reported by ACS Sustainable, 2020.

16.6.3 Fischer-Tropsch Diesel

F-T diesel is also called gas-to-liquid fuel. F-T diesel use biomass feedstocks such as wood, grass, forest wastes, manure and sludge, and agricultural wastes. 159 L of biodiesel fuel is produced by approximately 1 T of biomass having 30% moisture (Unruh et al. 2010). The gas is turned into a liquid which in turn is transformed into biofuel.

FTS reaction is as follows:

$$\mathrm{n\,CO} + (\mathrm{n} + \mathrm{m}/2) \rightarrow \mathrm{H_2C_nH_m} + \mathrm{H_2O}$$

here "n" indicates the mean length of the hydrocarbon chain and "m" represents the number of hydrogen atoms per carbon (Naik et al. 2010).

16.6.4 Biomethanol

Bioethanol is generated from the chemical process of converting methane (natural gas) to methanol. The process includes three steps: team reforming, synthesis, and distillation. Methanol can be recycled easily and hence, is preferred despite its toxicity as biodiesel (Demirbas 2009). Biomethanol production can also be done by partial oxidation and gasification of carbohydrates with water and oxygen (Demirbas 2009). It is also produced by either fermenting or distilling the crops constituting sugars and then converted to bioethanol and bioethanol.

16.6.5 Dimethyl Ether (DME)

The catalytic dehydration of biomethanol leads to the production of DME (CH_3OCH_3) which is the simplest ether. The reaction is as follows:

$$\mathrm{2CH_3OH} \rightarrow \mathrm{CH_3OCH_3} + \mathrm{H_2O}.$$

DME blended LPG has the following advantages:

1. No GHG emission.
2. DME can be supplied for DME-fueled vehicles by using LP infrastructure.
3. Compensates low cetane number of LPG and DME's low energy density.
4. LPG blended with DME has improved the fuel consumption of DME-fueled engines (Park and Lee 2014).

16.6.6 Waste Vegetable Oil

The oil is not valuable and hence can be used to reduce environmental pollution. This biomass-derived biofuel can be utilized directly by uniquely engineered diesel engines without any additional refining or blending (Raqeeb and Bhargavi 2015).

16.6.7 Pyrolysis Oils (Biocrude)

Biocrude is produced by ash pyrolysis by fast heating at about 1000 °F succeeded by fast cooling. The product is refined and upgraded and is used for transportation or stationary purposes like turbines, boilers, etc. (Wongkhorsub and Chindaprasert 2013).

16.6.8 Biosynthetic Natural Gas (Bio-SYN)

Renewable natural gas can be yielded by the purification of gasification and catalytic methanation product. Microorganisms can also produce biogas via anaerobic digestion which primarily consists of methane and carbon dioxide. The gas may be used in vehicles such as CNG and LNG, and it can also be injected into existing natural gas cylinders (Zhang et al. 2015).

16.6.9 Hydrotreated Vegetable Oil

It has well-known fuel properties like non-aromatic, and cetane and the absence of sulfur and can be employed as a diesel substitute (Datta et al. 2019).

16.6.10 Biomass to Liquid (BTL) Technology

The Fischer-Tropsch process is utilized to transform synthesis gas (Syn gas) into gasoline, jet fuel, and diesel (Datta et al. 2019).

Table 16.2 Prominent biofuel production countries and contribution to second-generation biofuel production

Country	Company	Product	Feedstock	Capacity	Percentage of the world's total second-gen biofuel production
USA	Coskata Renewable energy group, Inc. (Nasdaq: REGI) Verenium POET-DSM	Ethanol Biodiesel Ethanol Ethanol	Wood biomass Inedible corn oil Softwood and grass Corn crop residue	208 million L 972 million L 5.3 million L 3.8 billion L	35%
China	TMO renewables	Ethanol	Cassava stock	13.6 million L	24%
Canada	Enerkem	Ethanol	Municipal solid waste	40 million L	22%
Brazil	Raizen Petrobras GranBio	Ethanol Ethanol Biodiesel Ethanol	Sugarcane Sugarcane Plant oil/ animal fat Sugarcane bagasse	2.1 billion L 2 billion L 326,000 cubic meters 82 million L	9%

Source: UNCTAD report, 2016, Bhatia et al. 2017

16.7 Current Status, Production, and Marketing of Second-Gen Biofuels

Over the last decade, an increase in ethanol productivity in Brazil and the USA, as well as biodiesel production capacity in Europe, has contributed to biofuels rapidly gaining a significant position in the global liquid market (Table 16.2). According to Navigant Research (2014), in 2022, global biofuel consumption for road transportation might very well exceed 193.41 million liters per year, straightforwardly increasing the supply of biofuels.

Discussed below is the contribution of various countries in the production of second-generation biofuels.

16.7.1 United States

According to the United States Energy Information Agency (2015), biofuel accounted for 5% of total transportation fuel consumption in 2014. Various vehicle categories use different biofuel blends, such as light-duty vehicles that use E-10, a 10% ethanol blend of gasoline and ethanol. Furthermore, biodiesel blends, the most common of which are B-20 and B-5, can be easily used in heavy, medium, and light-duty vehicles designed for diesel use. Despite low biofuel production, the industry is making gradual progress in producing biofuel. In 2013, INEOS Bio commercialized cellulosic ethanol by completing the construction of its Vero Beach (Florida) facility, which used a variety of feedstocks including vegetable waste, municipal solid waste, and agricultural waste (Lane 2015).

16.7.2 *Brazil*

The PAISS program assisted in the establishment of two commercial and one pilot 2G ethanol plant with a total capacity of 140 million liters. The Brazilian Bioethanol Science and Technology Lab (CTBE) proposed various scenarios (14) for enhancing the yield of second-generation biofuel in the short and long term using various technologies. According to CTBE, the cost of producing second-generation biofuel will stay competitive with oil prices for a short-term US$ 123/barrel and a long-term US$ 44/barrel (Milanez et al. 2015). According to Milanez et al. (2015), mills can produce 105.6 kg of dry lignocellulosic material per tonne of sugarcane when steam consumption is approximately 360 kg/ton of sugarcane and 50% of sugarcane trash is harvested on 90% of the land area, yielding 216.9 L of ethanol from a single mill using existing technologies. As suggested by BNDES (2015), the expansion of the annual milling capacity of sugarcane by the sugarcane industry to up to 100 million tonnes of cane which is 80% viable will lead to the production of 2G ethanol in Brazil up to 1.5 billion liters by 2025.

16.7.3 *Africa*

In 2012, the estimation showed that Africa contributed less than 0.04% to global biofuel production which creates a challenge to producing second-generation biofuels. The present data show that the main source of first and second-generation biofuel production is Eastern and Southern African countries. The table below shows the African production of biofuel.

There are two main types of 2G biofuels found in the African continent namely—the transport industry and domestic use. Apart from these two-electricity generations from bagasse is also practiced in the continent.

- **Transportation:** It represents the economic hub for African biofuels. Countries such as Mozambique, Kenya, and Angola are well-known for producing first-generation biofuels, and several second-generation biofuel projects have been proposed. Project Solaris, for example, is being tested in South Africa. Solaris is a joint effort by South African Airways, Boeing, and various partners. Solaris is the energy-rich and nicotine-free tobacco variety that will be used in the project to produce bio-jet fuel.

- **Cooking:** Cookstove fuel based on sawdust and water hyacinths was produced by a private sector company named Green Energy Biofuels. The company claimed 200,000 customers in Ghana and Nigeria use its eco-friendly cookstoves and renewable biofuel coking gel.
- **Electricity generation using sugarcane bagasse:** The consumption of bagasse for electricity generation seems to have the ability to assist in the production of second-generation biofuels. Despite limited data, the data show a decrease in bagasse consumption for power generation.

16.8 Challenges for Biomass Utilization of Biofuels

Non-edible lignocellulosic biomass is gaining popularity as a renewable, cost-effective, and produce sufficient for relying heavily on petroleum and bringing down energy and material feedstock costs. Unlike fossil-based energy fuels like gasoline, natural gas, and coal, these resources emit no additional carbon dioxide into the atmosphere. Carbon dioxide captured during biomass growth is primarily used to offset carbon dioxide released during bioenergy/biofuel production. As a result, biomass systems are capable of lowering GHGs. The primary carbon sources for a wide range of fuels are biomass resources. The heterogeneous nature and rigid and complex structure of biomass components makes them difficult to fragment into simple fragments and/or transform into a diverse variety of real-worth components.

In general, the energy density of biomass is low, calling for the use of additional biomass feedstocks to generate energy as much as with conventional hydrocarbon fuels slightly higher O_2 content of biomass materials may also impede the conversion to other products such as fuels. For example, oxygen should be removed from biomass structures to yield hydrocarbon fuels that are nearly identical to those derived from petroleum. The dependability of processing methods can also vary according to the type of biomass used (Irmak 2019). The following challenges of conversion of biomass are:

16.8.1 Density

Lignocellulosic biomass substances have a low bulk density (80–150 kg/m^3 for grass biomass and 160–220 kg/m^3 for woody biomass). This complicates the handling of these massive volumes of feedstock as well as raises transportation and storage expenses (Sokhansanj and Fenton 2006). Biomass bulk density lies between

190 and 240 kg/m^3 for effective transport in variable sizes of trucks carrying 25-ton loads (Kaliyan et al. 2009). Depending on the biomass form, humidity levels, preparation methods, and so on, the concentration of biomass material can be greatly increased. This reduces the storage and transportation cost for densified material. The components are managed with basic machines or equipment due to their consistent shapes and dimensions (Kaliyan and Morey 2006).

16.8.2 Difficulties with Recalcitrance and Dissolution

Amidst the potential for biofuel manufacturing and chemical products as options available to petroleum, lignocellulosic materials' intricate and rigid structures restrict their use in such implementations. The physicochemical properties, pretreatment strategies, processing conditions, etc. aid in employing lignocellulosic biomass for biofuels production. Plant cell wall composition and cellular ultrastructure have a significant effect on the efficacy of biomass hydrolysis (dissolution). A biological process that converts biomass structures into sugars is known as hydrolysis. To be effectively transformed into a variety of products, biomass resources must first be fragmented into elements with lower molecular weights (e.g., oligo- and monosaccharides). Biomass hydrolysates can be used to make a variety of value-added products, such as biofuels (ethanol, hydrogen, and so on), essential chemicals (e.g., solvents), and food products (sugar and sugar alcohols, etc.) (Irmak 2019). Difficulties that occur during the hydrolysis of lignocellulosic biomass are:

- Hydrolysis methods that are currently available are both costly and time-consuming. The majority of them are harmful to the environment.
- In harsh hydrolysis conditions, released carbohydrates decompose.
- More steps are essential such as pretreatment, neutralization, etc.

Hazardous, toxic, or corrosive chemicals (e.g., alkali and acid treatments) and/or long processing durations (e.g., enzymatic hydrolysis) are required in the major hydrolysis processes commonly used for biomass solubilization, thereby rendering the procedure hazardous to the environment and/or costly Mineral acids are commonly used to dissolve hemicelluloses, whereas organosolv or alkaline pretreatments are used to dissolve lignin (Alvira et al. 2010; Galbe and Zacchi 2007). Because of the use of strong acids for biomass, hydrolysis has many additional disadvantages, such as increased energy consumption, equipment corrosion, the handling of hazardous chemicals, an additional crucial component of acid neutralization, the accumulation of byproducts that hinder fermentation (Galbe and Zacchi 2002; Sun and Cheng 2002), and other harmful emissions. As a result, current practices include undesirable processes and thus do not satisfy the requirements.

16.8.3 Complexity and Diversity

The three principal components of biomass are cellulose, hemicellulose, and lignin. The polymers are sorted in sophisticated non-uniform 3-D structures with varying degrees of polymerization. Depending on the biomass species, the polymerization rate or/and structure of these biopolymers could perhaps differ. Cellulose has a linear structure and is made up of (1–4) linked glucose subunits. Cellulose molecules determine the cell wall structure. Because of the structure's intra- and inter-chain H - bonding, the cellulose is crystalline, and this segment of the cellulose does not hydrolyze as easily as amorphous cellulose (Bali et al. 2016; Meng et al. 2017). Hemicellulose is an amorphous and random structure made up of a variety of heteropolymers including xyloglucan, arabinoxylan, glucomannan, xylan, and galactomannan. It has a lower degree of polymerization than cellulose. In contrast to cellulose, the monomer units of hemicellulose polysaccharides are arabinose, galactose, mannose, xylose, and rhamnose. Lignin is an aromatic compound that contains phenylpropane. The three phenyl propane foundations that makeup lignin are coniferyl alcohol, sinapyl alcohol, and p-coumaric alcohol. The configurations of lignin, cellulose, and hemicellulose in biomasses vary greatly between species. Some biomass materials, like hardwoods, have much more cellulose in their constructions than others, like straws. Softwood hemicellulose fractions are primarily composed of D-mannose-derived structures like galactoglucomannans, while hardwood hemicelluloses are composed of D-xylose-derived structures like arabinoglucuronoxylan (Puls 1997). The variety of biomass materials has the potential to have an influence on the transformation techniques used to generate biofuel or other valuable products from biomass materials (Fig. 16.4).

16.8.4 Moisture Content

Gasification and pyrolysis, for example, are traditional thermochemical conversion technologies that do not operate well with high moisture biomass materials. The quality of conversion processes can be reduced by high humidity. Water in raw biomasses is also undesirable seeing as moisture can be found in the gas emitted from these components. Fuels with an excessive amount of moisture are difficult to ignite and also lead to loss of energy in vaporization. The moisture content of biomass should preferably be below 20% to maximize the thermal value of the fuel generated. Because of the high cost, thoroughly rinsing the raw material before use in the transformation is not recommended. Some biomass conversion methods, on the other hand, make use of biomass with a significant amount of moisture. Hydrothermal conversion processes and biological reactions are suitable for high moisture biomass as compared to conventional drying. During storage, the high amount of water in biomass causes shortfalls in organic contents, biological degradation, and mold formation (e.g., carbohydrates) (Johansson et al. 2006), which may minimize the

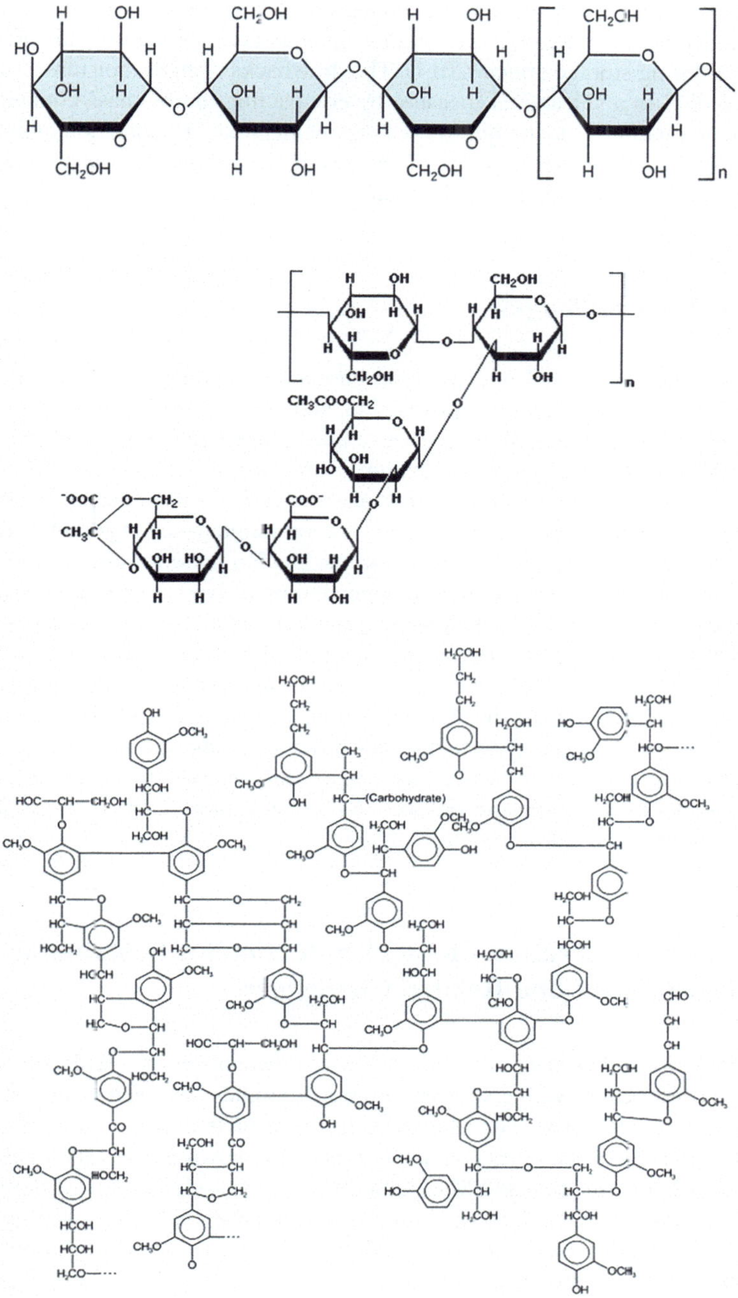

A

B

C

Fig. 16.4 Structure of (a) cellulose, (b) hemicellulose, and (c) lignin

outcome of the fuel generated from these materials. Maintaining biomass moisture at 10% can extend the material's preservation span and reduce sugar declination in the biomass during storage (Balan 2014). The drawbacks of high moisture content can be remedied to a substantial measure by compacting the biomass content for further unified qualities, a technique known as densification. Densification improves overall efficiency and storage capacity by increasing the volume of biomass materials. Moreover, this procedure raises the final product's price.

16.8.5 Other Challenges

Even though energy demand is consistent, biomass components are seasonal. In terms of transportation, cultivation, production, and storage, several biomasses outperform others. Kenaf, Miscanthus, and Switchgrass are energy crops that have economic benefits over crops (sweet sorghum, corn, sugar beet, sugarcane, etc.). Perennial energy crops, such as miscanthus and switchgrass, have the advantage of overcoming replantation every year, special care, or maintenance. Agricultural biomass residues (e.g., wheat straw, pulps, crop peels, etc.) are appealing to reduced feedstocks because they do not require a large amount of land for biomass production, and these biomass materials belong to agricultural land. Forest biomass is also an important supply of nutrients for the manufacture of biofuels and other important products. However, because of the high expenses of harvesting and transportation, their use is limited. Aside from the above-mentioned advantages and disadvantages, multiple sources of biomass feedstocks may differ in terms of constituents, standardized size, form, and so on, which has a major influence on the effectiveness of processing methods for a specific product. As a result, biorefinery biomass feedstocks must be streamlined.

16.9 Second-Generation Biofuel Production for Sustainable Development and Related Challenges

The sustainability and climate-related issues posed by conventional fuels are being addressed by biofuels as viable sources of energy. Studies have demonstrated that assuming biofuels are a superior alternative to fossil fuels in terms of combating climate change requires careful consideration because they are not carbon neutral, as believed earlier (Lundgren and Marklund 2013). T the sustainability of biofuels, on the other hand, depends on their reduced dependence on foreign oil imports, reduced GHG emissions, assisting rural and agricultural development through the use of plant and animal products, and also providing a reliable energy supply (Sims et al. 2009). Things, however, are not as reassuring as they seem. There may be a conflict between the biomass supply for biofuels and agricultural land for food crop cultivation (Lee 2011). Second-generation biofuels are very helpful in this situation because they utilize lignocellulosic plant material, food industry wastes, and even locally grown vegetable waste as their feedstock rather than food crops. Additionally, second-generation biofuels have higher environmental performance and use a wider variety of biomass as feedstock when compared with first-generation biofuels.

The second generation of biofuel, however, also faces difficulties such as high production costs, the need for new and improved production methods, logistical problems (production, storage, shipping, etc.), a lack of regulatory oversight of production quality, etc. (Alam et al. 2020). The current second-generation biofuel manufacturing processes are costly, which prevents their widespread production and commercialization. To assure the low-cost, efficient biochemical/thermochemical method of second-generation biofuel production from a variety of biomass feedstocks, concerted efforts in the direction of technological improvement are required.

Multiple barriers are encountered during biofuel production, such as production costs, environmental elements (deficit of soil and land area), and others. The first major impediment is the rising price of producing biofuel (bioethanol) and its economic viability in contrast to crude oil prices (Popp et al. 2016; Raud et al. 2019; Mizik and Gyarmati 2021; Sinitsyn and Sinitsyna 2021). Recently, Mizik and Gyarmati (2021) outlined the limiting factors, notably the economic aspect. The authors expressed numerous concerns regarding biofuel production, claiming that due to the expense and technological limitations, extremely high biofuels are not priced competitively. The second major impediment to overcome is operational constraints, which must be overcome in generating cost-effective and economically viable second-generation biofuels. The development of novel techniques is getting difficult, particularly when it comes to identifying the impediments associated with biofuel production (Raud et al. 2019; Shahid et al. 2021). Another significant impediment may be the availability of funds for ongoing and expanded research and advancement to bring biofuel to market, it can be achieved by the following:

(a) Specific and augmented support for increased energy crop yields.
(b) Sustainable biomass production.
(c) Lowering the supply chain's costs.
(d) Upgrading the conversion process.

16.10 Novel Technologies for Pretreatment

16.10.1 Ionic Liquid Pretreatment

Ionic liquid (IL) pretreatment of lignocellulose has lately garnered a great deal of attention. These compounds have been widely used owing to their environmentally responsible properties, low vapor pressures, and high thermal stability that prevent the release of toxic gases (Wang et al. 2018). As cations and anions, ILs are known to be beneficial for cellulose disintegration and typically contain morpholinium+, ammonium+, phosphonium+, imidazolium+, or pyridinium+. These IL groups are effective because they can form substantial H- bonds with hydroxyl groups. Using an IL-based biomass pretreatment, hemicellulose, and lignin are rapidly removed, while cellulose shape is significantly reduced (Li et al. 2010). The need for IL per co-solvents has also been recognized as an effective method for cellulose dissociation. According to reports, the co-solvent system has numerous benefits over parent ILs, including a high rate of dissociation, excellent thermal stability, and low viscosity (de Oliveira et al. 2018). Despite their attractive characteristics, ILs have limited industrial-scale consumption due to their flowability and the large quantities of expensive ILs required, resulting in significant operating costs (Cheah et al. 2020).

16.10.2 Hydrothermal Pretreatment

Hydrothermal mechanisms have played an important role in the production of fuels over the last decade. In hydrothermal pretreatments, subcritical water at high temperatures (374 °C) is used. These high temperatures deform plant cell walls of the plants, dissolve hemicelluloses and convert lignin to syngas/sugars (Zhang et al. 2012). Hydrothermal pretreatment of biomass could improve cellulose accessibility to enzymes by increasing biomass surface area and decreasing cellulose crystallinity (Binod et al. 2010). Equipment parameters, time of residence, temperature, pH, flow rate, and pressure all have an impact on the hydrolysis of lignocellulose biomass and the disintegration of lignin structure utilizing supercritical and subcritical water processes (Torres-Mayanga et al. 2019). Xiao and colleagues demonstrated in a recent study that incorporating ethanol into the hydrothermal liquefaction process increased permeation into the solid system of lignocellulose, resulting in high yields of bio-oil (Wu et al. 2019) (Fig. 16.5).

16.11 Conclusion

Second-generation biofuel uses lignocellulosic biomass which proves to be a more sustainable approach and more environmentally friendly. Lignocellulosic biomass is non-edible in nature hence, the competition with food crops is decreased. Even though lignocellulosic materials are by far the most readily available feedstock for the production of second-generation biofuels, significant research efforts have been made to improve the various underlying mechanisms (Yau and Easterling 2018; Raud et al. 2019). Although biomass is widely available, the cost of manufacturing second-generation biofuel remains high, owing to the pretreatment stage accounting for the highest difference in capital costs among the first and second generations. This is because reactor designs that can resist high pressures, extreme heat, and/or corrosive catalysts are required (Stephen et al. 2012). The majority of the current published studies were undertaken in a lab, so there is little data on the actual

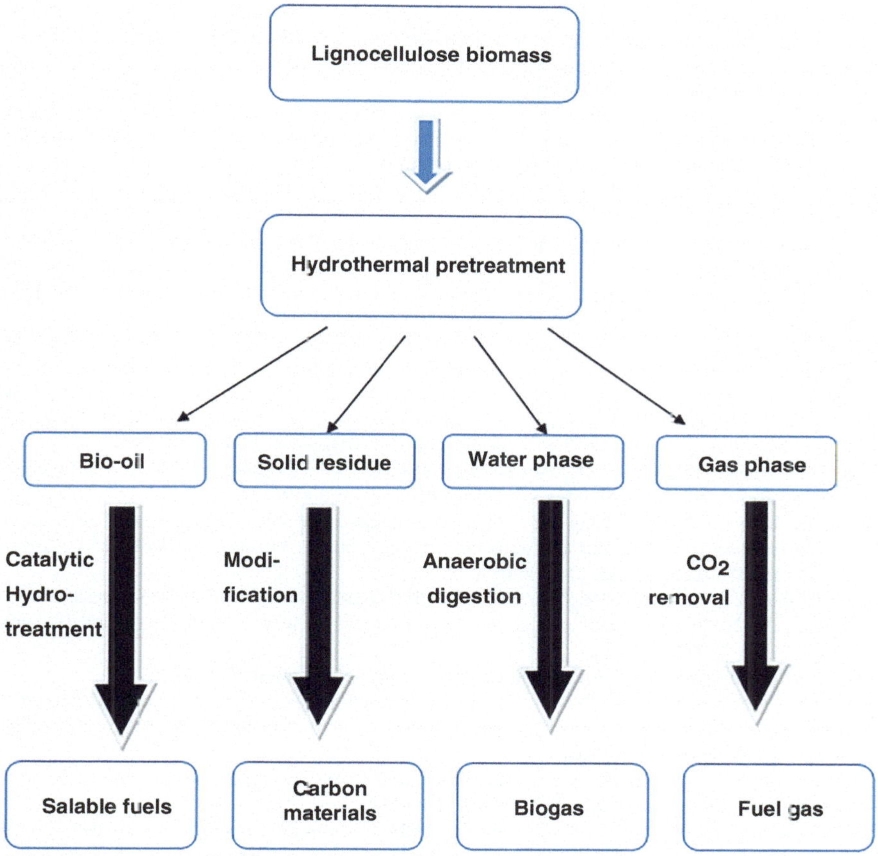

Fig. 16.5 Description of the hydrothermal pretreatment process (Cheah et al. 2020)

operating associated costs with all phases of lignocellulosic biomass biofuel production, namely delignification, hydrolysis, and fermentation. As a result, future research should concentrate on optimizing operating parameters and estimating the total cost of manufacturing biofuel on a large scale from lignocellulose biomass using various pretreatment methods. This understanding would open up the possibilities for large-scale lignocellulosic biofuel production.

References

Al Hatrooshi AS, Eze VC, Harvey AP (2020) Production of biodiesel from waste shark liver oil for biofuel applications. Renew Energy 145:99–105

Al Makishah NH, Rai AK, Neamatallah AA, Mabrouk AM (2019) Micrococcus luteus: a novel lipolytic bacterial strain isolated from local contaminated soil in Saudi Arabia. Sylwan 2019(163):132–152

Alam A, Wang Y, Liu F, Kang H, Tang S, Wang Y, Peng L (2020) Modeling of optimal green liquor pretreatment for enhanced biomass saccharification and delignification by distinct alteration of wall polymer features and biomass porosity in miscanthus. Renew Energy 159:1128. https://doi.org/10.1016/j.renene.2020.06.013

Alvira P, Tomás-Pejó E, Ballesteros M, Negro M (2010) Pretreatment technologies for an efficient bioethanol production process based on enzymatic hydrolysis: a review. Bioresour Technol 101:4851–4861. https://doi.org/10.1016/j.biotech.2009.11.093

Badawy T, Mansour MS, Daabo AM, Aziz MMA, Othman AA, Barsoum F, Basouni M, Hussien M, Ghareeb M, Hamza M et al (2021) Selection of second-generation crop for biodiesel extraction and testing its impact with nano additives on diesel engine performance and emissions. Energy 237:121605

Balan V (2014) Current challenges in commercially producing biofuels from lignocellulosic biomass. ISRN Biotechnol 2014:463074. https://doi.org/10.1155/2014/463074

Bali G, Khunsupat R, Akinosho H, Payyavula RS, Samuel R, Tuskan GA et al (2016) Characterization of cellulose structure of populus plants modified in candidate cellulose biosynthesis genes. Biomass Bioenergy 94:146–154. https://doi.org/10.1016/j.biombioe.2016.08.013

Barnwal BK, Sharma MP (2005) Prospects of biodiesel production from vegetable oils in India. Renew Sust Energ Rev 9:363–378. https://doi.org/10.1016/j.rser.2004.05.007

Bessou C, Ferchaud F, Gabrielle B, Mary B (2010) Biofuels, greenhouse gases and climate change. Agron Sustain Dev 31:1–79. https://doi.org/10.1051/agro/2009039

Bhatia SK, Kim SH, Yoon JJ, Yang YH (2017) Current status and strategies for second generation biofuel production using microbial systems. Energy Convers Manag 148:1142–1156. https://doi.org/10.1016/j.enconman.2017.06.07

Binod P, Sindhu R, Singhania RR, Vikram S, Devi L, Nagalakshmi S, Kurien N, Sukumaran RK, Pandey A (2010) Bioethanol production from rice straw: an overview. Bioresour Technol 101(13):4767–4774

BNDES (2015) Technical note on second-generation biofuel potentials in Brazil

Cadoux S, Riche AB, Yates NE, Machet JM (2012) Nutrient requirements of *Miscanthus x giganteus*: conclusions from a review of published studies. Biomass Bioenergy 38:14–22. https://doi.org/10.1016/j.biombioe.2011.01.015

Carriquiry Miguel A, Xiaodong D, Timilsina Govinda R (2011) Second generation biofuels: economics and policies. Energy Policy 39(7):4222–4234

Cheah WY, Sankaran R, Show PL, Tg Ibrahim TNB, Chew KW, Culaba A, Chang J-S (2020) Pretreatment methods for lignocellulosic biofuels production: current advances, challenges and future prospects. Biofuel Res J 25:1115–1127. https://doi.org/10.18331/BRJ2020.7.1.4

Das D, Veziroğlu TN (2001) Hydrogen production by biological processes: a survey of literature. Int J Hydrog Energy 26:13–28

Datta A, Hossain A, Roy S (2019) An overview on biofuels and their advantages and disadvantages. Asian J Chem 31(8):1851–1858. https://doi.org/10.14233/ajchem.2019.22098

De Cara S, Goussebaïle A, Grateau R et al (2012) Revue critique des études évaluant l'effet des changements d'affectation des sols sur les bilans environnementaux des biocarburants. Étude réalisée pour le compte de l'ADEME par l'INRA 1:1–96

De Oliveira MED, Vaughan BE, Rykiel EJJ (2005) Ethanol as fuel: energy, carbon dioxide balances, and ecological footprint. Bioscience 55:593–602

de Oliveira S, AMF GM, Ferraz A (2018) Role of hemicellulose removal during dilute acid pretreatment on the cellulose accessibility and enzymatic hydrolysis of compositionally diverse sugarcane hybrids. Ind Crop Prod 111:722–730

Deenanath ED, Iyuke S, Rumbold K (2012) The bioethanol industry in sub-Saharan Africa: history, challenges, and prospects. J Biomed Biotechnol 2012:1–11. https://doi.org/10.1155/2012/416491

Demirbas A (2009) Biofuels from agricultural biomass. Energ Sour A Recov Utiliz Environ Effect 31(17):1573–1582. https://doi.org/10.1080/15567030802094011

Demirbas MF (2011) Biofuels from algae for sustainable development. Appl Energy 88:3473–3480. https://doi.org/10.1016/j.apenergy.2011.01.059
Demirel Y (2017) Sugar versus lipid for sustainable biofuels. Int J Energy Res 42(3):881–884. https://doi.org/10.1002/er.3914
Donner SD, Kucharik CJ (2008) Corn-based ethanol production compromises goal of reducing nitrogen export by the Mississippi river. Proc Natl Acad Sci U S A 105:4513–4518
Dufosse K, Wassim Ben A, Benoit G (2017) Life-cycle assessment of agricultural feedstock for biorefineries. In: Life-cycle assessment of biorefineries. Elsevier, p 9780444635860
Elumalai PV, Nambiraj M, Parthasarathy M, Balasubramanian D, Hariharan V, Jayakar J (2021) Experimental investigation to reduce environmental pollutants using biofuel nano-water emulsion in thermal barrier coated engine. Fuel 285:119200
Fargione J, Hill J, Tilman D, Polasky S, Hawthorne P (2008) Land clearing and the biofuel carbon debt. Science 319(5867):1235–1238. https://doi.org/10.1126/science.1152747
Felizardo P, Neiva Correia MJ, Raposo I, Mendes JF, Berkemeier R, Bordado JM (2006) Production of biodiesel from waste frying oils. Waste Manag 26:487–494
Fu Y, Chen L, Zhang W (2016) Regulatory mechanisms related to biofuel tolerance in producing microbes. J Appl Microbiol 121:320–332
Galbe M, Zacchi G (2002) A review of the production of ethanol from softwood. Appl Microbiol Biotechnol 59:618–628. https://doi.org/10.1007/s00253-002-1058-9
Galbe M, Zacchi G (2007) Pretreatment of lignocellulosic materials for efficient bioethanol production. In: Scheper T, Olssen L (eds) Biofuels. Springer, Berlin, Heidelberg, pp 41–65. https://doi.org/10.1007/978-3-540-73651-6
Gasparatos A, Stromberg P, Takeuchi K (2013) Sustainability impacts of first-generation biofuels. Anim Front 3(2):12–26. https://doi.org/10.2527/af.2013-0011
Gavrilescu M, Chisti Y (2005) Biotechnology—a sustainable alternative for chemical industry. Biotechnol Adv 23:471–499. https://doi.org/10.1016/j.biotechadv.2005.03.004
Gnansounou E, Dauriat A (2010) Techno-economic analysis of lignocellulosic ethanol: a review. Bioresour Technol 101:4980–4991
Gomez LD, Clare GS, McQueen-Mason J (2008) Sustainable liquid biofuels from biomass: the writing's on the walls. New Phytol 178:473–485
Hays SG, Ducat DC (2014) Engineering cyanobacteria as photosynthetic feedstock factories. Photosynth Res 123:285–295
Huber GW, Corma A (2007) Synergies between bio- and oil refineries for the production of fuels from biomass. Angew Chem Int Ed 46:7184–7201
Ilmberger N, Streit WR (2017) Screening for cellulase encoding clones in metagenomic libraries. In: Streit W, Daniel R (eds) Metagenomics. Methods in molecular biology, vol 1539. Humana Press, New York, NY, pp 205–217
Irmak S (2019) Challenges of biomass utilization for biofuels. In: Abomohra AE (ed) Biomass for bioenergy—recent trends and future challenges. IntechOpen, London. https://doi.org/10.5772/intechopen.83752
Johansson J, Liss J, Gullberg T, Bjorheden R (2006) Transport and handling of forest energy bundles-advantages and problems. Biomass Bioenergy 30:334–341. https://doi.org/10.1016/j.biombioe.2005.07.012
Kaliyan N, Morey RV (2006) Factors affecting strength and durability of densified products. ASABE paper no. 066077. ASABE, St. Joseph, MI
Kaliyan N, Morey RV, White MD, Tiffany DG (2009) A tub-grinding/roll-press compaction system to increase biomass bulk density: preliminary study. In: 2009 ASABE annual international meeting, presentation paper number: 096658, Nevada
King ZA, Lloyd CJ, Feist AM, Palsson BO (2015) Next-generation genome-scale models for metabolic engineering. Curr Opin Biotechnol 35:23–29
Koh LP, Wilcove DS (2008) Is oil palm agriculture really destroying tropical biodiversity? Conserv Lett 1(2):60–64. https://doi.org/10.1111/j.1755-263x.2008.00011

Koh LP, Miettinen J, Liew SC, Ghazoul J (2011) Remotely sensed evidence of tropical peatland conversion to oil palm. Proc Natl Acad Sci 108(12):5127–5132. https://doi.org/10.1073/pnas.1018776108

Kulkarni MG, Dalai AK (2006) Waste cooking oil—an economical source for biodiesel, a review. Ind Eng Chem Res 45:2901–2913. https://doi.org/10.1021/ie0510526

Lane J (2015) INEOS bio: biofuels digest's 2015 5-minute guide; http://www.biofuelsdigest.com/bdigest/2015/02/17/ineos-bio-biofuels-digests-2015-5-minute-guide/

Lee H (2011) Chapter 21—mitigation: reducing greenhouse gas emissions, sinks, and solutions. In: Hannah L (ed) Climate change biology. Academic Press, pp 339–356. https://doi.org/10.1016/B978-0-12-374182-0.00016-9; ISBN 9780123741820

Li C, Knierim B, Manisseri C, Arora R, Scheller HV, Auer M, Vogel KP, Simmons BA, Singh S (2010) Comparison of dilute acid and ionic liquid pretreatment of switchgrass: biomass recalcitrance, delignification and enzymatic saccharification. Bioresour Technol 101:4900–4906

Liang J, Lin Y, Li T, Mo F (2017) Microbial consortium OEM1 cultivation for higher lignocellulose degradation and chlorophenol removal. RSC Adv 7:39011–39017

Liu G, Qin Y, Li Z, Qu Y (2013) Development of highly efficient, low-cost lignocellulolytic enzyme systems in the post-genomic era. Biotechnol Adv 31:962–975

Lundgren T, Marklund PO (2013) Assessing the welfare effects of promoting biomass growth and the use of bioenergy. Climate Chang Econom 4(01):1350003

Meng X, Pu Y, Yoo CG, Li M, Bali G, Park DY et al (2017) An in-depth understanding of biomass recalcitrance using natural poplar variants as the feedstock. ChemSusChem 10:139–150. https://doi.org/10.1002/cssc.201601303

Milanez A et al (2015) De promessa a realidade: como o etanol celulósico pode revolucionar a indústria da cana-de-açúcar – uma avaliação do potencial competitivo e sugestões de política pública. In: BNDES setorial 41. BNDES, Rio de Janeiro, pp 237–294

Millet DB, Apel E, Henze DK, Hill JD, Marshal JD, Singh HB, Tessum CW (2012) Natural and anthropogenic ethanol sources in North America and potential atmospheric impacts of ethanol fuel use. Environ Sci Technol 46:8484–8492

Mizik T, Gyarmati G (2021) Economic and sustainability of biodiesel production- a systematic literature review. Clean Technol 3:19–36

Muyibi SA, Ambali AR, Eissa GS (2008) Development-induced water pollution in Malaysia: policy implications from an econometric analysis. Water Policy 10:193–206

Nagle N, Lemke P (1990) Production of methyl ester from microalgae. Appl Biochem Biotechnol 24:355–361. https://doi.org/10.1007/BF02920259

Naik SN, Goud VV, Rout PK, Dalai AK (2010) Production of first and second generation biofuels: a comprehensive review. Renew Sust Energ Rev 14(2):578–597. https://doi.org/10.1016/j.rser.2009.10.003

Nimchua T, Thongaram T, Uengwetwanit T, Pongpattanakitshote S, Eurwilaichitr L (2012) Metagenomic analysis of novel lignocellulose-degrading enzymes from higher termite guts inhabiting microbes. J Microbiol Biotechnol 22:462–469

Nizami AS, Rehan M (2018) Towards nanotechnology-based biofuel industry. Biofuel Res J 5(2):798–799

Obidzinski K, Andriani R, Komarudin H, Andrianto A (2012) Environmental and social impacts of oil palm plantations and their implications for biofuel production in Indonesia. Ecol Soc 17(1):25. https://doi.org/10.5751/ES-04775-170125

OECD (2022). https://www.oecd.org/

Park SH, Lee CS (2014) Applicability of dimethyl ether (DME) in a compression ignition engine as an alternative fuel. Energy Convers Manag 86:848–863. https://doi.org/10.1016/j.enconman.2014.06.05

Patel A, Shah A (2021) Integrated lignocellulosic biorefinery: gateway for production of second generation ethanol and value added products. J Bioresour Bioprod 6:108–128

Pimentel D, Patzek TW (2005) Ethanol production using corn, switchgrass, and wood; biodiesel production using soybean and sunflower. Nat Resour Res 14:65–76. https://doi.org/10.1007/s11053-005-4679-8

Popp J, Harangi-Rakos M, Gabnai Z, Balogh P, Antal G, Bai A (2016) Biofuels and their co-products as livestock feed: global economic and environmental implications. Molecules 21:285

Puls J (1997) Chemistry and biochemistry of hemicelluloses. Relationship between hemicellulose structure and enzymes required for hydrolysis. Macromol Symp 120:183–196. https://doi.org/10.1002/masy.19971200119

Rai AK, Al Makishah NH, Wen Z, Gupta G, Pandit S, Prasad R (2022) Recent developments in lignocellulosic biofuels, a renewable source of bioenergy. Fermentation 8:161. https://doi.org/10.3390/fermentation8040161

Raqeeb MA, Bhargavi R (2015) J Chem Pharm Res 7:670

Rashamuse K, Tendai WS, Mathiba K, Ngcobo T, Mtimka S, Brady D (2017) Metagenomic mining of glycoside hydrolases from the hindgut bacterial symbionts of a termite (Trinervitermes trinervoides) and the characterization of a multimodular beta-1,4-xylanase (GH11). Biotechnol Appl Biochem 64:174–186

Raud M, Kikas T, Sippula O, Shurpali NJ (2019) Potentials and challenges in lignocellulosic biofuel production technology. Renew Sust Energ Rev 111:44–56

Rosson E, Sgarbossa P, Pedrielli F, Mozzon M, Bertani R (2021) Bioliquids from raw waste animal fats: an alternative renewable energy source. Biomass Convers Biorefin 11:1475–1490

Sarkar M, Sarkar B (2020) How does an industry reduce waste and consumed energy within a multi-stage smart sustainable biofuel production system? J Clean Prod 262:121200

Sawayama S, Inoue S, Dote Y, Yokoyama SY (1995) CO2 fixation and oil production through microalga. Energy Convers Manag 36:729–731

Scaife MA, Nguyen GT, Rico J, Lambert D, Helliwell KE, Smith AG (2015) Establishing Chlamydomonas reinhardtii as an industrial biotechnology host. Plant J 82:532–546

Searchinger T, Heimlich R, Houghton RA et al (2008) Use of U.S. croplands for biofuels increases greenhouse gases through emissions from land-use change. Science 319(5867):1238–1240. https://doi.org/10.1126/science.1151861

Sekoai PT, Ouma CNM, Du Preez SP, Modisha P, Engelbrecht N, Bessarabov DG, Ghimire A (2019) Application of nanoparticles in biofuels: an overview. Fuel 237:380–339

Shahid MK, Batool A, Kashif A, Nawaz MH, Aslam M, Iqbal N, Choi Y (2021) Biofuels and biorefineries: development, application and future perspectives emphasizing the environmental and economic aspects. J Environ Manag 297:113268

Sims RE, Mabee W, Saddler JN, Taylor M (2009) An overview of second generation biofuel technologies. Bioresour Technol 101(6):1570–1580. https://doi.org/10.1016/j.biortech

Singh N, Dhar D (2011) Microalgae as second generation biofuel a review: agronomy for sustainable development. Springer Verlag/EDP Sciences/INRA, pp 605–606

Sinitsyn AP, Sinitsyna OA (2021) Bioconversion of renewable plant biomass. Second-generation biofuels: raw materials, biomass pretreatment, enzymes, processes, and cost analysis. Biochemistry 86:S166–S195

Sokhansanj S, Fenton J (2006) Cost benefits of biomass supply and pre-processing. In: A BIOCAP research integration program synthesis paper. BIOCAP Foundation, Canada

Stephen JD, Mabee WE, Saddler JN (2012) Will second-generation ethanol be able to compete with first-generation ethanol? Opportunities for cost reduction. Biofuels Bioprod Biorefin 6(2):59–176

Strullu L, Cadoux S, Preudhomme M, Jeuffroy MH, Beaudoin N (2011) Biomass production and nitrogen accumulation and remobilization by *Miscanthus × giganteus* as influenced by nitrogen stocks in belowground organs. Field Crop Res 121:381–391. https://doi.org/10.1016/j.fcr.2011.01.005

Sun Y, Cheng J (2002) Hydrolysis of lignocellulosic materials for ethanol production: a review. Bioresour Technol 83:1–11. https://doi.org/10.1016/s0960-8524(01)00212-7

Tessum CW, Marshall JD, Hill JD (2012) A spatially and temporally explicit life cycle inventory of air pollutants from gasoline and ethanol in the United States. Environ Sci Technol 46:11408–11417

Torres-Mayanga PC, Lachos-Perez D, Mudhoo A, Kumar S, Brown AB, Tyufekchiev M, Dragone G, Mussatto SI, Rostagno MA, Timko M, Forster-Carneiro T (2019) Production of biofuel precursors and value-added chemicals from hydrolysates resulting from hydrothermal processing of biomass: a review. Biomass Bioenergy 130:105397

United States Energy Information Administration (EIA) (2015) Energy use for transportation; http://www.eia.gov/Energyexplained/?page=us_energy_transportation

Unruh D, Pabst K, Schaub G (2010) Fischer−Tropsch synfuels from biomass: maximizing carbon efficiency and hydrocarbon yield. Energy Fuel 24(4):2634–2641. https://doi.org/10.1021/ef9009185

van Dam J, Juginger M et al (2008) Overview of recent developments in sustainable biomass certification. Biomass Bioenergy 32(8):749–780

Wagner AO, Lackner N, Mutschlechner M, Prem EM, Markt R, Illmer P (2018) Biological pretreatment strategies for second generation lignocellulosic resources to enhance biogas production. Energies 11:17

Wang C, Dong D, Wang H, Mueller K, Qin Y, Wang H, Wu W (2016) Metagenomic analysis of microbial consortia enriched from compost: new insights into the role of actinobacteria in lignocellulose decomposition. Biotechnol Biofuels 9:22

Wang D, Shen F, Yang G, Zhang Y, Deng S, Zhang J, Zeng Y, Luo T, Mei Z (2018) Can hydrothermal pretreatment improve anaerobic digestion for biogas from lignocellulosic biomass? Bioresour Technol 249:117–124

Wongkhorsub C, Chindaprasert N (2013) A comparison of the use of pyrolysis oils in diesel engine. Energy Power Eng 5:350. https://doi.org/10.4236/epe.2013.54B068

Wu XF, Yin SS, Zhou Q, Li MF, Peng F, Xiao (2019) Subcritical liquefaction of lignocellulose for the production of biooils in ethanol/water system. Renew Energy 136:8

Yau YY, Easterling M (2018) Lignocellulosic feedstock improvement for biofuel production through conventional breeding and biotechnology. In: Biofuels: greenhouse gas mitigation and global warming. Springer, New Delhi, pp 107–140

Zhang Y, Fu X, Chen H (2012) Pretreatment based on two-step steam explosion combined with an intermediate separation of fiber cells-optimization of fermentation of corn straw hydrolysates. Bioresour Technol 121:100–104

Zhang W, He J, Engstrand P, Björkqvist O (2015) Energies 8:12795. https://doi.org/10.3390/en81112343

Zhu N, Yang J, Ji L, Liu J, Yang Y, Yuan H (2016) Metagenomic and metaproteomic analyses of a corn stover-adapted microbial consortium EMSD5 reveal its taxonomic and enzymatic basis for degrading lignocellulose. Biotechnol Biofuels 9:243

Chapter 17
Third-Generation Biofuels: A Promising Approach for Environmental Sustainability

Priyanka Thakur and Anjuvan Singh

Abstract Third-generation biofuels derived from microbial sources have recently emerged as viable fuel feedstocks, overcoming direct reliance on fossil fuels. Commercialising microbial technology for biofuel generation remains a complicated and contentious job, owing to a variety of problems related to life cycle evaluation and the technoeconomic feasibility of microorganism-derived biofuels. The long-term viability of first-generation biofuels derived from agricultural products is currently being questioned, as they are not seen as viable long-term energy sources. Similarly, second-generation biofuels derived from non-food feedstock have commercial hurdles due to high processing costs that render them commercially unviable. In contrast, because they overcome the drawbacks of previous generations of biofuels, third-generation microalgal biofuels are regarded as the most realistic and long-term response to emerging energy requirements of biofuel This brief chapter gives a summary of recent developments in economic research on the production of biofuels from first-, second-, third-, and fourth-generation microalgae.

Keywords Microalgae · Third-generation biofuel · Bioethanol · Biodiesel · First-generation biofuel · Second-generation biofuels · Transesterification · Biofuel

P. Thakur
Department of Microbiology, School of Bioengineering and Biosciences, Lovely Professional University, Phagwara, Punjab, India

A. Singh (✉)
Department of Biotechnology, School of Bioengineering and Biosciences, Lovely Professional University, Phagwara, Punjab, India

A. Karnwal, A. R. Mohammad Said Al-Tawaha (eds.), *Microbial Applications for Environmental Sustainability*, https://doi.org/10.1007/978-981-97-0676-1_17

17.1 Introduction

The third generation of biofuels was introduced as a replacement for conventional fossil fuels and chemical-based fuel for a number of reasons such as it will be reducing reliance on imported energy derived from depleting fossil fuel resources, utilising renewable biofuels in the transportation sector to combat global warming, and reducing emissions of particulates, sulphur, carbon monoxide, and hydrocarbons (Lene et al. 2009). Fuel defined as the volatile liquid that contains carbon and ignites to produce heat and energy. For example, artificial fuels like gasoline and diesel are used in cars while coal is utilised as a natural fuel in power plants to produce electricity. Natural gas is another fuel type that is used to produce energy which is considered as compatible to the environment while artificial fuels contribute to environmental degradation and are one of the factors in the greenhouse effect; therefore, sustainable fuels that are both affordable and environmentally friendly are needed in this present modern period. Therefore, in the modern period, biofuels are an alternate source that may effectively supply the current fuel demand in the energy sector. The depletion of fossil fuel supplies, the rising price of crude oil, worries about energy security, and the acceleration of global warming are other crucial factors for the sustainable production of biofuels.

These factors have generated widespread global interest in biofuels and other renewable energy alternatives. Various developed and rapidly developing nations are increasingly recognising biofuels as an essential energy source to accomplish objectives such as rural development, reducing dependence on foreign oil, and curbing greenhouse gas emissions, specifically Carbon Dioxide (CO_2) and Methane (Koh and Ghazoul 2008).

While secondary biofuels are available in three different varieties, primarily first-, second-, and third-generation biofuels, primary biofuels can be produced from common waste materials like animal waste and crop leftovers from forests, as well as traditional waste materials such as firewood, wood chips, and pellets (Dahlquist 2013). The initial phase of biofuel development comprises bioethanol and biodiesel, which involve the conversion of oils from crops like sugarcane, sugar beet, and oil crops, as well as starches, for example, wheat, barley, corn, and potatoes and sugars for bioethanol such as rapeseed, soybeans, sunflower, palm, coconut, used cooking oil, animal fats, etc. However, the lack of sustainability associated with the first generation of biofuels derived from these agricultural resources is presently a significant concern (Osman et al. 2021). The second generation (2G) of biofuels comprises lignocellulosic-based bioethanol, biobutanol, and syndiesel derived from non-food feedstock such as Jatropha, cassava, miscanthus, straw, wood, and grass. However, these second-generation biofuels face commercial challenges due to their high processing costs. Recognised as a practical and sustainable solution for future energy requirements, the third-generation of biofuels derived from microalgae overcomes the limitations of earlier generations (Sivakumar et al. 2010).

Third-generation biodiesel got from microalgae exhibits remarkable efficiency compared to other feedstock options. It demonstrates compatibility with various fuels, including diesel, gasoline, jet airways fuel, and aviation gasoline. Microalgae serve as a source of renewable and clean energy. According to studies, microalgae can produce more energy per hectare of land than conventional crops. The year-round availability of microalgae-based biodiesel is a huge benefit (Mofijur et al. 2019). *Cyanobacteria* which is commonly called as BGA (blue-green algae) are characterised by their prokaryotic cells, which have a blue-green colour due to the presence of the pigment phycocyanin. Eukaryotic microalgae encompass diatoms and green algae (Chlorophyta) like *Bacillariophyta*. It is estimated calculated that there are approximately 50,000 different species of algae, with around 30,000 of them studied and investigated. These extensively studied microalgae can be explored by researchers for various applications, including the production of diverse bioenergy products, pharmaceutical applications, human food production, and the creation of value-added bioproducts from food crops. Microalgae have the ability to thrive in various aquatic environments, including freshwater, brackish water, marine water, and wastewater. If sufficient nutrients are available in adequate quantities, municipal, industrial, agricultural, and domestic wastewater can all support the growth of microalgae. For instance, *Scenedesmus obliquus* has been successfully cultivated using a mixture of wastewater (WW) obtained from a secondary wastewater treatment facility and Bold's basal medium (BBM). The single-cell green microalgae were grown in this experiment in a 250-mL conical flask holding 150 mL of the WW and BBM mixture (Pal et al. 2019). Microalgae cultivation can be accomplished through two methods in either an open-type or closed-type photobioreactor. Reactor designs vary in different regions based on available space and resources, with adjustments made to optimise cultivation yields. While maintaining certain aspects of open-type reactors can pose challenges, precautionary measures can be implemented to mitigate their effects. In contrast, closed-type photobioreactors offer greater efficiency compared to open-type systems. They allow for precise control of variables such as light intensity, ensuring optimal growth conditions for microalgae (Jacob et al. 2021).

A potential replacement for conventional fossil fuels is biofuel. Four generations of biofuels have been created or suggested as a result of the rapid improvement in biofuel manufacturing technologies over the past several decades. The term "first-generation biofuel" describes fuels made from ingredients including grain, sugar, animal fats, and vegetable oils. Cellulosic biomass, primarily from agricultural non-edible matter, is used to create second-generation biofuel. Algal biofuel is referred to as "third-generation" biofuel. At every stage of manufacturing, the fourth generation of biofuel strives to absorb and store CO_2. Microalgae have huge industrial significance in terms of non-fuel products. Algal biomass may be used to make many number of valuable add or useful products or co-products for the industrial sector, including pigments, animal feed, organic fertiliser, nutraceuticals, cosmetics, chemicals, and food. It can even be burned to produce power. Proteins make up between 50% and 70% of the principal components of microalgae and are mostly used in human or animal feeding. Algal strains' proteins and pigments offer a lot of potential for usage in medicine and pharmaceuticals. Carotenoids, chlorophylls, and phycobiliproteins, natural pigments derived from algae, are used as colouring agents, biomaterials, cosmetics, and pharmaceuticals, as well as precursors to vitamins in food and animal feed (Mathimani and Pugazhendhi 2019).

17.2 Historical Perspective

In 2008, global energy consumption reached 533 EJ. However, due to escalating demand, it is projected to rise to 653 EJ by 2020 and further to 812 EJ by 2030. The top oil and liquid fuel producing countries are the United States (US), Saudi Arabia (SA), Russia, China, Canada, Iraq, the United Arab Emirates (UAE), and also Brazil. The utilisation of fuels from fossils has resulted in a significant surge in greenhouse gas emissions, contributing to global warming. In 2008 alone, approximately 32,083 million metric tonnes (MMT) of carbon dioxide were thrown into the atmosphere from burning of fossil fuel which is representing a substantial increase from 5977 MMT in 1950 and 1958 MMT in 1990 (Nanda et al. 2018). Non-OECD countries exhibited higher carbon dioxide emissions compared to OECD nations (Boden et al. 2009).

Utilising biofuels made from natural and renewable resources has a number of advantages, including the potential to reduce greenhouse gas emissions, fulfil rising energy needs, improve overall fuel efficiency, and boost employment in bio-based sectors. As the focus shifts towards biofuel development, there is a possibility that gasoline and other fossil fuels may become obsolete. Many developed and developing countries have implemented comprehensive policies to popularise the use of biofuels and expand their bioenergy markets, adopting a multimodal approach (Syrpas and Venskutonis 2020). The three generations of biofuels are the first generation, the second generation, and the third generation (Fig. 17.1).

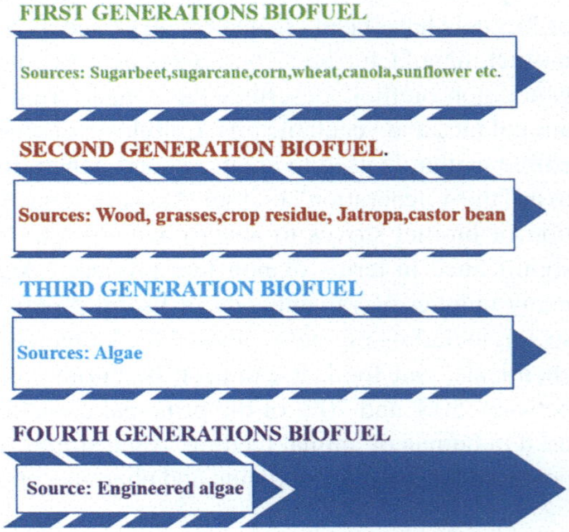

Fig. 17.1 Various generations of biofuels and its sources

17.3 First-Generation Biofuels

Agricultural and food processing wastes such as sugar, starch, vegetable oil, or animal fats are used to make first-generation biofuels using standard manufacturing techniques. Biodiesel, bioethanol, biooil, and biogas are popular first-generation biofuel examples. Vegetable oils are extracted, either with or without esterification, from the seeds of plants including soybean, oil palm, oilseed rape, and sunflower in order to make biodiesel, which is used as a fuel for internal combustion engines. By-products such as reduced animal fats are also used in the process (Naqvi and Yan 2015). Simple sugars from sugarcane or starchy plants like maize and wheat can be fermented to create bioethanol, which serves as a fuel for gasoline engines. Biomass is thermochemically converted into bio-oil, a process that is still in development. Biogas, an engine fuel, is produced by the anaerobic fermentation of organic waste, animal dung, agricultural waste, and energy crops (Singh et al. 2022; Martin 2010). First-generation biofuels made from edible biomass, such as starch (from potatoes, wheat, barley, and corn) or sugars (from sugarcane and sugar beetroot), were initially developed and showed a promising ability to reduce the combustion of fossil fuels and lower atmospheric CO_2 levels, which are absorbed by crops during growth. The effects of using edible crops as feedstocks on farming, biodiversity, and the food supply have been questioned. About 50 billion litres of first-generation biofuels, including biodiesel, bioethanol, and biogas, are produced commercially each year. Biogas, which is created by the anaerobic digestion of manure and other biomass sources, has fewer uses in transportation than the other two types of biofuels. Two key requirements must be met in order to maximise the advantages of reducing greenhouse gas (GHG) emissions from crops. These croplands and fertilisation techniques are great options for reducing nitrous oxide emissions and preventing the concentration of carbon in or on the ground. To maximise cost effectiveness, by-product manufacturing should also be prioritised and effectively utilised (Alalwan et al. 2019).

17.4 Second-Generation Biofuels

In addition to lowering carbon emissions and improving energy efficiency, second-generation biofuels made from lignocellulosic biomass can lessen a country's reliance on foreign energy imports. (2010); Naik et al. Non-food lignocellulosic substrates are a readily available and possibly inexpensive organic source for the synthesis of renewable chemicals and second-generation fuels. Lignocellulosic waste can come from a variety of sources, including agricultural waste (like straws, stover, and non-food seeds), industrial waste (like sawdust, paper mill by-products, and waste from the food industry), forestry waste (like hardwood and softwood), domestic waste (like kitchen waste, sewage, and waste papers), and municipal solid waste (Pabbathi et al. 2021; Mohtashami et al. 2021; Binod et al. 2019).

17.5 Third-Generation Biofuel: Microalgae

Microalgae are single-cell, microscopic microorganisms that naturally inhabit freshwater and marine environments. They exhibit a significantly higher level of diversity compared to plants, with over 300,000 species identified. Microalgae are known for their efficient utilisation of solar energy, surpassing higher plants in this aspect. Furthermore, the aquatic nature of microalgae allows them easier access to essential nutrients such as water and CO_2, as they thrive in a liquid environment (Chowdhury and Loganathan 2019; Ananthi et al. 2021).

A wide variety of chemical substances, including proteins, carbohydrates, fatty acids (lipids), antioxidants, enzymes, carotenoids, peptides, toxins, sterols, and polymers, are known to be produced by different types of microalgae. Numerous studies have been done to determine the essential components of different microalgae species. Some commonly studied microalgae species involved in biofuel production include *Botryococcus braunii, Chlorella* sp., *Crypthecodinium cohnii, Cylindrotheca* sp., *Dunaliella primolecta, Isochrysis* sp., and many others (Powar et al. 2022; Goswami et al. 2022).

17.6 Fourth-Generation Biofuels

Engineered algae are used in fourth-generation biofuels to produce ethanol from oxygenic photosynthetic organisms and this technology can be used to make gaseous biofuels, algal ethanol, algal butanol, four different types of carbon alcohols, and algal biodiesel (Satari et al. 2019; Lü et al. 2011).

17.7 Production of Microalgal Biomass

Microalgae cultivation for biomass frequently requires high costs and demands more sophisticated technologies. Microalgae must have a source of light, CO_2, water, and inorganic salts in order to develop their photosynthetic activities. The temperature needs to be carefully controlled between 20 °C and 30 °C because the majority of microalgae thrive best at these temperatures (Alishah Aratboni et al. 2019; Yadav and Sen 2017). The generation of biofuel must rely on the free sunshine despite frequent and yearly variations in natural light intensity in order to save costs. Nitrogen (N), phosphorus (P), iron (Fe), and occasionally silicon are necessary elements that must be given by the growth medium for the inorganic components that make up the algal cell (Kaplan et al. 2017; Radzun et al. 2015). Two varieties of microalgae can be grown, as illustrated in Fig. 17.2.

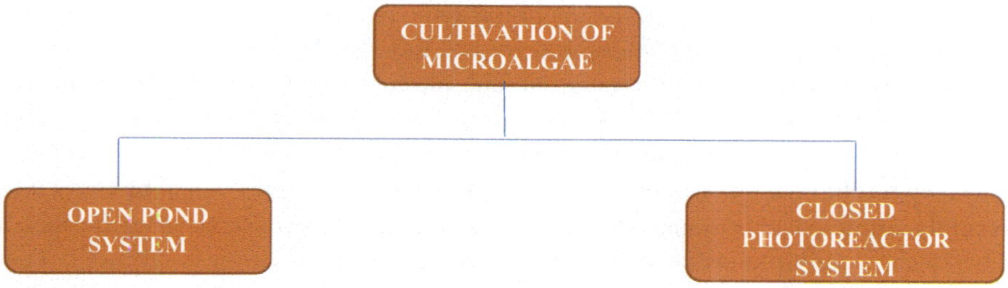

Fig. 17.2 Cultivation of microalgae

17.7.1 Open Pond System

The simplest and traditional approach for algae cultivation involves placing algae samples in an open pond filled with water. In this method, the algae thrive through natural photosynthesis and the provision of inorganic fertilisers, enabling their growth. The only challenge is that other microorganisms might easily pollute the pond, potentially causing the algae to die or get damaged (Shen et al. 2009; Dragone et al. 2010; Borowitzka and Moheimani 2013; Kowthaman et al. 2022). The large-scale commercial systems used now are largely open-air systems, although the first research on microalgal mass cultivation mostly focused on closed-culture systems. This inclination is motivated by financial factors because closed-culture systems are frequently expensive and difficult to scale up. Additionally, closed systems frequently need artificial lighting indoors, which results in high energy expenditures as opposed to open-air systems that may use natural sunlight. However, only a few kinds of algae can be successfully grown in open-air settings. Open-air systems typically come in one of four major forms, each with specific benefits and drawbacks: shallow large ponds, tanks, circular ponds, and raceway ponds. The actual production in open-air pond systems frequently falls short of the theoretical potential, and it might be difficult to regulate the external environment. A key element that strikes a compromise between the necessity for sufficient water depth for mixing and avoiding major changes in ionic composition owing to evaporation is the pond's depth. Shallower ponds provide greater water accessibility for the cells. For example, paddle-wheel raceway ponds in Australia typically have depths of 20–30 cm, while large unmixed ponds used for cultivating Dunaliella can reach depths of up to 50 cm, with sizes up to approximately 250 hectares (Tredici 2004).

17.7.2 Close Photobioreactors System

This device provides a suitable habitat for growing algae in a laboratory environment. The algae get the sunlight, water, nutrients, and CO_2 that they require to grow, just like in an open pond system. There are numerous types of photobioreactors

(bioreactors that use light to cultivate phototrophic microorganisms) included in the system, including the plate photobioreactor, the most challenging and sophisticated system currently available, fermentation tank photobioreactors, and tubular photobioreactor. (Schipper et al. 2021; Rehman et al. 2022; Peter et al. 2022). Every year, massive amounts of microalgal biomass are produced, primarily in open ponds. Closed photobioreactors can only produce a few hundred tonnes. Algal biomass has recently attracted enormous interest for the manufacture of low-value biofuels. Therefore, a multiple-order of-magnitude increase in production is required. The benefits of using photobioreactors in this procedure are obvious: they provide cultivation under a wide range of conditions or, to a certain extent, prevent the production strain from being outcompeted by other algae or contaminated by undesired microorganisms or grasses. Higher areal productivities and the reduction of water loss by evaporation are the two key advantages of closed bioreactor systems. A typical photobioreactor contains three phases: a liquid phase that serves as the medium, a solid phase made up of the cells, and a gas phase. A superimposed radiation field that contains light is what makes photoreactors special, but it is occasionally referred to as the "fourth phase." Understanding the relationship between environmental factors and the biological response is necessary for designing an adaptive PBR (Posten 2009). The development of affordable large-scale culture methods for the algae, which has been and still is a gradual process, is one of the many aspects that determine the viability of commercial large-scale production of microalgae. Cultural system to employ depends on a number of factors. The biology of the algae, the cost of labour, energy, water, nutrients, climate (if the culture is outdoors), and the sort of finished product are all important factors to take into account. A comparison of the fundamental characteristics of the various large-scale cultural systems is also necessary. These characteristics include light utilisation efficiency, temperature controllability, hydrodynamic stress on the algae, the ability to keep the culture unialgal and/or axenic, and ease of scalability from the laboratory to large-scale (Tredici 2004).

17.8 Microalgae and Its Characteristics

Microalgae can be of both types, i.e., autotrophic and heterotrophic microalgae. If they are autotrophic, they obtain their carbon from inorganic substances. Autotrophs can be photoautotrophic, which oxidises inorganic substances for energy, or chemoautotrophic, which uses light as a source of energy (Ananthi et al. 2021; Barone et al. 2020). Microalgae employ organic substances for growth if they are heterotrophic. Heterotrophs can be either photoheterotrophs, which oxidise organic substances for energy, or chemoheterotrophs, which use light as a source of energy. Certain microalgae species exhibit a mixotrophic nature, which combines both photosynthesis and heterotrophy. Photosynthesis plays a vital role for autotrophic algae, as it involves the conversion of solar radiation and CO_2 through chloroplasts, resulting in the production of adenosine triphosphate (ATP) and oxygen (O_2). ATP serves

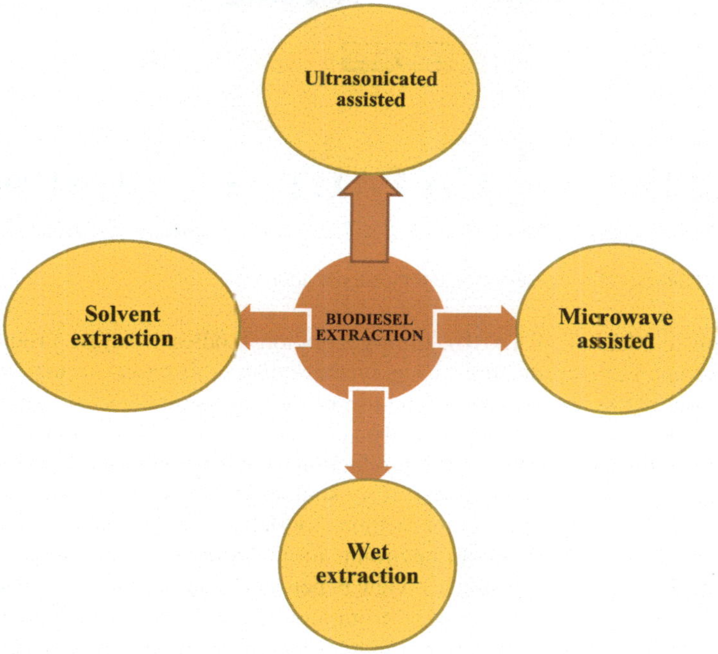

Fig. 17.3 Various extraction techniques for biodiesel extraction

as the usable energy currency at the cellular level, while O_2 is utilised in the process of respiration to generate energy for growth (Chaleshtori et al. 2022). Fig. 17.3 illustrates various extraction techniques employed for biodiesel extraction.

17.9 Transesterification of Third-Generation Biofuels

A triglyceride (fat or oil) reacting with an alcohol to produce esters and glycerol is known as transesterification. The triglyceride and alcohol combine during the esterification process in the presence of a catalyst, often sodium hydroxide or potassium hydroxide. To achieve complete conversion of the fat or oil to its esters, excessive alcohol is typically used (Fukuda et al. 2001). Alcohol and fat or oil are combined to produce mono-alkyl ester, the primary component of biodiesel, as well as glycerol (Fig. 17.4). Transesterification is the process by which non-edible oil seeds, vegetable oils, animal fats used in cooking oil, and microbial lipids or single-cell oil (from algae, oleaginous yeast, filamentous fungi, and bacteria) are converted into fatty acid methyl esters (biodiesel). In addition to acids, bases, and other liquid or heterogeneous materials, free or immobilised enzymes are also used as transesterification catalysts. (Bajaj et al. 2010; Amini et al. 2017; Moazeni et al. 2019).

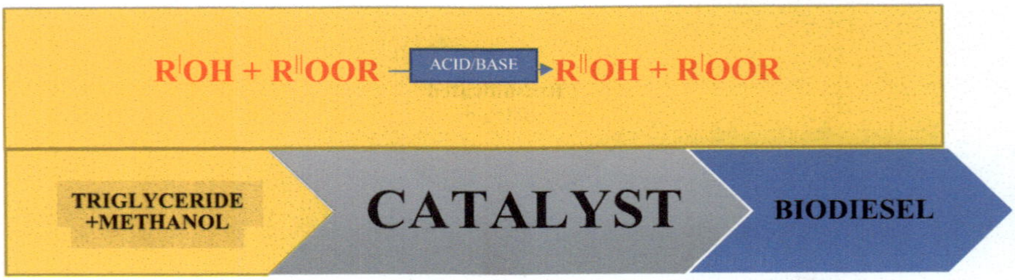

Fig. 17.4 Synthesis of biodiesel through transesterification

According to Plata et al. (2010), third-generation biodiesel can be produced by homogeneous transesterification using synthetic algae oil based on the fatty acid composition of *Chorella vulgaris.* The use of this synthetic algae oil provides the opportunity to assess the impact of catalyst concentration, reaction temperature, and the methanol/oil molar ratio on the production of third-generation biodiesel. The demand for using clean, alternative energy sources instead of fossil fuels is rising. Researchers are looking at different methods for delivering affordable clean energy from the plentiful biomass in our environment, including biodiesel, in an effort to meet these demands. As a result, a review paper on the nanocatalysed transesterification process for producing biodiesel from microalgae is presented. Only a few of the economic applications include using microalgae for fuel production, food supplement extraction, and CO_2 capture for biorefineries that produce biomethane, biohydrogen, bioethanol, and other by-products. Simplified equation for transesterification reaction is given below in Fig. 17.3 where R^{I} is the alcohol which is exchanged with R^{II} group of the ester.

Because they can be recycled, are more tolerant of variations in the quality of the raw material, can produce biodiesel with fewer steps, less energy, and significantly less wastewater, can improve product separation, and can create higher-quality glycerol, enzymes are potentially more advantageous than alkaline or acid catalysts. However, the usage of enzymes is restricted because of their slow rate of reaction, high cost, and reduced enzymatic activity (Aarthy et al. 2014).

17.10 Microalgae and Macroalgae Cultivation

Specific environmental factors, such as variations in temperature, brightness of light, mixing conditions, nutritional makeup, and gas exchange, are required for successful microalgae cultivation. In addition to open and closed cultivation systems, microalgae can be produced through various metabolic routes, such as photoautotrophic, heterotrophic, and mixotrophic pathways. For the growth and productivity of microalgae, each of these cultivation techniques offers certain benefits and considerations (Zuccaro et al. 2020). The production of macroalgae on a large scale started decades ago, mostly in Asian nations, but Hughes et al. (2012) highlighted that it is currently growing in Europe and America as a result of considerable technical transfer (Hughes et al. 2012). Finding an appropriate site location for the environmentally responsible production of macroalgal biomass demands meeting a number of social, legal, financial, and other constraints (such as depth, extent of shipping, and accessibility to ports). Coastal areas are often where large-scale macroalgal farming occurs (Thomas et al. 2019). Microorganisms have a simple multicellular or unicellular structure that allows them to grow swiftly and endure harsh conditions. Microalgae have a major advantage over terrestrial plants in the conversion of solar energy into biomass because of their straightforward cellular structure and submerged aquatic environment. With easy access to water, CO_2, and other crucial nutrients, they go through a photosynthetic process akin to that of terrestrial plants. Diatoms (Bacillariophyceae), green algae (Chlorophyceae), blue-green algae (Cyanophyceae), and golden algae (Chrysophyceae) are the four main subgroups of microalgae species. Factors including photoperiod, light intensity, temperature, and salinity have a substantial impact on the growth rate and biomass accumulation of microalgae Similar to terrestrial plants, microalgae utilise light energy for biomass development through photosynthesis. It is important to consider factors like photoperiod, temperature, and salinity to optimise and maximise microalgal growth and biomass yield. In order to maintain maximum photosynthetic performance, it is vital to comprehend and control environmental elements including light, photoperiod, temperature, and salt content in the culture. Microalgal photosynthesis also plays a significant role in carbon dioxide absorption and oxygen production. The growth of microalgae is significantly influenced by temperature, which has an impact on many different elements of their development. Up until a certain point, microalgal development grows linearly with temperature. Beyond this point, cell growth gradually declines, similar to the effect of light intensity. Temperature not only impacts growth but also affects the size and biochemistry of microalgae. High temperatures can be detrimental to microalgae, potentially leading to cell death. Conversely, temperatures above or below the optimal range may render microalgae inactive for growth without causing their destruction. It is important to maintain the appropriate temperature conditions to support optimal microalgal growth and avoid detrimental effects on their viability (Gani et al. 2019). Due to their efficiency and low resource requirements, closed-type photobioreactor cultivation methods using biofilms have grown in interest recently. Biofilms are structured collections of bacteria that develop over solid surfaces submerged in extracellular polymeric materials. In microalgae biofilms, a symbiotic relationship exists between specific microalgae and bacterial species. If the right amount of nutrients, light, and humidity are present, these biofilms can develop on flat or tubular bioreactors. The microalgal biomass paste can be readily gathered by scraping it off the surface, negating the need for time-consuming dewatering methods (Hirani et al. 2018).

17.11 Application of Third Generation of Biofuels

The production of third-generation biofuels comes from a variety of materials, including algae, sewage sludge, and municipal solid waste. A wide variety of biofuels, including liquid fuels such as ethanol, methanol, biodiesel, and Fischer-Tropsch diesel, as well as gaseous fuels like hydrogen and methane, can be produced using renewable resources. Various resources, including forestry and agricultural waste, waste from industrial processing, municipal solid waste, and urban wood waste, are used as biomass feedstock for the generation of biofuels. These feedstocks offer a wealth of opportunities for environmentally friendly and sustainable biofuel production. The extraction of gasoline from third-generation (3G) microalgal feedstocks has favourable economic and environmental consequences despite their tiny size. To overcome obstacles that limit production, the growing, harvesting, and extraction processes used to make microalgal biodiesel can all be modified. The development of optimised microalgal biodiesel blends allows for the creation of engine maps that calibrate engine properties through stationary or dynamic engine tests. Extensive research has been conducted on the synthesis of biodiesel from microalgae, its application in internal combustion (IC) engines. Third-generation biofuels, including microalgal biodiesel, have a range of potential applications, some of which include:

17.11.1 Biohydrogen for Fuel Cell Vehicles

Due to its great conversion efficiency, sustainability, and environmental friendliness, hydrogen is considered as the fuel for the future. Hydrogen is created by the processes of pyrolysis, electrolysis, gasification, and steam reforming. However, the development of a renewable energy source and a workable replacement for fossil fuels both rely on the biological processes that produce hydrogen. Future energy sources that could significantly outperform present ones include biohydrogen (Nagarajan et al. 2021). Microalgae can be considered as more dependable and reasonably priced source of biohydrogen. For the synthesis of hydrogen, Rashid et al. (2013) offered a number of techniques, including pyrolysis, electrolysis gasification, and steam reforming. The two-step process of producing hydrogen from microalgae starts with carbon fixation in stage one and progresses to anaerobic digestion in stage two. The main problems of a two-stage procedure include inefficient anaerobic digestion, limited hydrogen generation, and high expense. A fresh focus on algal biomass has emerged in numerous green systems that aim to produce energy from living stuff. The method of photosynthetic energy storage and utilisation in microalgae supports metabolism and cell growth by utilising readily accessible natural resources, including water and carbon dioxide. These organisms have a high capacity for CO_2 concentration and can obtain the nutrients required for growth from substrates or fluids derived from industrial waste. Algal culture systems, which are straightforward and inexpensive, guarantee good growth rates. Some microbes, such as bacteria and microalgae, are the fundamental components of processes that can manufacture hydrogen in a way that is completely environmentally sustainable. Microalgal hydrogen can be produced through biological processes that are either directly or indirectly light-dependent or through fermentation and thermochemical conversion of biomass. Among microalgae, there is a great deal of research on the hydrogenases of *Chlamydomonas reinhardtii*. Two well-known Fe-hydrogenase genes are present in this microalgal model organism: HydA1, which largely acts in anoxic situations, and hydrogenase activity, which is tightly regulated and one of the many reactions that microalgae display under stressful or anoxic environments. The protein is carried to the chloroplasts via a transit peptide at the N-terminal end after it has been translated in the cytoplasm. Modifying the culture conditions is a straightforward approach to instantly influence the growth or synthesis of specific metabolites. Regarding the medium components, the mentioned sulphur (S) deprivation leads to the desired outcomes only when it is followed by a phase of photosynthetic growth that provides all the necessary components for growth. Along with this method, a different modality that simulates starvation using modest amounts of S has also recently been investigated. This modality predicts distinct growth and a deprivation stage with a number of operational issues. It is also important to carefully consider the carbon source that is selected, as it has been found that storing starch reserves before a deficiency may be essential to sustaining productivity. Mixotrophic ecosystems contain both organic and inorganic carbon sources.

17.11.2 Biofuels from Algae

Algae biofuels might offer a realistic substitute for fossil fuels, but before they can compete on the fuel market and be widely adopted, they must first overcome a lot of barriers (Hannon et al. 2010). The world's energy resources, which are currently based mostly on non-renewable fuels, are undergoing a gradual depletion. Global population expansion and a rising economy have also led to higher energy consumption, which has increased waste production. Petroleum is still the main source of energy and materials in modern society, which has had negative consequences for the environment and energy security (Singh and Gu 2010). Microalgae are photosynthesis carrying microorganisms that can rapidly and in large quantities synthesise lipids, proteins, and carbohydrates. Both biofuels and valuable chemicals can be made from these products through purification. For the goal of generating biofuel, two samples of algae *Cladophora fracta* and *Chlorella protothecoid* were examined by Demirbas 2011.

The cost reduction of algae culture systems, harvesting methods, and downstream processes are the primary issues that must be addressed, in addition to the variable price of fossil-based fuels. The use of renewable feedstocks for the synthesis of bioenergy and bio-based products presents a promising solution to meet the growing global demand for energy and materials. Extensive research is underway to develop efficient technologies for converting biomass into fuels and chemicals. Among the various options, the potential of producing ethanol from algae is particularly intriguing. However, there are still significant barriers, both technological and financial, that hinder the commercialisation of algal biomass processes. The economic viability of such procedures is a critical factor for successful commercialisation, considering the existing knowledge and available literature in this field. Overcoming these challenges is essential to realise the full potential of algae-based fuels (Uduman et al. 2010; Leite et al. 2013; Singh and Gu 2010).

17.11.3 Biogas

Biogas, a biofuel generated through anaerobic digestion of organic waste, consists mainly of methane (57–75%) and carbon dioxide (25–45%). Recent studies examining the feasibility of utilising microalgae and macroalgae as substrates for biogas production have yielded diverse outcomes (Debowski et al. 2013). Since the 1950s, there has been considerable interest in producing biogas from microalgae due to their rapid growth potential in various climates and the high concentration of digestible macromolecules they can achieve. However, commercial-scale production of microalgae-based biogas is still in its early stages due to several technoeconomic challenges, including the recalcitrance of components, low biomass loading rates, interference from operational factors, and the costly and energy-intensive biomass harvesting process (Zabed et al. 2020). The challenges of using algal biomass as a substrate are mostly to blame for the lack of interest in algal biomass as a source of biomass for biogas production systems (Debowski et al. 2013). Since the 1950s, there has been a great deal of interest in producing biogas from microalgae due to the fact that they can grow quickly on non-arable land and amass a large number of digestible macromolecules in the biomass. However, commercial production of microalgae-based biogas is still in its early stages due to some current technoeconomic challenges, such as the requirement of a costly and energy-intensive step in biomass harvesting, the recalcitrance of the components, the low biomass loading rate, and the interference of various operational factors. The use of microalgae as a biomass feedstock is now being investigated as a way to lessen our reliance on fossil fuels for transportation, heating, and electricity generation. Research on microalgae as a source of chemicals, cosmetics, health products, and food for both people and animals has also been conducted in depth. Anaerobic digestion, one of the most advanced techniques for utilising microalgal biomass, has produced biomethane with encouraging results. The possible integration of this process into the microalgae biorefinery concept, which offers a source of bioenergy on the one hand and nutrients (nitrogen, phosphorus, and CO_2) and water for microalgae growth on the other, is something that makes it attractive. Biogas is created by the anaerobic decomposition of organic wastes such as sewage sludge, leftover food from farming, and industrial residues. Recently, this method has been applied to microalgae that were left over after lipid extraction. The release of long-chain fatty acids and alcohols from lipids, sugars from carbohydrates, and amino acids from proteins all occur as a result of the hydrolysis of complex chemical structures. Then, through the processes of acetogenesis and acidogenesis, simple organic molecules undergo fermentation to make acetic acid, which is used to manufacture other volatile fatty acids, including propionic, butyric, and valeric acids. Last but not least, biomethane is produced from hydrogen and acetate through processes called hydrogenotrophic and acetoclastic methanogenesis, respectively. Faster growth rates and the capacity to be grown in non-arable land areas, lakes, or the ocean are only two of the possible advantages algae offer over higher plants. As a result, there is less competition for food and feed (Limongi et al. 2021).

17.11.4 Biodiesel

Microalgae have the potential to produce a variety of sustainable biofuels. Examples include methane produced by the anaerobic digestion of algal biomass and biodiesel manufactured from microalgal oil (Chisti 2007). In place of fossil fuels, biodiesel is a natural energy source that is effective, long-lasting, environmentally friendly, and soon biodegradable. By trans esterifying organic fats or oils, a variety of methyl esters are produced (Bozbas 2008). Bioethanol and biodiesel are the two potential renewable fuels that have gotten the most attention. Current agricultural crop production techniques for biodiesel and bioethanol cannot replace fossil-based transportation fuels in a sustainable manner. However, there is an alternative. Microalgal biodiesel is the sole renewable biofuel that seems to be able to completely replace petroleum-derived transportation fuels without having an adverse effect on the supply of food and other crop products. Microalgae can be far more productive than most oil crops, including oil palm, in their ability to sustainably produce the required quantities of biodiesel (Chisti 2008). The manufacture of biodiesel from microalgae has a lot of potential but is still far from becoming carbon neutral or economically viable. The lack of established background knowledge in the field contributes to the issue in some respects. To find future paradigm shifts, we should search for both dramatic improvements and deeper understandings of algae. Integration of engineering and biology will be crucial in achieving this (Scott et al. 2010).

These problems could be resolved by using microalgae as a source of raw materials for the production of biofuel since their growth does not interfere with the production of food for people or animals. This method of development also eliminates the need for huge apparatus for spreading chemical products because nutrients and water are given directly to the bioreactor. One of the main benefits of microalgae is their high lipid content. Examples include microalgae, which can produce a maximum of 137 m3 of lipids per hectare per year and may include up to 75% (dry weight) of lipids, and palm oil, which can produce a maximum of 5.4 m3 of lipids per hectare per year and can contain up to 36% dry weight) of lipids triglycerides

(Alishah Aratboni et al. 2019). Biodiesel fuels derived from various vegetable oils have viscosities similar to diesel. These fuels have high flash points and cetane numbers despite having slightly lower volumetric heating values. Biodiesel is an excellent option to replace diesel if the necessity arises because its properties are typically comparable to those of diesel. Numerous studies have documented the kinetics of transesterification processes that are both acid- and alkali-catalyzed. Dufek et al. (1972) investigated the esterification and transesterification of 9(lO)-carboxystearic acid and its mono- and di-methyl esters in the presence of acids. In their investigation of the transesterification of soybean oil and other vegetable oils with alcohols, Freedam et al. investigated the rate constants and reaction order in relation to the kind of alcohol, molar ratio, type and amount of catalyst, and reaction temperature. Microalgae are photosynthetic microorganisms that convert sunlight, water, and carbon dioxide into algal biomass. Many microalgae contain an abundance of oil that can be processed into biodiesel using current technologies. This article addresses the possibility of microalgae to completely replace petroleum-derived transportation fuels like gasoline, jet fuel, and diesel by producing biodiesel in a sustainable manner.

17.11.5 Bioethanol

The requirement for an alternative to petroleum fuel may be fulfilled by bioethanol. The Earth's natural resources are under threat due to ongoing exploitation of natural resources, particularly fossil fuels, to meet market demand for fuel. Bioethanol generated from microalgae is one of the possible biofuels for energy security and environmental sustainability (Phwan et al. 2018). The current classification of bioethanol production into first (saccharine or starch-based), second (lignocellulosic materials), and third (microalgal biomass) generations differs depending on the technology and raw materials employed. Using genetically altered cyanobacteria to collect light, water, and nutrients and convert them into ethanol through a process known as photofermentation is referred to as "fourth-generation" technology in the context of emerging technology (Silva and Bertucco 2019). Due to the effectiveness of bioethanol when paired with gasoline and its low toxicity, it can replace petroleum oil without requiring any engine modifications. The existence of a high-octane number, which minimises engine knocking, a higher oxygen content, which lowers the emission of polluting gases into the environment, and the ability to be used directly as pure alcohol in sectors that use gasoline are all further benefits of bioethanol. In order to increase bioethanol production, there are a number of key obstacles that must be taken into account, one of which is the search for suitable feed material. According to the feedstock and the methods used for processing it, the production of bioethanol has also been classified into three generations: the first, second, and third generations (Maity and Mallick 2022). Microalgal biomass could be transported to biorefineries, where the residual biomass from the extraction procedures will be used to make high-value goods which result in the lowering production costs and waste accumulation.

17.11.6 Utilisation of Microalgae Biodiesel in Diesel Engines

Haik et al. (2011) looked into the effects of engine load and speed on the burning of algae-based biodiesel in a diesel engine. They determined that even though algal biodiesel produces less torque than diesel fuel, it generates more heat. They asserted that increasing power output might be possible by postponing the period of algal biodiesel injection. Islam et al. (2015) assessed the combustion behaviour of microalgal biodiesel in a single-cylinder engine running under varied load situations. The researchers found that when mixing microalgae biodiesel with diesel fuel, the in-cylinder pressure and output torque of diesel engines are lowered by a maximum of 4.5% while the engine is operating at full power. Thus, it may be said that microalgae are effective enough to replace currently available chemical fuel.

17.11.7 Developments in the Cultivation of Microalgae

Using biofilm in liquid suspension is a viable method for cultivating microalgae in open ponds and photobioreactors. The characteristics of the biofilm's growth and associated influence variables require further study. According to Ji et al., *Pseudochlorococcum*, an oleaginous microalga, prefers to accumulate on hydrophilic substrata as opposed to hydrophobic substrata when grown in an attached biofilm (Pal et al. 2019). Production of industrial algal biomass is significantly affected economically by harvesting and dewatering. To lessen these effects, energy-efficient processes for these process steps must be created. When the pH was altered or when two biological flocculants, chitosan and the synthetic tannin CFL-PT, were added, the microalga *Scenedesmus obtusiusculus* flocculated in a salt-based medium (Bracharz et al. 2018). Table 17.1 describes a few microalgae species and their environments.

The current direction of research seems to be the industrialisation of lab-scale microalgae culture using existing technology. Working towards large-scale production requires consideration of a number of aspects, starting with the scaling-up challenges that are now being encountered. Since planting and harvesting account for around 30% of the entire production expenses, simple, inexpensive technologies are

Table 17.1 Some species of the microalgae and their habitat

Sr	Name of the Species	Nutrient composition	Water habitat
1.	*Chlorella emersonii*	Lipids, carbohydrates	Freshwater
2.	*Botyrococcus braunii*	High Lipid contents and varying amount of carbohydrates	Freshwater
3.	*Tetraselmus suscia*	High fatty acid	Marine
4.	*Isochysis galbana*	High carbohydrates	Marine
5.	*Nanochloropsis oculata*	High omega 3 fatty acids	Marine

urgently needed. To reduce production costs and boost downstream revenues, a biorefinery strategy might be used. Numerous elements should be given inside the account in order to further develop this idea in order to make scale-up profitable. Different product extraction procedures are typically more feasible when the process volume is reduced. Additionally, choosing the right dewatering techniques based on the type of algae can significantly reduce the cost. However, as the outcomes of the modelling literature could be different from the actual procedure, the actual test must be conducted. It is possible to replicate the perfect microalgae growing conditions using mathematical models (Chew et al. 2018).

17.12 Status of the Third-Generation Biofuels After Post Pandemic

The COVID-19 pandemic has had a significant negative influence on the world economy, which has resulted in a sharp decrease in both fuel consumption and price. Worldwide preventative measures have also had an impact on agricultural techniques and the transportation of feedstocks, which has limited the production of biofuels. A number of sustainable organisations have already begun developing strategies to take advantage of this circumstance and convert more energy and fuel to renewable sources. Effective measures are therefore urgently required to advance the production of third-generation biofuels and gradually encourage their replacement of previous-generation feedstocks. Many novelists have previously examined the significance and application of biofuels made from algal biomass. Before and after COVID-19, there are a number of obstacles to the synthesis more amount of third-generation biofuels. The main barrier to large-scale production and, consequently, the inadequacy to satisfy the world's demand for this fuel, is its high cost of manufacturing and distribution. Third-generation biofuel has a lot to offer, such as several integrated uses and cutting-edge applications in the future fuel sector. Thus, it is critical to manage the current situation while highlighting algae biofuel's potential as a long-term supply. Third-generation feedstocks have emerged as highly favourable alternatives to 1G (first generation) and 2G (second generation) feedstocks in the production of biofuels. However, it is evident from various literature sources that there is still a need for advancements in algal biofuel technologies, particularly in the areas of algae culture and harvesting operations. With careful analysis of past and present data, it becomes evident that algae biofuels hold substantial potential in replacing fossil fuels, especially with their increased utilisation in the power sector. Although a rise in biofuel production and consumption is predicted by the COVID-19 scenario, it is yet unknown what kind of algae can be farmed for third-generation biofuels. It is therefore highly recommended that technology be created to make the manufacturing of algal biofuels effective and inexpensive (Cervera and Villalba 2021; Maliha and Abu-Hijleh 2022).

17.13 Current Review About Algal Production Technology

Transesterification, the creation of biohydrogen, and biomass hydrolysis can all be accelerated by biochar. The most effective method to successfully gather and harvest the complete biomass is to co-cultivate filamentous fungus with specific microalgae. Although the screening procedure can be labour-intensive and time-consuming, UV mutagenesis can be used to increase the lipid content of microalgae. However, the development of molecular biology methods like guided RNA and CRISPR/Cas9 has opened up new opportunities for genetically altering algae. These developments present promising ways to satisfy the expanding energy demands of the world (Godbole et al. 2021).

Due to their exceptional structural properties, biodegradability, adaptability to different microalgae species, and environmental friendliness, chitosan-based flocculants have attracted a lot of attention for application in microalgae harvesting. In addition to having a faster growth rate and creating a variety of goods, algal biomass has the ability to remove carbon from the air and clean wastewater. Despite this, a number of difficulties arise in their growing, harvesting, and production procedures. It is now possible to make the entire system carbon negative by reducing CO_2 emissions, using less energy, and cutting production costs, however their widespread application is still inadequate (Yin et al. 2021).

17.14 Conclusion

For the third generation of biofuels, microalgae have been investigated as a potential source that could lessen reliance on scarce resources. The feasibility of microalgae in the production of biofuels is acknowledged at the end of this chapter, which also emphasises the need for more research. The potential for manufacturing biofuels, biogas, dietary supplements, and other important items from various microalgae species has been investigated. It is essential to optimise cultivation methods in both open pond systems and closed photobioreactor systems, ensuring optimal output of biofuels, to meet the rising demand for biofuels. The development and scalability of microalgae-based biofuel production will be aided by continued study in this field. Technologies such as biomass growing, dewatering, harvesting, conversion to biofuels and bioproducts, biorefining, nutrient recycling, and more are used to produce biofuels from third-generation feedstocks. Due to their high yields when compared to other types of feedstocks, microalgae have attracted a lot of attention in recent years for the purpose of making biofuels. However, it is essential to optimise the most effective processes and appropriate extraction conditions in order to make the production of biofuels from microalgae cost-effective. In order to cut costs, research should concentrate on combining multiple ideas and creating efficient production techniques. The transition from crude fossil fuels to third-generation biofuels made from algae will be considerably aided by this development. The creation of tailored microalgae is currently under way, opening the door for the subsequent generation of biofuels.

Acknowledgement The Department of Microbiology, Department of Biotechnology, School of Bioengineering and Biosciences at Lovel Professional University Punjab, India, are gratefully acknowledged by the authors.

Statements and Declarations According to the authors, there are no known financial or personal conflicts between them that would have looked to have an impact on the research described in this paper.

Conflict of Interest No conflict of interest has been disclosed by the authors.

References

Aarthy M, Saravanan P, Gowthaman MK, Rose C, Kamini NR (2014) Enzymatic transesterification for production of biodiesel using yeast lipases: an overview. Chem Eng Res Des 92(8):1591–1601

Alalwan HA, Alminshid AH, Aljaafari HA (2019) Promising evolution of biofuel generations. Subject review. Renew Energy Focus 28:127–139

Alishah Aratboni H, Rafiei N, Garcia-Granados R, Alemzadeh A, Morones-Ramírez JR (2019) Biomass and lipid induction strategies in microalgae for biofuel production and other applications. Microb Cell Factories 18(1):1–17

Amini Z, Ilham Z, Ong HC, Mazaheri H, Chen WH (2017) State of the art and prospective of lipase-catalyzed transesterification reaction for biodiesel production. Energy Convers Manag 141:339–353

Ananthi V, Raja R, Carvalho IS, Brindhadevi K, Pugazhendhi A, Arun A (2021) A realistic scenario on microalgae based biodiesel production: third generation biofuel. Fuel 284:118965

Bajaj A, Lohan P, Jha PN, Mehrotra R (2010) Biodiesel production through lipase catalyzed transesterification: an overview. J Mol Catal B Enzym 62(1):9–14

Barone R, De Napoli L, Mayol L, Paolucci M, Volpe MG, D'Elia L et al (2020) Autotrophic and heterotrophic growth conditions modify biomolecole production in the microalga Galdieria sulphuraria (Cyanidiophyceae, Rhodophyta). Mar Drugs 18(3):169

Binod P, Gnansounou E, Sindhu R, Pandey A (2019) Enzymes for second generation biofuels: recent developments and future perspectives. Bioresou Technol Rep 5:317–325

Boden TA, Marland G, Andres RJ (2009) Global, regional, and national fossil-fuel CO2 emissions. Carbon dioxide information analysis center, oak ridge national laboratory. US Department of Energy, Oak Ridge

Borowitzka MA, Moheimani NR (2013) Open pond culture systems. In: Algae for biofuels and energy. Springer, Dordrecht, pp 133–152

Bozbas K (2008) Biodiesel as an alternative motor fuel: production and policies in the European Union. Renew Sust Energ Rev 12(2):542–552

Bracharz F, Helmdach D, Aschenbrenner I, Funck N, Wibberg D, Winkler A et al (2018) Harvest of the oleaginous microalgae Scenedesmus obtusiusculus by flocculation from culture based on natural water sources. Front Bioeng Biotechnol 6:200

Cervera R, Villalba MR (2021) Will the Covid-19 pandemic transform our urban habitat? A microalgae Photobioreactors labyrinth-garden as an answer for the post-Covid Sustainable City. In:

IOP conference series: earth and environmental science, vol 754. IOP Publishing, Philadelphia, PA, p 012027
Chaleshtori SN, Shamskilani M, Babaei A, Behrang M (2022) Municipal wastewater treatment and fouling in microalgal-activated sludge membrane bioreactor: cultivation in raw and treated wastewater. J Water Process Eng 49:103069
Chew KW, Reen CS, Loke SP, Jiun YY, Chuan LT, Jo-Shu C (2018) Effects of water culture medium, cultivation systems and growth modes for microalgae cultivation: a review. J Taiwan Inst Chem Eng 91:S1876107C18303134. https://doi.org/10.1016/j.jtice.2018.05.039
Chisti Y (2007) Biodiesel from microalgae. Biotechnol Adv 25(3):294–306
Chisti Y (2008) Biodiesel from microalgae beats bioethanol. Trends Biotechnol 26(3):126–131
Chowdhury H, Loganathan B (2019) Third-generation biofuels from microalgae: a review. Curr Opin Green Sustain Chem 20:39–44
Dahlquist E (ed) (2013) Biomass as energy source: resources, systems and applications. CRC, New York
Debowski M, Zielinski M, Grala A, Dudek M (2013) Algae biomass as an alternative substrate in biogas production technologies. Rev Renew Sustain Energ Rev 27:596e604. https://doi.org/10.1016/J.RSER.2013.07.029
Demirbas MF (2011) Biofuels from algae for sustainable development. Appl Energy 88(10):3473–3480
Dragone G, Fernandes BD, Vicente AA, and Teixeira JA (2010) Third generation biofuels from microalgae
Dufek EJ, Butterfield RO, Frankel EN (1972) Esterification and transesterification of 9 (10)-carboxystearic acid and its methyl esters. Kinetic studies. JAOCS 49:302
Fukuda H, Kondo A, Noda H (2001) Biodiesel fuel production by transesterification of oils. J Biosci Bioeng 92(5):405–416
Gani P, Sunar NM, Matias-Peralta HM, Apandi N (2019) An overview of environmental factor's effect on the growth of microalgae. J Appl Chem Nat Resour 1(2):1
Godbole V, Pal MK, Gautam P (2021) A critical perspective on the scope of interdisciplinary approaches used in fourth-generation biofuel production. Algal Res 58:102436
Goswami L, Kayalvizhi R, Dikshit PK, Sherpa KC, Roy S, Kushwaha A et al (2022) A critical review on prospects of bio-refinery products from second and third generation biomasses. Chem Eng J 137677:137677
Haik Y, Selim MY, Abdulrehman T (2011) Combustion of algae oil methyl ester in an indirect injection diesel engine. Energy 36(3):1827–1835
Hannon M, Gimpel J, Tran M, Rasala B, Mayfield S (2010) Biofuels from algae: challenges and potential. Biofuels 1(5):763–784
Hirani AH, Javed N, Asif M, Basu SK, Kumar A (2018) A review on first-and second-generation biofuel productions. In: Biofuels: greenhouse gas mitigation and global warming. Springer, Cham, pp 141–154
Hughes AD, Kelly MS, Black KD, Stanley MS (2012) Biogas from macroalgae: is it time to revisit the idea? Biotechnol Biofuels 5(1):1–7
Islam MA, Rahman MM, Heimann K, Nabi MN, Ristovski ZD, Dowell A et al (2015) Combustion analysis of microalgae methyl ester in a common rail direct injection diesel engine. Fuel 143:351–360
Jacob A, Ashok B, Alagumalai A, Chyuan OH, Le PTK (2021) Critical review on third generation micro algae biodiesel production and its feasibility as future bioenergy for IC engine applications. Energy Convers Manag 228:113655
Kaplan D, Richmond AE, Dubinsky Z, Aaronson S (2017) Algal nutrition. In: CRC handbook of microalgal mass culture. CRC Press, Boca Raton, FL, pp 147–198
Koh LP, Ghazoul J (2008) Biofuels, biodiversity, and people: understanding the conflicts and finding opportunities. Biol Conserv 141(10):2450–2460. https://doi.org/10.1016/j.biocon.2008.08.005

Kowthaman CN, Kumar PS, Selvan VAM, Ganesh D (2022) A comprehensive insight from microalgae production process to characterization of biofuel for the sustainable energy. Fuel 310:122320

Leite GB, Abdelaziz AEM, Hallenbeck PC (2013) Algal biofuels: challenges and opportunities. Bioresour Technol 145:134e141. https://doi.org/10.1016/J.BIORTECH.2013.02.007

Lene F, Christensen KV, Norddahl B (2009) A review of the current state of biodiesel production using enzymatic transesterification. Biotechnol Bioeng 102(5):1298–1315. https://doi.org/10.1002/bit.22256

Limongi AR, Viviano E, De Luca M, Radice RP, Bianco G, Martelli G (2021) Biohydrogen from microalgae: production and applications. Appl Sci 11(4):1616

Lü J, Sheahan C, Fu P (2011) Metabolic engineering of algae for fourth generation biofuels production. Energy Environ Sci 4(7):2451–2466

Maity S, Mallick N (2022) Trends and advances in sustainable bioethanol production by marine microalgae: a critical review. J Clean Prod 131153:131153

Maliha A, Abu-Hijleh B (2022) A review on the current status and post-pandemic prospects of third-generation biofuels. Energy Syst 14:1–32

Martin MA (2010) First generation biofuels compete. New Biotechnol 27(5):596–608

Mathimani T, Pugazhendhi A (2019) Utilization of algae for biofuel, bio-products and bio-remediation. Biocatal Agric Biotechnol 17:326–330

Moazeni F, Chen YC, Zhang G (2019) Enzymatic transesterification for biodiesel production from used cooking oil, a review. J Clean Prod 216:117–128

Mofijur M, Rasul MG, Hassan NMS, Nabi MN (2019) Recent development in the production of third generation biodiesel from microalgae. Energy Procedia 156:53–58

Mohtashami Z, Bozorgi-Amiri A, Tavakkoli-Moghaddam R (2021) A two-stage multi-objective second generation biodiesel supply chain design considering social sustainability: a case study. Energy 233:121020

Nagarajan D, Dong CD, Chen CY, Lee DJ, Chang JS (2021) Biohydrogen production from microalgae—major bottlenecks and future research perspectives. Biotechnol J 16(5):2000124

Naik SN, Goud VV, Rout PK, Dalai AK (2010) Production of first and second generation biofuels: a comprehensive review. Renew Sust Energ Rev 14(2):578–597

Nanda S, Rana R, Sarangi PK, Dalai AK, Kozinski JA (2018) A broad introduction to first-, second-, and third-generation biofuels. In: Recent advancements in biofuels and bioenergy utilization. Springer, Singapore, pp 1–25

Naqvi M, Yan J (2015) First-generation biofuels. In: Handbook of clean energy systems. Wiley, Hoboken, NJ, pp 1–18

Osman AI, Qasim U, Jamil F, Ala'a H, Jrai AA, Al-Riyami M et al (2021) Bioethanol and biodiesel: bibliometric mapping, policies and future needs. Renew Sust Energ Rev 152:111677

Pabbathi NPP, Velidandi A, Tavarna T, Gupta S, Raj RS, Gandam PK, Baadhe RR (2021) Role of metagenomics in prospecting novel endoglucanases, accentuating functional metagenomics approach in second-generation biofuel production: a review. Biomass Convers Biorefin 13:1371–1398

Pal P, Chew KW, Yen HW, Lim JW, Lam MK, Show PL (2019) Cultivation of oily microalgae for the production of third-generation biofuels. Sustainability 11(19):5424

Peter AP, Tan X, Lim JY, Chew KW, Koyande AK, Show PL (2022) Environmental analysis of Chlorella vulgaris cultivation in large scale closed system under waste nutrient source. Chem Eng J 433:134254

Phwan CK, Ong HC, Chen WH, Ling TC, Ng EP, Show PL (2018) Overview: comparison of pretreatment technologies and fermentation processes of bioethanol from microalgae. Energy Convers Manag 173:81–94

Plata V, Kafarov V, Moreno N (2010) Optimization of third generation biofuels production: biodiesel from microalgae oil by homogeneous transesterification. Chem Eng 21

Posten C (2009) Design principles of photo-bioreactors for cultivation of microalgae. Eng Life Sci 9(3):165–177

Powar RS, Yadav AS, Ramakrishna CS, Patel S, Mohan M, Sakharwade SG et al (2022) Algae: a potential feedstock for third generation biofuel. Materials Today: Proceedings 63:A27–A33
Radzun KA, Wolf J, Jakob G, Zhang E, Stephens E, Ross I, Hankamer B (2015) Automated nutrient screening system enables high-throughput optimisation of microalgae production conditions. Biotechnol Biofuels 8(1):1–17
Rashid N, Rehman MSU, Memon S, Ur Rahman Z, Lee K, Han J-I (2013) Current status, barriers and developments in biohydrogen production by microalgae. Renew Sust Energ Rev 22:571e579. https://doi.org/10.1016/J.RSER.2013.01.051
Rehman M, Kesharvani S, Dwivedi G, Suneja KG (2022) Impact of cultivation conditions on microalgae biomass productivity and lipid content. Mater Today Proceedings 56:282–290
Satari B, Karimi K, Kumar R (2019) Cellulose solvent-based pretreatment for enhanced second-generation biofuel production: a review. Sustain Energy Fuels 3(1):11–62
Schipper K, Al-Jabri HMS, Wijffels RH, Barbosa MJ (2021) Techno-economics of algae production in the Arabian peninsula. Bioresour Technol 331:125043
Scott SA, Davey MP, Dennis JS, Horst I, Howe CJ, Lea-Smith DJ, Smith AG (2010) Biodiesel from algae: challenges and prospects. Curr Opin Biotechnol 21(3):277–286
Shen Y, Yuan W, Pei ZJ, Wu Q, Mao E (2009) Microalgae mass production methods. Trans ASABE 52(4):1275–1287
Silva CEDF, Bertucco A (2019) Bioethanol from microalgal biomass: a promising approach in biorefinery. Braz Arch Biol Technol 62:1516
Singh J, Gu S (2010) Commercialization potential of microalgae for biofuels production. Renew Sust Energ Rev 14:2596e2610. https://doi.org/10.1016/j.rser.2010.06.014
Singh A, Prajapati P, Vyas S, Gaur VK, Sindhu R, Binod P et al (2022) A comprehensive review of feedstocks as sustainable substrates for next-generation biofuels. Bioenerg Res 16:105–122
Sivakumar G, Vail DR, Xu J, Burner DM, Lay JO, Ge X, Weathers PJ (2010) Bioethanol and biodiesel: alternative liquid fuels for future generations. Eng Life Sci 10(1):8–18
Syrpas M, Venskutonis PR (2020) Algae for the production of bio-based products. In: Biobased products and industries. Elsevier, Amsterdam, pp 203–243
Thomas J-BE, Ramos FS, Grondahl F (2019) Identifying suitable sites for macroalgae cultivation on the Swedish west coast. Coast Manag 47:88e106. https://doi.org/10.1080/08920753.2019.1540906
Tredici MR (2004) Mass production of microalgae: photobioreactors. In: Handbook of microalgal culture: Biotechnology and applied phycology, vol 1. Blackwell Publishing, pp 178–214
Uduman N, Qi Y, Danquah MK, Forde GM, Hoadley A (2010) Dewatering of microalgal cultures: a major bottleneck to algae-based fuels. J Renew Sustain Energy 2:012701. https://doi.org/10.1063/1.3294480
Yadav G, Sen R (2017) Microalgal green refinery concept for bio sequestration of carbon-dioxide Vis-à-Vis wastewater remediation and bioenergy production: recent technological advances in climate research. J CO2 Util 17:188–206
Yin Z, Chu R, Zhu L, Li S, Mo F, Hu D, Liu C (2021) Application of chitosan-based flocculants to harvest microalgal biomass for biofuel production: a review. Renew Sust Energ Rev 145:111159
Zabed HM, Akter S, Yun J, Zhang G, Zhang Y, Qi X (2020) Biogas from microalgae: technologies, challenges and opportunities. Renew Sust Energ Rev 117:109503
Zuccaro G, Yousuf A, Pollio A, Steyer J-P (2020) Microalgae cultivation systems. Microalgae cultivation for biofuels production. Elsevier, Amsterdam, pp 11–29. https://doi.org/10.1016/b978-0-12-817536-1.00002-3

Chapter 18
The Utilization of Microorganisms for the Sustainable Development of Environmental Issues

Rahul Kanaoujiya, Abdel Rahman Mohammad Said Al-Tawaha, Ghulam Mustafa, Ruhul Amin, Shekhar Srivastava, and Abdel Razzaq Mohammad Al-Tawaha

Abstract The future of food security is becoming an increasingly significant concern as a result of exponential human population growth. Challenges such as land degradation, reduced agricultural productivity, and vulnerability to both abiotic and biotic pressures are becoming increasingly prevalent within the agricultural sector. Furthermore, human activities like urbanization and industrialization are contributing to the production of large amounts of waste, which poses a substantial risk to ecosystems worldwide. Beneficial soil microorganisms (BSMs) are proposed as potential tools for achieving sustainable environmental management. These microorganisms exhibit various mechanisms that hold promising commercial potential for addressing major environmental concerns. Currently, products utilizing beneficial microbes are being successfully employed in agroecosystems. In this review, we highlight the significance of beneficial soil microbes, particularly bacteria and fungi that enhance plant growth, in the context of environmentally responsible agriculture and the mitigation of environmental issues.

Keywords Microbes · Cyanobacteria · Biofertilizers · Bioherbicides · Biopesticides

---

R. Kanaoujiya (✉) · G. Mustafa · S. Srivastava
Department of Chemistry, Central University of Allahabad, Prayagraj, Uttar Pradesh, India
e-mail: rahul.k@allduniv.ac.in

A. R. Mohammad Said Al-Tawaha
Department of Biological Sciences, Al Hussein Bin Talal University, Ma'an, Jordan

R. Amin
Faculty of Pharmaceutical Science, Assam Down Town University, Guwahati, Assam, India

A. R. Mohammad Al-Tawaha
Department of Crop Science, Faculty of Agriculture, Universiti Putra Malaysia, Serdang, Selangor, Malaysia

National Agricultural Research Center, Baqa, Jordan

A. Karnwal, A. R. Mohammad Said Al-Tawaha (eds.), *Microbial Applications for Environmental Sustainability*, https://doi.org/10.1007/978-981-97-0676-1_18

18.1 Introduction

The expansion of the world population remains a significant factor contributing to the increased strain on available land resources, which are crucial for the production of food, fuel, and raw materials (Fróna et al. 2019). In order to address the increasing need, agricultural practitioners largely depend on the use of chemical fertilizers and pesticides, leading to the degradation of land and the decline of biodiversity (Tilman et al. 2002; Arora et al. 2018; Timsina 2018; Kumar et al. 2019b). The aforementioned method has become prevalent as a result of the escalating need for agricultural goods (Popp et al. 2013; Srivastav 2020), although it has adverse impacts on both soil health and biodiversity. Land degradation results in a significant yearly agricultural value loss of no less than $10 billion (Mishra et al. 2016; Hasnat et al. 2018).

Researchers have identified several factors contributing to soil degradation, with anthropogenic activities playing a direct role (Lehman et al. 2015; Zucca et al. 2021). Land degradation can result in soil exhaustion, salinization, and desertification (İmamoglu and Dengiz 2019). Although numerous technological innovations are being implemented to increase crop yield, they must still address the global food demand (Khan et al. 2021). The use of plant-associated microorganisms is a possible avenue for addressing the many difficulties connected with salinity, fertility, degradation, and habitat loss. These microbial entities have unique abilities, directly enhancing soil fertility, promoting plant growth, and mitigating biotic and abiotic stresses (Mishra et al. 2016; Arora et al. 2018; Pandey et al. 2019; Sahu et al. 2019; Timmis and Ramos 2021). Beneficial soil microbes facilitate plant growth by nutrient absorption, the formation of complex soil matrices, and the enhancement of plant defense responses through the release of various metabolites (Etesami and Maheshwari 2018; Pascale et al. 2020; Sharma 2022).

In addition, it has been shown that soil bacteria with advantageous characteristics has the capacity to adapt to many environmental situations such as high salt levels, limited water availability, invasion by unwanted plant species, insufficient nutrient supply, and the presence of toxic heavy metals (Nadeem et al. 2014; Egamberdieva et al. 2017; Pathania et al. 2020). Significantly, there is a growing recognition of the importance of beneficial soil microbes, not only in facilitating plant growth but also in their involvement in the decomposition of organic waste, the detoxification of harmful substances like pesticides, and the alleviation of stress (Sundari et al. 2019; Halim et al. 2020; Xiang et al. 2022). The present study explores the discourse around plant growth-promoting rhizobacteria (PGPR), plant growth-promoting fungi (PGPF), mycorrhiza, and cyanobacteria, and their significant contribution to the field of sustainable agriculture (Arora et al. 2016; Dubey et al. 2017; Tijjani and Khairulmazmi 2021).

18.2 The Importance of Beneficial Microbes in Soil Ecosystems

Soil, a natural physical covering, is a vital interface between the solid, liquid, and gaseous phases of the Earth's surface. It encompasses rocks, decomposed organic matter, water, and air in soil pores (Reichardt and Timm 2014; Al-Kaisi et al. 2017). As the foundation of all terrestrial ecosystems, soil harbors diverse microorganisms such as bacteria, archaea, and fungi, which play integral roles in the functioning of ecosystems. The interdependency among these microbial communities is crucial for ecosystem stability. Remarkably, even a single gram of soil contains an estimated amount of 10^{10} to 10^{11} bacteria, 6000 to 50,000 distinct bacterial species, and up to 200 million fungal hyphae (Ma 2019). Significantly, these beneficial microbes not only support plant growth but also contribute to the enhancement of soil health (Das and Pradhan 2016; Meena et al. 2020). However, soil also harbors pathogenic bacteria that can infiltrate or parasitize plants, leading to decreased agricultural productivity (Jones et al. 2009; Alegbeleye et al. 2018). Therefore, understanding the role of beneficial soil microbes is crucial, as they contribute to the overall health of soil ecosystems while inhibiting the growth of destructive microorganisms.

18.3 Role of Cyanobacteria

Cyanobacteria, prokaryotic organisms employing photosynthesis, are ubiquitous in nature as depicted (Subashchandrabose et al. 2013; Singh et al. 2016). They can be found in a variety of natural habitats, including lakes, ponds, springs, wetlands, streams, and rivers. In addition to these aquatic environments, cyanobacteria also play a vital role in soil ecosystems (Ananya and Ahmad 2014; Zahra et al. 2020).

Cyanobacteria, particularly those capable of nitrogen fixation, have been instrumental in maintaining the fertility of rice fields, earning them the designation of natural biofertilizers (Syiem et al. 2017; Kuraganti et al. 2020). Free-living and symbiotic cyanobacteria have found applications in sustainable agriculture (Prasanna and Kaushik 2007; Prasanna et al. 2014). It is worth mentioning that many nitrogen-fixing cyanobacteria, including *Nostoc linkia, Anabaena variabilis, Aulosira fertilissima, Calothrix* sp., and *Scytonema* sp., have been detected across a wide range of agro-ecological zones and used for the purpose of rice growth (Roy 2021). While the utilization of cyanobacteria in rice fields has a long history, their application with other crops has only recently garnered attention (Singh 2014; Venkatachalam et al. 2016).

Cyanobacteria have been used in rice fields for a long time, but they have only recently been tried with other crops (Kaushik et al. 2019; Bergi and Trivedi 2020). Cyanobacteria have long been utilized in rice fields but have only recently been tested on other crops (Nain et al. 2010; Bergi and Trivedi 2020). Extracellular polysaccharides, peptides, and lipids that are excreted by cyanobacteria as they mature in the soil are also important to the production of healthy soil (Garlapati et al. 2019; Mukherjee and Sen 2021). Cyanobacteria, according to some accounts, can thrive in saline soil, where most plants (apart from halophytes) cannot, and can even help make the soil more productive (Saad and Atia 2014).

18.4 Role of Plant Growth-Promoting Rhizobacteria (PGPR)

Promoting plant growth rhizobacteria are microorganisms that colonize the plant root and either directly or indirectly supports plant development (Arora et al. 2017). This type of soil bacteria can colonize plant roots and promote plant development. Stimulating the growth of plants rhizobacterial species can be classified into a number of different genera, some of which are shown in Table 18.1. These genera include *Azoarcus, Azospirillum, Rhizobium,* and *Serratia* (Arora et al. 2017). In the past few years, extensive research has been conducted to elucidate the symbiotic relationship between plants and microorganisms. The promotion of growth in plants encompasses various mechanisms, such as the synthesis of phytohormones, the fixation of atmospheric nitrogen (N_2), the production of siderophores as iron chelators, and the solubilization of inorganic minerals such as phosphorus (P), potassium (K), and zinc (Zn), thereby enhancing their accessibility for plant growth (Etesami and Adl 2020; Arıkan and Karakoyun 2022). Phytopathogens' growth can be effectively inhibited through various mechanisms employed by plants. These mechanisms include the production of antibiotic or antifungal metabolites, sequestration

Table 18.1 Some of the biopesticides used in agroecosystems have beneficial soil microbes in them (plant growth-promoting rhizobacteria)

S. no	Target organism	Type of beneficial soil microbes (plant growth-promoting rhizobacteria)	General name	References
1	Lepidopteran pests	*B. thuringiensis* subspecies israelensis	VectoBac	Belousova et al. (2021)
2	Turf fungal diseases	*Pseudomonas aureofaciens*	Turf fungal diseases	Ünal et al. (2019)
3	Bacterial speck	*Pseudomonas syringae*	AgriPhage	Hernandez et al. (2020)
4	Fire blight	*Pantoea agglomerans*	BlightBan C9–1	Mikiciński et al. (2016)
5	Crown gall disease	*Agrobacterium radiobacter*	NoGall	Escobar et al. (2003)
6	Bacterial wilt, root rot	*Pseudomonas fluorescens*	Biomonas, Esvin	Pandya and Saraf (2014)
7	Seedling diseases	*Bacillus pumilus*	Sonata	Sansinenea (2019)

of iron from the rhizosphere, induction of systemic resistance (ISR), synthesis of enzymes that degrade fungal cell walls, and competition with neighboring plants for root binding sites (Kaur et al. 2016; Saeed et al. 2021).

Plant growth-promoting rhizobacteria are also thought to be possible microbes that protect plants from different environmental stresses in both standard and stressed environments (Ambrosini and Passaglia 2017; Fan and Smith 2021). The valuable role of plant growth-promoting rhizobacteria in sustainable agriculture is well-described in a wealth of literature (Mustafa et al. 2019). According to research, they can also be employed in rehabilitating degraded land, improving soil quality, lowering environmental contaminants from soils, and preventing climate change (Agegnehu et al. 2017).

18.5 Fighting Climate Change

Climate change is now recognized as a significant environmental issue, which has a substantial impact on the composition and functioning of soil microbial communities (Classen et al. 2015; Jansson and Hofmockel 2020). The soil is widely acknowledged as a significant reservoir of atmospheric CO_2 and organic carbon on a worldwide scale, with an estimated carbon content that is almost double the amount present in the Earth's atmosphere (Lal 2004; Smith 2008). The involvement of soil microbes in the regulation of the global carbon pool has been well acknowledged in the scientific literature (Hartman and Richardson 2013; Gougoulias et al. 2014). Hence, even little variations in the global carbon stream may have a significant influence on the amounts of atmospheric carbon dioxide. To exemplify, it is widely thought that soils have released an estimated range of 40–90 billion metric tons of organic carbon, out of the total stored amount of around 2000 metric tons, as a result of anthropogenic environmental alterations. Researchers globally have extensively studied beneficial soil microorganisms (BSMs) in light of the consequences of changing climates. These microbes possess distinct characteristics that allow them to flourish under fluctuating climatic circumstances (Mishra et al. 2016).

The PGPR presents a promising strategy to mitigate the effects of climate change, with particular emphasis on the development of strains adaptable to stress-ridden agroecosystems (Ojuederie et al. 2019; Naamala and Smith 2020; French et al. 2021). Mycorrhizal fungi, in a similar vein, represent an effective tool in climate change mitigation, as they have evolved over time in response to shifts in the global climate (Pritchard 2011). These fungi have developed various mechanisms to meet the potential increased nutrient requirements of their host plants when faced with changing environmental conditions. The utilization of beneficial soil microbes offers a sustainable approach to preserving soil organic matter and facilitating carbon credit generation (Das and Ghosh 2020).

In conventional agricultural practices, the application of chemical fertilizers leads to increased emissions of greenhouse gases such as CO_2, CH_4, and nitric oxide (NO). In agriculture, however, using beneficial soil microbes as inoculants has shown that it can cut down on the need for chemical fertilizers, which can help fight climate change (Ahirwar et al. 2020; Sammauria et al. 2020).

18.6 Biofertilizers

Biofertilizers include compounds including of viable microorganisms that colonize the rhizosphere of plants. The fundamental objective of this study is to improve the delivery or availability of essential nutrients and/or growth-promoting substances to the specific crop, as stated by (Ritika and Utpal 2014). Table 18.2 presents the use of alternative fertilizers in agricultural contexts as a viable replacement for conventional fertilizers (Arora 2018). The rise in the utilization of biofertilizers, as may be ascribed to their capacity to uphold soil health, alleviate pollution, and reduce dependence on chemical inputs within agricultural methodologies (Singh and Yadav 2022). Based on the calculations presented by Li et al. (2010), it is generally recommended for farmers to apply a minimum of 100 kg of N_2 per hectare. In contrast to the often observed consumption efficiency of around 40%, a considerable proportion of the applied fertilizer is lost via processes such as leaching or volatilization (Mishra et al. 2016). According to Mishra et al. (2016), it has been estimated that symbiotic cyanobacteria provide a range of 7–80 kg N_2 per hectare per year, free-living bacteria contribute 15 kg N_2 per hectare per year, and associative (endophytic) bacteria contribute 36 kg N_2 per hectare per year. According to Martins-Loução et al. (2022), it has been shown by scientists that cereal crops may get up to 30% of their nitrogen via biological nitrogen fixation (BNF) when subjected to high levels of phosphorus, potassium, and microelement fertilization. For almost a century, rhizobial inoculants have been used as biofertilizers in the cultivation of legume crops. According to Pathak et al. (2021), several studies have shown the potential of

Table 18.2 Some characteristics of the biofertilizers that are based on beneficial soil microbes

S. no	Microbes used	Biofertilizers	Application	References
1	Cyanobacteria	Nitrogen fixers	Due to their presence, nitrogen uptake in rice fields was facilitated	Priya et al. (2022)
2	Azospirillum species	Nitrogen fixers	It is a nonsymbiotic, free-living nitrogen fixer	Flora et al. (2021)
3	Bacillus species, Pseudomonas species	Zinc solubilizers	Zinc sulfide (ZnS) may reduce the rate of zinc sulfate production	Kumar et al. (2019a)
4	Pseudomonas species Rhizobium species Mycorrhiza	Phosphate solubilizers	Mycorrhiza can aid in increasing plant absorption of an insoluble form of phosphate in a number of ways	Ordoñez et al. (2016)
5	Bacillus species Pseudomonas species	Potassium solubilizers	Potassium is found in the insoluble form in rocks and silicate minerals and must be dissolved	Etesami et al. (2017)

rhizobium inoculation in mitigating the detrimental effects of soil-borne fungi on plants. The genus Azospirillum has been intensively researched as a prominent group of nonsymbiotic nitrogen-fixing bacteria. Many studies have shown that increasing plant nitrogen content and making phytohormones such as auxins, cytokinins, and gibberellins are two ways that plants can grow faster and better (Kurepin et al. 2014; Nawaz et al. 2020). An increasing amount of research shows that legumes, along with the symbiotic fungus rhizobia and acid thiobacillus meliloti (AMF), may help plants absorb nitrogen and phosphate better (Mishra et al. 2016). People know that PGPR can take up iron in low-iron conditions and make plants more productive. In some agroecosystems, biofertilizers are also used to dissolve "nonexchangeable" K. In vitro and in vivo studies of PGPF for the solubilization of insoluble zinc compounds have been conducted in great detail (Kaur and Rani 2023). However, it has been reported that some PGPR genera, including Acinetobacter, Bacillus, Gluconacetobacter, and Pseudomonas, can solubilize insoluble zinc (Gandhi and Muralidharan 2016; Kumar et al. 2019a).

18.7 Bioherbicides

The use of bioherbicides allows for the sustainable management of weeds and herbs that are damaging to crops (Thakur et al. 2020; Hasan et al. 2021). Most bioherbicides that worked were based on fungi, since bacteria are not used much in weed control and viruses are hard to deal with because they need a specific host and spread through vectors (Shi and Bode 2018; Hajek and Eilenberg 2018).

Various *Puccinia canaliculata* and *Chondrostereum purpureum* rust fungus-based bioherbicides have been marketed to the general public under brand names (Mishra et al. 2016). Bioherbicides have also been made with the bacteria *Xanthomonas campestris* pv. poannua (Xcp) and *P. syringae* pv. tagetis (Pst). Both of these bacteria are harmful to plants (Mishra et al. 2016).

18.8 Biopesticides

Pests (invertebrates, plant pathogens, and weeds) decimate the potential yield of crops by about 40% before harvest and another 20% after harvest (Barreto et al. 2012). However, chemical-based pesticides should be avoided for sustainable agriculture due to their negative environmental impact (Arora et al. 2022; Elnahal et al. 2022). Biocontrol agents are another name for microbial pesticides (BCAs) (Verma et al. 2019; Abd-Elgawad 2020). According to Abdollahdokht et al. (2022), the use of biopesticides derived from microorganisms is advantageous compared to conventional chemical pesticides. This is primarily attributed to their enhanced selectivity and reduced or negligible toxicity, as indicated in Table 18.3. Bacterial, fungal, and viral biological control agents (BCAs) may be categorized into three overarching

Table 18.3 Some of the biopesticides used in agroecosystems have beneficial soil microbes in them (plant growth-promoting fungi)

S. no	Target organism	Type of beneficial soil microbes (plant growth-promoting fungi)	General name	References
1	Aspergillus flavus producing aflatoxin	*Aspergillus flavus*	Afla-guard	Alshannaq et al. (2018)
2	Soil and foliar diseases	*Trichoderma polysporum*	Binab T	Woo et al. (2014)
3	Soil-borne pathogens	*Trichoderma viride*	Monitor, Trichoguard	Kandula et al. (2015)
4	Coleoptera and lepidopteran insect	*Metarhizium anisopliae*	Biomagic, Biomet	Rajula et al. (2021)
5	Herbicide—Dodder	*Alternaria destruens*	Smolder	Boyetchko and Peng (2004)
6	Root rots	*Pythium oligandrum*	Polyversum	Pisarčik et al. (2020)

classifications. Multiple species of Pseudomonas have shown the ability to effectively control phytopathogens via biological means. *Trichoderma harzianum*, a fungal biopesticide, has been used for the purpose of eradicating plant diseases (Sehrawat and Sindhu 2019; Chandra et al. 2020). According to D'Ambrosio et al. (2022), this substance has antagonistic properties against rhizoctonia, pythium, fusarium, and several other soil-borne phytopathogens. The baculovirus is widely used as a viral biological control agent (BCA) in the commercial production of phage-based insecticides.

At present, microbial biopesticides such as *Trichoderma, Pseudomonas*, and *Bacillus* are extensively used as biofungicides, while phytophthora serves as a bioherbicide. Additionally, Bacillus thuringiensis is a commonly employed bioinsecticide (Mishra et al. 2016). Based on the available data, it has been observed that bacterial biopesticides hold a dominant position in the biopesticide market, constituting approximately 74% of the market share across various crop types. In contrast, fungal biopesticides represent a smaller proportion at 10%, while viral biopesticides account for 5%. Predator biopesticides contribute 8% to the market, whereas the remaining 3% is attributed to "other" biopesticides (Mishra et al. 2016; Seenivasagan and Babalola 2021; Šunjka and Mechora 2022).

18.9 Challenges and Prospects for the Future

Agroecosystems and the delicate balance of our world are under intentional assault due to the excessive use of chemical fertilizers and pesticides. As a direct consequence, agriculturalists have grown increasingly aware of the declining soil fertility and reduced crop yields. Their embrace of beneficial soil microorganisms (BSMs) in agroecosystems has gained momentum as a means to enhance plant productivity and restore soil quality in their natural state. Soil microbes possess a diverse array of mechanisms that render them ideal candidates for sustainable agriculture and environmental management.

Soil bacteria that are beneficial and supplemented with inoculants might be considered as a feasible substitute for synthetic chemicals. These compounds are responsible for several environmental problems and provide a substantial toxicological risk to ecosystems. Beneficial microorganisms are often used in biocontrol strategies due to their environmentally favorable characteristics and ability to effectively target certain organisms. The potential for enhancing the involvement of biotic soil crusts (BSMs) in promoting environmental sustainability is contingent upon our capacity to discover novel knowledge pertaining to their ecological characteristics, population dynamics, and functional attributes in many ecological settings.

As the global population continues to rise, the agricultural sector will increasingly rely on BSMs to bolster agricultural production while upholding environmentally friendly practices.

18.10 Conclusion

Using BSMs to manage sustainable agriculture and the environment has a lot of benefits. Their use as biofertilizers and biopesticides is becoming more common, and it helps agroecosystems in a big way. They are well suited for a variety of stress management techniques due to their resilience in dangerous environments. As we gain a better understanding of how beneficial soil microbes respond in agroecosystems and their potentially important effects on environmental restoration, we collectively advance the cause of sustainable development. More research is needed to determine how effectively BSMs can be used in light of climate change. It has been hypothesized that better chances for their widespread application in environmental sustainability could be achieved by clarifying the mechanisms involved in their interaction with plants under extreme conditions.

References

Abd-Elgawad MM (2020) Plant-parasitic nematodes and their biocontrol agents: current status and future vistas. In: Management of phytonematodes: recent advances and future challenges. Springer, Singapore, pp 171–203

Abdollahdokht D, Gao Y, Faramarz S, Poustforoosh A, Abbasi M, Asadikaram G, Nematollahi MH (2022) Conventional agrochemicals towards nano-biopesticides: an overview on recent advances. Chem Biol Technol Agric 9(1):1–19

Agegnehu G, Srivastava AK, Bird MI (2017) The role of biochar and biochar-compost in improving soil quality and crop performance: a review. Appl Soil Ecol 119:156–170

Ahirwar NK, Singh R, Chaurasia S, Chandra R, Ramana S (2020) Effective role of beneficial microbes in achieving the sustainable agriculture and eco-friendly environment development goals: a review. Front Microbiol 5:111–123

Alegbeleye OO, Singleton I, Sant'Ana AS (2018) Sources and contamination routes of microbial pathogens to fresh produce during field cultivation: a review. Food Microbiol 73:177–208

Al-Kaisi MM, Lal R, Olson KR, Lowery B (2017) Fundamentals and functions of soil environment. In: Soil health and intensification of agroecosytems. Academic Press, Cambridge, MA, pp 1–23

Alshannaq AF, Gibbons JG, Lee MK, Han KH, Hong SB, Yu JH (2018) Controlling aflatoxin contamination and propagation of aspergillus flavus by a soy-fermenting *Aspergillus oryzae* strain. Sci Rep 8(1):1–14

Ambrosini A, Passaglia LM (2017) Plant growth–promoting bacteria (PGPB): isolation and screening of PGP activities. Curr Protocol Plant Biol 2(3):190–209

Ananya AK, Ahmad IZ (2014) Cyanobacteria "the blue green algae" and its novel applications: a brief review. Int J Innov Appl Stud 7(1):251

Arıkan Ş, Karakoyun M (2022) Nutrient availability in temperate fruit species: new approaches in bacteria and mycorrhizae. In: Sustainable horticulture. Academic Press, Cambridge, MA, pp 39–54

Arora NK (2018) Agricultural sustainability and food security. Environ Sustain 1(3):217–219

Arora NK, Mehnaz S, Balestrini R (eds) (2016) Bioformulations: for sustainable agriculture. Springer, Berlin, pp 1–283

Arora S, Singh AK, Sahni D (2017) Bioremediation of salt-affected soils: challenges and opportunities. In: Bioremediation of salt affected soils: an Indian perspective, pp 275–301

Arora NK, Fatima T, Mishra I, Verma M, Mishra J, Mishra V (2018) Environmental sustainability: challenges and viable solutions. Environ Sustain 1(4):309–340

Arora S, Murmu G, Mukherjee K, Saha S, Maity D (2022) A comprehensive overview of nanotechnology in sustainable agriculture. J Biotechnol 355:21–41

Barreto RW, Ellison CA, Seier MK, Evans HC (2012) Biological control of weeds with plant pathogens: four decades on. In: Integrated pest management: principles and practice, p 299

Belousova ME, Malovichko YV, Shikov AE, Nizhnikov AA, Antonets KS (2021) Dissecting the environmental consequences of bacillus thuringiensis application for natural ecosystems. Toxins 13(5):355

Bergi J, Trivedi R (2020) Bioremediation of saline soil by cyanobacteria. In: Microbial bioremediation & biodegradation. Springer, Singapore, pp 447–465

Boyetchko SM, Peng G (2004) Challenges and strategies for development of mycoherbicides. In: Fungal biotechnology in agricultural, food, and environmental applications, vol 21, pp 111–121

Chandra H, Kumari P, Bisht R, Prasad R, Yadav S (2020) Plant growth promoting *Pseudomonas aeruginosa* from *Valeriana wallichii* displays antagonistic potential against three phytopathogenic fungi. Mol Biol Rep 47(8):6015–6026

Classen AT, Sundqvist MK, Henning JA, Newman GS, Moore JA, Cregger MA, Patterson CM et al (2015) Direct and indirect effects of climate change on soil microbial and soil microbial-plant interactions: what lies ahead? Ecosphere 6(8):1–21

D'Ambrosio G, Cariddi C, Mannerucci F, Bruno GL (2022) In vitro screening of new biological limiters against some of the main soil-borne phytopathogens. Sustainability 14(5):2693

Das SK, Ghosh GK (2020) Soil health management through low cost biochar technology. In: Biochar applications in agriculture and environment management. Springer, Cham, pp 193–206

Das I, Pradhan M (2016) Potassium-solubilizing microorganisms and their role in enhancing soil fertility and health. In: Potassium solubilizing microorganisms for sustainable agriculture. Springer, New Delhi, pp 281–291

Dubey RK, Tripathi V, Edrisi SA, Bakshi M, Dubey PK, Singh A et al (2017) Role of plant growth-promoting microorganisms in sustainable agriculture and environmental remediation. In: Advances in PGPR research, pp 75–124

Egamberdieva D, Wirth SJ, Alqarawi AA, Abd-Allah EF, Hashem A (2017) Phytohormones and beneficial microbes: essential components for plants to balance stress and fitness. Front Microbiol 8:2104

Elnahal AS, El-Saadony MT, Saad AM, Desoky ESM, El-Tahan AM, Rady MM, El-Tarabily KA et al (2022) The use of microbial inoculants for biological control, plant growth promotion, and sustainable agriculture: a review. Eur J Plant Pathol 162:1–34

Escobar MA, Dandekar AM (2003) Agrobacterium tumefaciens as an agent of disease. Trends Plant Sci 8(8):380–386

Etesami H, Adl SM (2020) Plant growth-promoting rhizobacteria (PGPR) and their action mechanisms in availability of nutrients to plants. In: Phyto-microbiome in stress regulation, pp 147–203

Etesami H, Maheshwari DK (2018) Use of plant growth promoting rhizobacteria (PGPRs) with multiple plant growth promoting traits in stress agriculture: action mechanisms and future prospects. Ecotoxicol Environ Saf 156:225–246

Etesami H, Emami S, Alikhani HA (2017) Potassium solubilizing bacteria (KSB): mechanisms, promotion of plant growth, and future prospects. A review. J Soil Sci Plant Nutr 17(4):897–911

Fan D, Smith DL (2021) Characterization of selected plant growth-promoting rhizobacteria and their non-host growth promotion effects. Microbiol Spectr 9(1):e00279–e00221

Flora Y, Rabha P, Shinde A, Jha P, Jobby R (2021) Non-symbiotic bacteria for soil nitrogen fortification. In: Sustainable agriculture reviews, vol 52. Springer, Cham, pp 417–435

French E, Kaplan I, Iyer-Pascuzzi A, Nakatsu CH, Enders L (2021) Emerging strategies for precision microbiome management in diverse agroecosystems. Nat Plants 7(3):256–267

Fróna D, Szenderák J, Harangi-Rákos M (2019) The challenge of feeding the world. Sustainability 11(20):5816

Gandhi A, Muralidharan G (2016) Assessment of zinc solubilizing potentiality of *Acinetobacter* sp. isolated from rice rhizosphere. Eur J Soil Biol 76:1–8

Garlapati D, Chandrasekaran M, Devanesan A, Mathimani T, Pugazhendhi A (2019) Role of cyanobacteria in agricultural and industrial sectors: an outlook on economically important byproducts. Appl Microbiol Biotechnol 103(12):4709–4721

Gougoulias C, Clark JM, Shaw LJ (2014) The role of soil microbes in the global carbon cycle: tracking the below-ground microbial processing of plant-derived carbon for manipulating carbon dynamics in agricultural systems. J Sci Food Agric 94(12):2362–2371

Hajek AE, Eilenberg J (2018) Natural enemies: an introduction to biological control. Cambridge University Press, Cambridge

Halim MA, Rahman MM, Megharaj M, Naidu R (2020) Cadmium immobilization in the rhizosphere and plant cellular detoxification: role of plant-growth-promoting rhizobacteria as a sustainable solution. J Agric Food Chem 68(47):13497–13529

Hartman WH, Richardson CJ (2013) Differential nutrient limitation of soil microbial biomass and metabolic quotients (q CO_2): is there a biological stoichiometry of soil microbes? PLoS One 8(3):e57127

Hasan M, Ahmad-Hamdani MS, Rosli AM, Hamdan H (2021) Bioherbicides: an eco-friendly tool for sustainable weed management. Plan Theory 10(6):1212

Hasnat GT, Kabir MA, Hossain MA (2018) Major environmental issues and problems of South Asia, particularly Bangladesh. In: Handbook of environmental materials management, pp 1–40

Hernandez CA, Salazar AJ, Koskella B (2020) Bacteriophage-mediated reduction of bacterial speck on tomato seedlings. Phage 1(4):205–212

İmamoglu A, Dengiz O (2019) Evaluation of soil quality index to assess the influence of soil degradation and desertification process in sub-arid terrestrial ecosystem. Rendiconti Lincei Scienze Fisiche e Naturali 30(4):723–734

Jansson JK, Hofmockel KS (2020) Soil microbiomes and climate change. Nat Rev Microbiol 18(1):35–46

Jones DL, Nguyen C, Finlay RD (2009) Carbon flow in the rhizosphere: carbon trading at the soil–root interface. Plant and Soil 321(1):5–33

Kandula DR, Jones EE, Stewart A, McLean KL, Hampton JG (2015) Trichoderma species for biocontrol of soil-borne plant pathogens of pasture species. Biocontrol Sci Tech 25(9):1052–1069
Kaur A, Rani R (2023) Role of soil microorganisms in sustainable crop production. In: Microbial based land restoration handbook, vol 2. CRC, Boca Raton, FL, pp 1–30
Kaur H, Kaur J, Gera R (2016) Plant growth promoting rhizobacteria: a boon to agriculture. Int J Cell Sci Biotechnol 5:17–22
Kaushik MS, Kumar A, Abraham G, Dash NP, Singh PK (2019) Field evaluations of agrochemical toxicity to cyanobacteria in rice field ecosystem: a review. J Appl Phycol 31(1):471–489
Khan N, Ray RL, Kassem HS, Hussain S, Zhang S, Khayyam M, Ihtisham M, Asongu SA (2021) Potential role of technology innovation in transformation of sustainable food systems: a review. Agriculture 11(10):984
Kumar A, Dewangan S, Lawate P, Bahadur I, Prajapati S (2019a) Zinc-solubilizing bacteria: a boon for sustainable agriculture. In: Plant growth promoting rhizobacteria for sustainable stress management. Springer, Singapore, pp 139–155
Kumar R, Kumar R, Prakash O (2019b) The impact of chemical fertilizers on our environment and ecosystem. Chief Ed 35:69
Kuraganti G, Edla S, Pallaval VB (2020) Cyanobacteria as biofertilizers: current research, commercial aspects, and future challenges. In: Advances in plant microbiome and sustainable agriculture, pp 259–278
Kurepin LV, Zaman M, Pharis RP (2014) Phytohormonal basis for the plant growth promoting action of naturally occurring biostimulators. J Sci Food Agric 94(9):1715–1722
Lal R (2004) Soil carbon sequestration impacts on global climate change and food security. Science 304(5677):1623–1627
Lehman RM, Cambardella CA, Stott DE, Acosta-Martinez V, Manter DK, Buyer JS, Maul JE, Smith JL, Collins HP, Halvorson JJ, Kremer RJ, Lundgren JG, Ducey TF, Jin VL, Karlen DL (2015) Understanding and enhancing soil biological health: the solution for reversing soil degradation. Sustainability 7(1):988–1027
Li F, Miao Y, Hennig SD, Gnyp ML, Chen X, Jia L, Bareth G (2010) Evaluating hyperspectral vegetation indices for estimating nitrogen concentration of winter wheat at different growth stages. Precis Agric 11(4):335–357
Ma S (2019) Soil microbe communities and their interactions with understorey plants in temperate forests. Doctoral dissertation, Ghent University
Martins-Loução MA, Dias T, Cruz C (2022) Integrating ecological principles for addressing plant production security and move beyond the dichotomy 'good or Bad'for nitrogen inputs choice. Agronomy 12(7):1632
Meena RS, Kumar S, Datta R, Lal R, Vijayakumar V, Brtnicky M, Marfo TD et al (2020) Impact of agrochemicals on soil microbiota and management: a review. Land 9(2):34
Mikiciński A, Sobiczewski P, Puławska J, Maciorowski R (2016) Control of fire blight (*Erwinia amylovora*) by a novel strain 49M of *Pseudomonas graminis* from the phyllosphere of apple (*Malus* spp.). Eur J Plant Pathol 145(2):265–276
Mishra J, Prakash J, Arora NK (2016) Role of beneficial soil microbes in sustainable agriculture and environmental management. Clim Change Environ Sustain 4(2):137–149
Mukherjee R, Sen S (2021) Agricultural sustainability through nitrogen fixation: approaches and techniques. Harvest 6(1):48–55
Mustafa S, Kabir S, Shabbir U, Batool R (2019) Plant growth promoting rhizobacteria in sustainable agriculture: from theoretical to pragmatic approach. Symbiosis 78(2):115–123
Naamala J, Smith DL (2020) Relevance of plant growth promoting microorganisms and their derived compounds, in the face of climate change. Agronomy 10(8):1179
Nadeem SM, Ahmad M, Zahir ZA, Javaid A, Ashraf M (2014) The role of mycorrhizae and plant growth promoting rhizobacteria (PGPR) in improving crop productivity under stressful environments. Biotechnol Adv 32(2):429–448

Nain L, Rana A, Joshi M, Jadhav SD, Kumar D, Shivay YS, Prasanna R et al (2010) Evaluation of synergistic effects of bacterial and cyanobacterial strains as biofertilizers for wheat. Plant and Soil 331(1):217–230

Nawaz K, Chaudhary R, Sarwar A, Ahmad B, Gul A, Hano C, Anjum S et al (2020) Melatonin as master regulator in plant growth, development and stress alleviator for sustainable agricultural production: current status and future perspectives. Sustainability 13(1):294

Ojuederie OB, Olanrewaju OS, Babalola OO (2019) Plant growth promoting rhizobacterial mitigation of drought stress in crop plants: implications for sustainable agriculture. Agronomy 9(11):712

Ordoñez YM, Fernandez BR, Lara LS, Rodriguez A, Uribe-Velez D, Sanders IR (2016) Bacteria with phosphate solubilizing capacity alter mycorrhizal fungal growth both inside and outside the root and in the presence of native microbial communities. PLoS One 11(6):e0154438

Pandey A, Tripathi A, Srivastava P, Choudhary KK, Dikshit A (2019) Plant growth-promoting microorganisms in sustainable agriculture. In: Role of plant growth promoting microorganisms in sustainable agriculture and nanotechnology. Woodhead Publishing, Sawston, pp 1–19

Pandya U, Saraf M (2014) In vitro evaluation of PGPR strains for their biocontrol potential against fungal pathogens. In: Microbial diversity and biotechnology in food security. Springer, New Delhi, pp 293–305

Pascale A, Proietti S, Pantelides IS, Stringlis IA (2020) Modulation of the root microbiome by plant molecules: the basis for targeted disease suppression and plant growth promotion. Front Plant Sci 10:1741

Pathak AP, Rathod MG, Devarshe AM, Hundekar MR, Tengse SA, Kamble GT (2021) Entomopathogenic microorganisms as biopesticides: a review. Front Life Sci 3:95

Pathania P, Rajta A, Singh PC, Bhatia R (2020) Role of plant growth-promoting bacteria in sustainable agriculture. Biocatal Agric Biotechnol 30:101842

Pisarčik M, Hakl J, Hrevušová Z (2020) Effect of *Pythium oligandrum* and poly-beta-hydroxy butyric acid application on root growth, forage yield and root diseases of red clover under field conditions. Crop Prot 127:104958

Popp J, Pető K, Nagy J (2013) Pesticide productivity and food security. A review. Agron Sustain Dev 33(1):243–255

Prasanna R, Kaushik BD (2007) Cyanobacteria in soil health and sustainable agriculture. Health Environ 3:91–105

Prasanna R, Sood A, Ratha SK, Singh PK (2014) Cyanobacteria as a green option for sustainable agriculture. Cyanobacteria 9:145–166

Pritchard SG (2011) Soil organisms and global climate change. Plant Pathol 60(1):82–99

Priya H, Dhar DW, Singh R, Kumar S, Dhandapani R, Pandey R, Kumar A et al (2022) Co-cultivation approach to decipher the influence of nitrogen-fixing cyanobacterium on growth and N uptake in rice crop. Curr Microbiol 79(2):1–16

Rajula J, Karthi S, Mumba S, Pittarate S, Thungrabeab M, Krutmuang P (2021) Current status and future prospects of entomopathogenic fungi: a potential source of biopesticides. In: Recent advancement in microbial biotechnology, pp 71–98

Reichardt K, Timm LC (2014) Soil, plant and atmosphere. Springer Nature, Cham

Ritika B, Utpal D (2014) Biofertilizer, a way towards organic agriculture: a review. Afr J Microbiol Res 8(24):2332–2343

Roy A (2021) Biofertilizers for agricultural sustainability: current status and future challenges. In: Current trends in microbial biotechnology for sustainable agriculture. Springer, Singapore, pp 525–553

Saad A, Atia A (2014) Review on freshwater blue-green algae (cyanobacteria): occurrence, classification and toxicology. Biosci Biotechnol Res Asia 11(3):1319–1325

Saeed Q, Xiukang W, Haider FU, Kučerik J, Mumtaz MZ, Holatko J, Mustafa A et al (2021) Rhizosphere bacteria in plant growth promotion, biocontrol, and bioremediation of contaminated sites: a comprehensive review of effects and mechanisms. Int J Mol Sci 22(19):10529

Sahu PK, Singh DP, Prabha R, Meena KK, Abhilash PC (2019) Connecting microbial capabilities with the soil and plant health: options for agricultural sustainability. Ecol Indic 105:601–612
Sammauria R, Kumawat S, Kumawat P, Singh J, Jatwa TK (2020) Microbial inoculants: potential tool for sustainability of agricultural production systems. Arch Microbiol 202(4):677–693
Sansinenea E (2019) *Bacillus* spp.: as plant growth-promoting bacteria. In: Secondary metabolites of plant growth promoting rhizomicroorganisms. Springer Nature, Singapore, pp 225–237
Seenivasagan R, Babalola OO (2021) Utilization of microbial consortia as biofertilizers and biopesticides for the production of feasible agricultural product. Biology 10(11):1111
Sehrawat A, Sindhu SS (2019) Potential of biocontrol agents in plant disease control for improving food safety. Def Life Sci J 4(4):220–225
Sharma P (2022) Role and significance of biofilm-forming microbes in phytoremediation—a review. Environ Technol Innov 25:102182
Shi YM, Bode HB (2018) Chemical language and warfare of bacterial natural products in bacteria–nematode–insect interactions. Nat Prod Rep 35(4):309–335
Singh S (2014) A review on possible elicitor molecules of cyanobacteria: their role in improving plant growth and providing tolerance against biotic or abiotic stress. J Appl Microbiol 117(5):1221–1244
Singh JS, Kumar A, Rai AN, Singh DP (2016) Cyanobacteria: a precious bio-resource in agriculture, ecosystem, and environmental sustainability. Frontiers in microbiology 7:529
Singh B, Yadav JP (2022) Sustainable agriculture development through organic farming in India. ECS Trans 107(1):14919
Smith P (2008) Land use change and soil organic carbon dynamics. Nutr Cycl Agroecosyst 81(2):169–178
Srivastav AL (2020) Chemical fertilizers and pesticides: role in groundwater contamination. In: Agrochemicals detection, treatment and remediation. Butterworth-Heinemann, Oxford, pp 143–159
Subashchandrabose SR, Ramakrishnan B, Megharaj M, Venkateswarlu K, Naidu R (2013) Mixotrophic cyanobacteria and microalgae as distinctive biological agents for organic pollutant degradation. Environ Int 51:59–72
Sundari SK, Prakash A, Yadav P, Kumari A (2019) Plant growth-promoting microbes as front-runners for on-site remediation of organophosphate pesticide residues in agriculture soils. In: Phyto and Rhizo remediation. Springer, Singapore, pp 249–285
Šunjka D, Mechora Š (2022) An alternative source of biopesticides and improvement in their formulation—recent advances. Plan Theory 11(22):3172
Syiem MB, Singh AK, Rai AN (2017) N2-fixing cyanobacterial systems as biofertilizer. In: Agro-environmental sustainability. Springer, Cham, pp 43–61
Thakur N, Kaur S, Tomar P, Thakur S, Yadav AN (2020) Microbial biopesticides: current status and advancement for sustainable agriculture and environment. In: New and future developments in microbial biotechnology and bioengineering. Elsevier, Amsterdam, pp 243–282
Tijjani A, Khairulmazmi A (2021) Global food demand and the roles of microbial communities in sustainable crop protection and food security: an overview. In: Role of microbial communities for sustainability, pp 81–107
Tilman D, Cassman KG, Matson PA, Naylor R, Polasky S (2002) Agricultural sustainability and intensive production practices. Nature 418(6898):671–677
Timmis K, Ramos JL (2021) The soil crisis: the need to treat as a global health problem and the pivotal role of microbes in prophylaxis and therapy. J Microbial Biotechnol 14(3):769–797
Timsina J (2018) Can organic sources of nutrients increase crop yields to meet global food demand? Agronomy 8(10):214
Ünal F, Aşkın A, Koca E, Yıldırır M, Bingöl MÜ (2019) Mycelial compatibility groups, pathogenic diversity and biological control of *Sclerotium rolfsii* on turfgrass. Egypt J Biol Pest Control 29(1):1–7

Venkatachalam S, Ranjan K, Prasanna R, Ramakrishnan B, Thapa S, Kanchan A (2016) Diversity and functional traits of culturable microbiome members, including cyanobacteria in the rice phyllosphere. Plant Biol 18(4):627–637
Verma PP, Shelake RM, Das S, Sharma P, Kim JY (2019) Plant growth-promoting rhizobacteria (PGPR) and fungi (PGPF): potential biological control agents of diseases and pests. In: Microbial interventions in agriculture and environment. Springer, Singapore, pp 281–311
Woo SL, Ruocco M, Vinale F, Nigro M, Marra R, Lombardi N, Lorito M et al (2014) Trichoderma-based products and their widespread use in agriculture. Open Mycol J 8(1):71–126
Xiang L, Harindintwali JD, Wang F, Redmile-Gordon M, Chang SX, Fu Y, He C, Muhoza B, Brahushi F, Bolan N, Jiang X, Ok YS, Rinklebe J, Schaeffer A, Zhu Y-G, Tiedje JM, Xing B (2022) Integrating biochar, bacteria, and plants for sustainable remediation of soils contaminated with organic pollutants. Environ Sci Technol 56(23):16546–16566
Zahra Z, Choo DH, Lee H, Parveen A (2020) Cyanobacteria: review of current potentials and applications. Environments 7(2):13
Zucca C, Middleton N, Kang U, Liniger H (2021) Shrinking water bodies as hotspots of sand and dust storms: the role of land degradation and sustainable soil and water management. Catena 207:105669

SPRINGER NATURE

GPSR Compliance

The European Union's (EU) General Product Safety Regulation (GPSR) is a set of rules that requires consumer products to be safe and our obligations to ensure this.

If you have any concerns about our products, you can contact us on ProductSafety@springernature.com

In case Publisher is established outside the EU, the EU authorized representative is:

Springer Nature Customer Service Center GmbH
Europaplatz 3
69115 Heidelberg, Germany

The manufacturer's authorised representative in the EU is Springer Nature Customer Service Centre GmbH, Europaplatz 3, 69115 Heidelberg, Germany. If you have any concerns regarding our products, please contact ProductSafety@springernature.com

Printed and bound by CPI Group (UK) Ltd, Croydon, CR0 4YY
01/10/2026
02225864-0013